THE AGE OF OIL

*The Mythology, History,
and Future of the World's Most
Controversial Resource*

LEONARDO MAUGERI

THE LYONS PRESS
Guilford, Connecticut
An imprint of The Globe Pequot Press

The Lyons Press is an imprint of The Globe Pequot Press.

10 9 8 7 6 5 4 3 2

Printed in the United States of America

ISBN 978-1-59921-118-3

First published in 2006 by Praeger Publishers
First Lyons Press edition, 2008

The Library of Congress has previously cataloged an earlier hardcover edition as follows:

Maugeri, Leonardo, 1964–
  The age of oil : the mythology, history, and future of the world's most controversial resource / Leonardo Maugeri.
    p. cm.
  Includes bibliographical references and index.
  ISBN 0-275-99008-7 (alk. paper)
  1. Petroleum—History—19th century. 2. Petroleum—History—20th century. 3. Petroleum—History—21st century. 4. Petroleum industry and trade—History—19th century. 5. Petroleum industry and trade—History—20th century. 6. Petroleum industry and trade—History—21st century. 7. Petroleum reserves. I. Title.
TN870.M339 2006
553.2'82—dc22   2006006632

To Enzo Viscusi,
who more than anyone else accompanied me on my journey through
the oil industry and the complexities of international affairs.
His unstinting generosity and wholehearted friendship
have made this book possible.

Contents

PART TWO
Misperceptions and Problems Ahead

Appendixes

Note by the Author

In order to help the reader, I resorted to some simplifications. In dealing with the history of oil companies, I used their current names instead of the original ones they held for part of their existence. So, for instance, the reader will find BP instead of Anglo-Persian Oil Company and—later—Anglo-Iranian Oil Company; Chevron instead of Standard Oil Company of California; Exxon instead of Standard Oil of New Jersey; Mobil instead of Standard Oil New York-Vacuum Oil; Total rather than Compagnie Française du Pétrole. Although some may argue against this choice, I think it will allow the reader to avoid being confused by too many names and too many confusing changes that add no value to the tale.

As to the transliteration of foreign names and words (Arabic, Iranian, Russian, etc.) I simply made the choice to use the forms prevailing in current American usage, avoiding those phonetic symbols that—in my opinion—have no sense in the American language.

Personally, I dislike abbreviations. However, I must admit that it is quite boring to constantly repeat "million barrels per day" instead of a simple "mbd," or "Organization of Petroleum Exporting Countries" in place of its OPEC acronym. So I abdicated to some essential abbreviations.

Preface

Oil has played a unique role in the economy and history of modern times. No other raw material has been so critical in shaping the destiny of nations, the development of military and global trade strategies, and relationships between countries. No other raw material has offered such great promises for improving the well-being of entire nations, promises which sadly remained unrealized, and which often turned into curses looming over their future. No other resource has had such a huge impact on the geography of our world, and the way our societies interact and are organized. More than any other raw material, then, petroleum has shaped our lives, and inevitably such a prominent role has made it the world's most controversial resource.

Throughout its history, "black gold" has given rise to myths and obsessions, fears and misperceptions of reality, and ill-advised policies that have weighed heavily on the world's collective psyche. Even today the vast majority of public opinion tends to think of oil as a kind of "witch's brew" identified with wars, greed, and unspeakable power plays orchestrated by transnational elites engaged in schemes worthy of spy novels. At the same time, ever since it burst into human life during the second half of the nineteenth century, oil has always been subject to unpredictable changes that have caught to world by surprise. Those who thought they could control it for their own benefit have been thwarted time and again by oil's boom and bust production cycles, its frequent market crises and often uncontrollable price fluctuations, as well as the political explosions in which it has played such a crucial role.

It should therefore surprise no one that most attempts at predicting the future behavior of oil have yielded such grossly inaccurate results.

As we look at the oil industry today, we find that these soothsaying efforts have not died. Once again, pseudoscience is used to spread fear through doom-and-gloom scenarios predicting catastrophic shortages, while ignoring the lessons of history and rejecting the cold logic of reality. Once again, the comforting slogan of "independence from oil" resounds in the world arena, proclaimed by most Western governments and even by the U.S. President George Bush, whose administration yet continues to favor the more irresponsible habits of oil consumption that has fed America'a addiction with "black gold."

This overdramatization and sterile overreaction to oil's cyclical behavior draws its strength from the gap between the realities of the oil market and the perceptions of the casual observer.

Unfortunately, a true understanding of oil can be reached only by penetrating complex technical elements, the abstruse prose of experts, and esoteric indicators, coupled with a deep knowledge of past events, economics, and geopolitics. Public perception, on the other hand, is shaped mainly by the more simplistic language of the media, which is more at ease with extremes.

The lack of clear information is chronic and generalized, compounded by a decline in public interest since the 1990s, in sharp contrast with the obsessive concern of the previous two decades. Underlining the public's declining interest was the belief that oil had become "just another commodity," a resource facing an irreversible decline in its importance. This attitude took root after the petroleum "countershock" of 1986, and became more deeply ingrained in the last decade of the century as the world floated atop a sea of crude while demand grew only fitfully. At the same time, the more developed economies were reducing their dependence on raw materials and heavy industry, building their wealth on intelligence and inventiveness, on microchips and services. For the first time in modern history, their dependence on oil was no longer the subject of strategic concern or a matter of existential struggles for access to energy sources, as had been predicted during the 1970s and the early 1980s. In this context, the history of oil seemed to become a tale of past struggles with seemingly no future.

But by the start of the new millennium, black gold aggressively reclaimed center stage, surprising all pundits, and once again became the object of old-fashioned fears and obsessions, sometimes in new disguise. The fear of an inevitable exhaustion of reserves came back, along with the fear of an apocalyptic clash of civilizations, pitting Islam against the rest of the world and threatening access to the largest global oil deposits.

It is not easy to escape the trap of catastrophism, particularly when so many elements seem to justify a grim picture of the future. Yet we are not on the verge of any catastrophe. What is needed to fully grasp the issue is a sober and deeper approach, free of sensationalism. This approach must examine the history of the industry and the ways in which it has endowed its participants with a unique DNA that has led to extraordinarily similar reactions, behavior, and thought patterns over many decades. Such an approach is needed to sweep away the myths and distortions that underpin so many current assessments of the industry, lingering like evil spirits impervious to every kind of exorcism. They appear at the first sign of crisis and conspire to exacerbate it.

Only by retracing the history of the industry can we grasp its current reality and come to understand the economic, technical, social, and geopolitical variables that make it at once more complex and less dramatic than it is generally portrayed.

At the end of 2002, as I was beginning to draft a pair of articles that were ultimately published in *Foreign Affairs*[1] and *Science*,[2] I came to realize that only a book-length analysis could do justice to the depth and complexity of the oil business. And I decided to undertake the challenge even though my day-to-day responsibilities as an oil industry executive seemed to present a nearly insurmountable obstacle to such a project.

The more I went into it, the more I became convinced that a comprehensive book was needed, one that could take into account the entire history of oil and not just current problems. The last great work on the subject, Daniel Yergin's masterpiece *The Prize*, had been published in 1991.[3] Since then a few good books had appeared, but all of them addressed a specialized audience. Making matters worse, starting in 2000, a flurry of "eat-and-run" books began to hit the market, all exploiting a growing hysteria about oil by pointing to a dire as well as superficial forecast of the energy future.

As I stole time from holidays, weekends, and vacations and cut back on sleep, I was able to draw on a great deal of material I had been archiving in electronic form since my days as a graduate student preparing for my doctorate. Without this extensive documentation, supported by the lessons learned in my professional life, I would never have been able to put this book together in a little more than two years.

As you surely will have understood by now, this book is devoted to the history, economics, and geopolitics of oil.

The history of oil is covered in Part One, which is the book's longest. It begins at the very dawn of the industry, with the unexpected consequences of "Colonel" Drake's first modern oil well in 1859. It ends with the dramatic situation of post-Saddam Iraq, the public fear of a world seemingly running out of oil, the breathtaking growth of Chinese energy consumption, and the threat of radical Islam to the largest global oil producers.

The reader will find tales of peculiar characters and extravagant missions, the designs of great powers and the aspirations of emerging countries, together with the events that marked the rise of oil as a vital factor in modern society and its influence on shaping of the world we live in. The thread that runs throughout this history is the unpredictable succession of booms and busts that made it unique, often leading to the failure of ill-conceived policies and the destruction of entrepreneurial undertakings.

In reconstructing this tale, I also attempt to debunk most of the myths and obsessions generated by each cycle, and which to this day remain a part of the collective perception about oil. Probably the most enduring among them is the link between the fear of oil shortages and the global quest for control of reserves and the security of supplies.

It was the fear of oil shortages that moved the great powers to develop their first oil-driven foreign policies at the dawn of the twentieth century, leading to British control of what was then Persia (today's Iran) and to the establishment of today's Iraq (then Mesopotamia) in the 1920s. Later on, it was the perception of dwindling oil resources in the United States that inspired the close links between the American government and Saudi Arabia's. And it was the fear that Arab oil would fall under the influence of the Soviet Union that largely shaped American foreign policy in the Middle East after World War II.

Yet, the obsessive fear of a world short of oil and the political analyses and responses it produced always proved to be inconsistent with reality. Over almost 150 years, the dominant characteristics of the oil market have been oversupply and low prices, sometimes temporarily interrupted by shocking reversals. Each period of dramatic expectation that the end of oil was near concluded with a major oil glut. A prominent oilman saw this trend as early as 1925, at the peak of a groundless wave of fear about the future of oil, when he remarked:

My father was one of the pioneers in the oil industry. Periodically ever since I was a small boy, there has been an agitation predicting an oil shortage, and always in the succeeding years production has been greater than ever before.[4]

Five years later, the world was plunged into one of the worst over-production crises ever, leading to a major and prolonged collapse in prices. Similarly, in the 1970s an outpouring of gloomy forecasts predicted a day of doom that never materialized. The Central Intelligence Agency (CIA) was among the most persistently negative analysts, producing forecasts like this:

We believe that world oil production is probably at or near its peak. Simply put, the expected decline in oil production is the result of a rapid exhaustion of accessible deposits of conventional crude oil. Politically, the cardinal issue is how vicious the struggle for energy supply will become.[5]

In 1986, another huge wave of overproduction led to a new oil price collapse. Understanding the origin and cultural consequences of most oil myths and obsessions is essential to penetrate the issues facing the world of oil today. This is addressed in Part Two, which is much shorter than Part One. Here we explore questions such as *Are we running out of oil? Will China upset the world's oil consumption trends? Are there real alternatives to oil? What is the real impact of Islamic radicalism on oil-rich areas?*—and many others—in an attempt to give the reader a comprehensive insight into the complexities of the industry, and the wrongheaded interpretations to which they are usually subjected.

Naturally, understanding some of the most debated issues of our times requires some technical background, which I have tried to make as simple as possible for the casual reader. Particularly, the familiar ghost of a world short of oil—that has once again reappeared on the global stage—requires the reader to know the basic rules governing the discovery, extraction, production, and consumption of hydrocarbons. In spite of all doomsayers, those rules make it possible to understand some basic elements. First, we are very far from having an acceptable knowledge of subsurface oil resources, and there is plenty of evidence that a huge potential for future oil production still exists. Catastrophists of all kinds usually tend to underestimate the latter by cranking into their pseudo-scientific models data which assume a quasi-perfect knowledge of the ultimate level of existing resources—and this is the first major flaw in their predictions. Second, those same advocates of doom always tend to overestimate consumption, making it a price-indifferent function—while in the long term demand always responds to prices. The problem is that in a prolonged high-price scenario, as in a low-price scenario, all actors involved—including oil companies—tend to see the future as the result of "adaptive expectations about current conditions,"[6] and in so doing they indefinitely prolong the situation in which they find themselves. So it is no surprise that in times like ours, a pessimistic view about the future of oil overwhelms the debate. Yet the final message of this book—if I may anticipate it—is that nothing we are experiencing today is a major departure from the historical cycles of the oil market.

This does not mean we can expect stability or avoid prolonged periods of tension in the future. Uncertainty and volatility are characteristics of all human activity, and they have been a constant throughout the history of the oil industry. But focusing only on negative concerns, using them as a platform to build visions of a dark future is like looking at a single tree, and missing the forest. Thus, for example, even as China's oil consumption is growing exponentially, consumption is dropping in other areas and new technology is providing less energy-demanding solutions to support our lifestyle. At the same time oil production and reserves worldwide are silently growing, irrespective of any catastrophic assumptions to the contrary.

Oil supplies will not be exhausted when the Oil Age ends, because it is overabundant and because more intelligent and ethical ways to use it are emerging. By the same token, even if Islamic terrorists were to gain control over an oil-rich country, they would still have to cope with the elusive laws governing oil.

The conclusion line is that the wolf is not at the door. And even though dramatization is an unavoidable by-product of everything that concerns oil, there is nothing that dooms us to a vicious struggle for securing our future oil needs in the face of strangling shortages and geopolitical turmoil. Only the inability of decision makers to grasp this reality and to act accordingly may push us to that brink.

Everyone pursues the hobby that gives him or her the most pleasure. In my case I was fortunate in choosing one that largely coincided with my professional work. Nonetheless, writing a book is always a difficult experience. Trying to write a serious book, double or triple checking all sources, is even more difficult. Doing this in another language, and while engaged in a job which potentially consumes all of one's available time, is something that can stretch intellectual and physical strengths to their limits.

Beyond my personal fascination and daily engagement with the issues of this book, what helped me in overcoming many obstacles of this undertaking was the support of several people.

To start with, I have had the enormous good fortune to be blessed with the most lovable and understanding wife, Carmen, who has patiently supported and encouraged me in this project. Without her at my side, most things I managed to accomplish in my life would have never materialized.

I have an enduring obligation to my dear friends Claude Erbsen and Teddy Jefferson, the first priceless editors of the manuscript, who carefully read it while it was coming out of my pen (or rather, my personal computer) and made important textual suggestions and corrections. Naturally, I am wholeheartedly grateful to Hilary Claggett, Senior Editor for politics and current events at Praeger, who first—and without the help of a literary agent—demonstrated strong enthusiasm for my manuscript and supported its publication through a process involving three levels of approval by three different boards. Her personal dedication and support were something writers pray for when dealing with a publisher. I would also like to thank Carla Talmadge, project manager of my book, for her excellent and careful supervision of the editing process.

Many other persons helped me by double checking figures and statistics, or in discussing some of the issues I was dealing with. Most of them work with me at the department of strategy at Eni: I hope I remember them all, and I apologize if I omit someone. In rigid alphabetic order, they are: Marco Aversa, Paola Dagnino, Simonetta De Bartolo, Alvaro Donadelli, Fabio Ercoli, Alessandro Lanza, Sabina Manca, Giampiero Marcello, Maurizio Maugeri, Alberto Navarretta, Sandro Osvaldella, Cristiano Pattumelli, Salvatore Pino, Andrea Quarta, Manuela Rondoni, Mario Salustri, Giuseppe Sammarco, Lorenzo Siciliano, Maria Antonietta Solinas, Claudia Squeglia, Claudia Tenaglia, Antonella Tolentino, and Anna Maria Tibuzzi.

The enchanting hospitality of Ginesta and Alessandro Guerrera at their beautiful vacation home in the Turkish sea village of Kas over several summer holidays has been another important element in allowing me to write and correct parts of the book. The breathtaking and inspiring beauty of the place, along with the warmth of their friendship, has been a catalyst for ideas and the determination to work.

My brother Alessandro also helped me through long discussions about economics, finance, and history. His brilliant insights and appropriate advice have been decisive at critical points of this book. I also owe grateful appreciation and enduring affection to my two secretaries, Anna Laura De Francisci and Nadia Sturmann, who did not hesitate to devote part of their free time (and even Saturdays) organizing all the documents and even the electronic version of the manuscript.

Other friends, scholars, analysts, and several personalities of the oil world has helped me to carry out my project, through their advices, but also with their critics. In this case, however, I would risk losing some good friend by omitting his name in a long but incomplete list. And because I made the mistake not to take an orderly file of all the persons who somehow contributed to my work, I prefer to omit that list at all.

Finally, I have endless gratitude for Enzo Viscusi, to whom this book is dedicated. As always in his life, and as always when devoting himself to a cause or to a friend, he spared no effort in convincing me to go through with this project, acting as the real coordinator behind the different stages that brought it into being. Not surprising for a man of his talents, he served as literary agent, promoter, taskmaster, and adviser, depending on the circumstances. More than anything else, however, he helped me with the support of his friendship.

I still cannot decide if the final result has been worth the effort, and I leave that to the judgment of the readers, while inviting them to put the blame uniquely on me for any errors or deficiencies encountered.

List of Abbreviations

AGIP   Azienda Generale Italiana Petroli
API   American Petroleum Institute
Aramco Arabian American Oil Company
bcm   billion cubic meters
bpd   barrels per day
BNOC   British National Oil Company
BP   British Petroleum
Caltex California Texas Oil Company
CERA   Cambridge Energy Research Associates
CFP   Compagnie Française de Pétroles
CNOCs   Consumers' National Oil Companies
EIA   Energy Information Administration
ENI   Ente Nazionale Idrocarburi (until 1992)
GDP   Gross Domestic Product
GTL   Gas-To-Liquids
IEA   International Energy Agency
IOCs   International Oil Companies
IPC   Iraq Petroleum Company
IPE   International Petroleum Exchange
mbd   million barrels per day
MEES   Middle East Economic Survey
NIOC   National Iranian Oil Company
Nymex   New York Mercantile Exchange
OAPEC   Organization of Arab Petroleum Exporting Countries
OGJ   *Oil&Gas Journal*

OOP	Original Oil in Place
OPEC	Organization of Petroleum Exporting Countries
PDVSA	Petroleos de Venezuela
PEMEX	Petróleos Mexicanos
PIW	Petroleum Intelligence Weekly
PNOCs	Producers' National Oil Companies
PSA	Production Sharing Agreement
ROCE	Return on Capital Employed
SUVs	Sport-Utility Vehicles
TPC	Turkish Petroleum Company
TRC	Texas Railroad Commission
UAE	United Arab Emirates
UAR	United Arab Republic
URR	Ultimate Recoverable Resources
U.S.	United States
WACC	Weighted Average Cost of Capital
WTI	Western Texas Intermediate
3-D	Three-dimensional (seismic)

PART ONE

A History of an Unreliable Market (and the Bad Policies It Prompted)

CHAPTER 1

John D. Rockefeller's Cursed Legacy

Oil slipped abruptly into modern life by a back door. Long before its rise to prominence in energy production, its entry into daily use was brought about by people's need for a cheaper and more flexible source of illumination.

Indeed, petroleum derivatives have been exploited since the emergence of human civilization, particularly in ancient Mesopotamia and elsewhere in the Middle East, where a primitive but significant oil industry supplied asphalt for building roads, mastic for waterproofing ships, architecture, and hydraulics, as well as essential components for many medicines and treatments. Bitumen was used in warfare and many other fields. However, paradoxically, after having been widely used in ancient times, its eventual applications throughout the centuries were marginal and mainly confined to those places where oil was easily available through surface seepage.

After a long plunge into obscurity, oil partially reemerged in the mid-1850s, when parallel experiments by amateur and professional chemists were undertaken in Europe and the United States to refine oil to obtain an illuminating fuel. Among the many claiming to be the modern inventor of oil distillation, a Canadian scientist is due a special note of praise: Abraham Gesner, who first patented in the United States in 1854 a new oil product, Kerosene, to be used for "illuminating or other purposes."[1] Since it was cheaper, safer, and better than any existing illuminant, its use spread in Western Pennsylvania and New York City,[2] partly because of a favorable circumstance. Whale oil, until then the illuminating fuel preferred by the wealthy (the only ones who could afford artificial light), was running out as a result of intensive overfishing of whales in the Atlantic

Ocean. Procuring additional supplies involved traveling to faraway seas, like South Africa's, which brought an immediate jump in the product's price. Yet the most serious obstacle to petroleum's penetration of the market was producing it in sufficient volumes. All the extraction techniques applied since ancient times involved the collection of surface crude seepage with primitive instruments and amateurish devices; in most cases, oil was still picked up by hand. There were scattered examples of subsurface drilling in places like France, Japan, and some other Asian countries—particularly Azerbaijan—but this never was adopted as a common practice worldwide. The great revolution occurred in Pennsylvania in 1859, when Edwin Drake first succeeded in extracting oil from its rocky underground prison with a drilling machine.

A would-be "Colonel" with no skill at all in geology or engineering, Drake seemed doomed to a life of disappointments. He had undertaken many activities, dreaming of heroic achievement, only to fail at each, eventually resigning himself to much more modest jobs such as steamboat night clerk, farm laborer, and many others.[3] While on leave in New Haven, Connecticut, for a painful form of arthritis, Drake became acquainted with George Bissell, a local banker who had established with a few partners a small company for extracting oil on a commercial scale on the muddy hills around the small village of Titusville, Pennsylvania. Drake's quixotic proclivity for hopeless missions, along with his forced condition as a convalescent, made him the right person to carry out the curious adventure Bissell had in mind, and he accepted operational leadership of the undertaking. To render it more appealing to new investors, Bissell and his partners dignified Drake with the title of Colonel although the only uniform he had ever worn was that of a railroad conductor.

Unexpectedly enough for a man with no record but failure, on August 28, 1859, Drake succeeded in striking oil with the new method he had devised at the suggestion of Bissell. Working with a small team of local workers, he had erected a wooden tower housing a large steam-driven wheel around which was coiled a cable with an iron bit attached at one end. The wheel rotated, raising the cable and its armament by pulley, and then letting it fall to the ground, thereby excavating a hole. Used for drilling salt domes, this technique had already been tried for oil exploration in Azerbaijan in 1847, but Drake added to it something of his own that proved decisive. He drove a pipe down the hole and ordered his men to drill inside it, so that water and loose material from the sides of the hole did not impede the iron bit from going farther.[4] Thus the

Colonel established the drilling prototype for the modern oil industry, which was eventually improved in Texas in 1901, thanks to the adoption of rotary drilling.[5]

He also introduced one of the most durable landmarks of the oil world, by casually resorting to Pennsylvania's forty-two-gallon (around 159 liters) wooden barrel to gather and transport crude: mainly used in the whiskey business, the barrel would become the fundamental measure of production and consumption still in use today in the oil market. The Drake venture's initial production was thirty-five barrels per day, sold at the staggering price of forty dollars each, the equivalent of around six to seven hundred dollars today.

Colonel Drake's epoch-making experiment is considered the birth date of the oil industry. Rumors and articles about its success aroused a dreamlike infatuation with this substitute for whale oil that promised to be an elixir of prosperity—"black gold," as it was dubbed by newspapers and popular songs. All of a sudden, the fields of Western Pennsylvania were invaded by thousands of amateur petroleum seekers—nicknamed *wildcatters**—along with transporters, refiners, traders, dealers, bankers, speculators, and the ever-present swindlers. In 1861, the first oil refinery came onstream, and the first cargo of oil exported from the United States sailed for London from Philadelphia, with the oil loaded in barrels. In 1865, the first successful pipeline was completed, with a capacity of 800 barrels per day and a length of five miles.[6] Thus began what could be called the "Black Gold Rush." Oil production soared, with kerosene making its way onto the American market and soon to Europe as well. But the dawning industry soon became a potential nightmare for many of its irrational pioneers.

Not only were discoveries of new sources of crude erratic and unpredictable, but once a discovery was made, the novice producers' ignorance of the elusive features of oil deposits, coupled with a legal framework that gave owners full rights to the minerals beneath the surface of their land encouraged a foolish overexploitation of the new fields. As a result, recurring gluts flooded the market and pushed oil prices down, bankrupting many operators who had spent all their savings, and borrowings, in their quest for fortune. Wild fluctuations in the price of oil became a

---

*The expression "wildcatters" originated from the fact that the first oil wells were drilled in isolated and hostile places where the drillers could hear the cries of wildcats. Today the term "wildcat" is widely used to indicate an exploration well.

common feature of the business. In 1860, the oil price precipitated to ten cents, then in 1861 it rebounded to ten dollars; in 1862 the price fluctuated between 10 cents and $2.25 per barrel, averaging $1.5 per barrel. Eventually, the average price of an oil barrel at the wellhead was $3.5 in 1863, $8 in 1864, $4 in 1866, $2.8 in 1867, $5.8 in 1869, $4.2 in 1871, and less than $2 in 1873.[7] The arithmetic average, however, hides dramatic ups and downs within each single year that gave the U.S. oil market a rollercoaster shape during its formative years. Paradoxically, for long periods of time the cost of the wooden barrel itself—which could fluctuate between $2.50 and $3.50—far exceeded the value of its contents.

As for Colonel Drake, he escaped neither the whiplash undulations of the inscrutable fledging industry nor the scornful destiny of repeated disappointment that had pursued him throughout his life. In 1861, a well he drilled burst into flames, destroying all the machinery and infrastructure of his company. With his savings, the Colonel threw himself into new businesses, such as oil and stock trading, only to meet with more failure. Drake ended his life poor, kept from falling into abject misery only by the help of pitying friends and, eventually, by a small pension from the State of Pennsylvania.

It was in this landscape that John D. Rockefeller (1839–1937) emerged as one of the most genial and merciless fathers not only of the oil industry but of modern industry as a whole.[8]

A bookkeeper by training and then a trader in various goods who entered the oil refining business by chance in Ohio in 1863, Rockefeller realized within a few years that the burgeoning oil market was doomed to permanent chaos if left to the blind appetites of hundreds of improvised fortune seekers. The calamitous results produced by foolish and chaotic competition reinforced Rockefeller's innate distrust of the supposed virtues of the free market, particularly its thaumaturgic capability for self-adjustment.

His world was not Adam Smith's. In Smith's world, each person contributed to the overall progress of society by embarking on and competing in economic activities, while the steady working of an invisible hand corrected all imbalances. But as soon as Rockefeller shifted his gaze from the theoretical framework of the British father of modern economics, he saw only the world as it was: a brutal blind struggle fueled by rapacity and greed. To his mind, there was no "invisible hand" at work behind this world, which—in the case of oil—was moved by irrational people, whose addiction to building castles in the air brought disaster on themselves and on the whole oil business. It was difficult not to agree with such a view. In

1869, for instance, refining capacity was three times higher than crude production, and 90 percent of refiners worked at a structural loss.[9] As Rockefeller himself once observed, "[O]ftentimes, the most difficult competition comes, not from the strong, the intelligent, the conservative competitor, but from the man who is holding on by the eyelids and is ignorant of his costs, and anyway he's got to keep running or bust."[10] According to his line of reasoning, open competition was by no means the best solution to the woes of the budding American industry; on the contrary, it was its main evil.

A look beyond the worrisome situation of the oil sector could not but confirm Rockefeller's pessimism. Indeed, the entire American economy seemed prey to a demon that blew upon the fire of irrational speculation, industrial chaos, and unethical behavior, leaving no safe haven for anyone. It was the "Gilded Age" described by Mark Twain, a frantic and highly risky laboratory where upcoming entrepreneurs, stubborn pioneers, charlatans, and "robber barons" mixed in a sea of sweeping corruption, spectacular swindles, and steady distortion of market rules. In 1869, for instance, an amazing wave of speculation on gold provoked a dramatic crash of the Stock Exchange—the second in six years— triggering a huge chain reaction of bankruptcies and opening to the path to a long depression. Watching this world with profound distaste, the devout Baptist Rockefeller poured real religious fervor into his dealings with what he considered the evil of his times.

Surprisingly gifted with figures and quick calculation and a master at penetrating the intricacies of business and rationalizing them in an orderly framework, Rockefeller began to conceive of a great rational architecture that could be superimposed on the oil industry to put an end to its boom-and-bust cycles and their deadly consequences. His final solution was at once simple, grandiose, and awful: to suppress competition altogether.

Rockefeller's design aimed at taking over all of the so-called downstream structures of the oil market, such as refineries, transportation routes, pipelines, ships, and so on, which he saw as the manageable bottlenecks between producers and consumers. Conversely, he always considered too erratic and thus unmanageable the control of oil production, which he left to adventurous wildcatters. Never keen on improvisation or gambling, he started by laying the foundations of his own war machine, establishing in 1870 a new corporation, Standard Oil. He then embarked on a comprehensive plan to cure the embolism that would have otherwise destroyed the vital circulation of the oil business. Accordingly, he first moved to consolidate the entire oil refining business in Standard's home town, Cleveland, Ohio, then one of the main producing centers of the United States. Between February and March 1872, he bought twenty-two out of twenty-six refining companies in what would come to be known as the "Cleveland Massacre." Eventually, in a Rossinian crescendo he launched an impressive nationwide acquisition campaign, bringing nearly all American oil refining and service companies under his control.

Most of those all he won over by persuasion, sparing no effort to appeal to his rivals' common interest in avoiding self-destructive competition. Those who accepted were well rewarded. Rockefeller offered them positions as shareholders and even top managers at Standard Oil, options that would have made them millionaires as well as optimal long-term allies of Rockefeller himself. In other cases, he concluded running arrangements with independent refiners, guaranteeing them a certain profit if they accepted a ceiling on their output, and acting as a "swing producer"—i.e., curtailing Standard Oil's own production to maintain an adequate level of prices—in periods of overproduction. However, those refiners and traders who stood against him were relentlessly squeezed out of the market and saw their hopes turn to ashes.

Never was such an ambitious design so perfectly and cynically realized. But Rockefeller did not make it all on his own. The key architect of the most audacious and controversial Standard Oil's moves was Henry Flagler, Rockefeller's most brilliant and unscrupulous partner.

The man who in his later years developed Florida and transformed it into the "American Riviera," founding Miami and Palm Beach, Flagler was as buoyant and aggressive as Rockefeller was taciturn and patient. A risk taker by nature, he did not hesitate to display his life motto written on a small plaque on his desk—"Do unto others as they would to you—and do it first."[11] It was Flagler who probably suggested and eventually negotiated the decisive deals that permitted Standard Oil to destroy competition, notably secret agreements with the main American railroads to obtain large discounts ("rebates") on oil transport fees in return for guaranteeing railroads large volumes of business transporting petroleum.

With pipelines still in their infancy, railroads were critical to the transport of any product—oil included. Thus Standard's deals with them turned a major ingredient of the company's rapid success, as well as the ghost that later persecuted Rockefeller and finally led to the destruction of his empire. Thanks to those deals, on average Standard Oil was granted a 20–30 percent discount with respect to its competitors; adding insult to injury, Rockefeller and his partners even received a relevant fee—about 25 cents for every dollar—for every barrel the railroad transported for shippers other than Standard Oil, thereby securing an astonishing advantage to the latter.[12]

To escape the mortal embrace of Rockefeller's monster-creation, its rivals had to embark on the daunting undertaking of building up the first long-distance pipeline ever. At that time, such a project was considered a foolish hazard because no technical guidance or feasibility study existed proving it was possible to transport oil on long distances. Yet Rockefeller's rival effectively succeeded in carrying out a 110-mile-long pipeline connecting Western Pennsylvania's oilfields with the Pennsylvania and Reading Railroad. Named Tidewater Pipeline, the project was completed in 1879, and in the same year oil began flowing though that major technological achievement, something "comparable to the Brooklyn Bridge four years later"—in the words of Daniel Yergin.[13] It took a few years, however, before Standard Oil jumped into the new technological frontier and assumed control of brand-new long-distance pipelines.

When in the early 1880s Rockefeller finally implemented his monopolistic plan, he controlled 90 percent of U.S. refineries and pipelines, owned the vast majority of tank cars used for both road and rail transport, and controlled the entire production of high-grade railroad lubricants, along with the largest tanker fleet for exporting oil worldwide; all this at a time when the United States accounted for 85 percent of world crude oil production and refining. Moreover, in an age before real-time information transmission, a vast network of Standard Oil agents monitored every corner of the country, carefully tracking all retail prices, all kerosene sales, and the behavior of all local competitors, as well as any hint of new oil discoveries that could affect the value of crude. This information soon reached Standard Oil headquarters in New York and the executive committee of the Group, which made all strategic decisions related to each specific issue. So, for instance, if some intrepid competitor tried to lower its kerosene price in one state, Rockefeller's men ordered its local company to go even lower, while at the same time ordering a price increase in another state to compensate for the loss. Of course, small independent rivals of Standard Oil survived, particularly if they presented no possible threat to the absolute predominance of the former. But on the whole, the multi-tentacled war machine created by Rockefeller could wipe anyone off the market, anytime, at his pleasure. It was the founder's dream come true, the perfect instrument for suppressing hated competition and regulating the steady growth of the oil market.

Rockefeller's talent in fulfilling this dream proved unique. He combined great vision and sophisticated thinking with an obsessive drive for mastering details—numbers—as well as controlling costs and efficiency. He was the first modern industrialist to apply central planning as an essential tool for putting a strategy into action and carefully controlling its implementation. The only problem with Rockefeller's huge empire was its fragmentation, which was not an easy problem to resolve.

In the late 1880s the United States still had no law allowing for federal incorporation, a situation that made "every corporation created by a state foreign to every other state," in Rockefeller's judgment.[14] Simply put, a company could not hold stock in another state's company, which officially prevented the oil tycoon from gaining formal and centralized control of his vast holdings. But he was not a man to be stopped by laws he did not deem right. So he asked his chief legal adviser to find a solution to the problem which, without technically violating existing rules, simply circumvented them. The response came in the form of a device that would soon become a distinguishing feature of the late nineteenth century: the *Trust*.

To take advantage of this legal loophole, Rockefeller and his associates had to establish one or more companies in each state where Standard Oil had industrial activities. Eventually, the major shareholders of those companies had to transfer their shares to a Board of Trustees based in New York; the latter calculated the value of each individual share package against the overall value of the Group's companies' shares, and then assigned each shareholder a proportional quantity of trustee certificates. Through this intricate mechanism, Rockefeller ended up holding the controlling stake of Standard Trust, with a 27 percent quota of its certificates; with the shares held by his brother William and other members of his family, his control rose above 40 percent. An informal executive committee presided over by John D. Rockefeller himself ruled the whole system, a sort of shadow board whose legal and effective powers were not written down in any official document or bylaw. Nonetheless, it was the quasi-supernatural presence of Rockefeller that oversaw the silent management of all of the provinces of the empire, the coordination and integration of their activities, the strategic goals and step-by-step actions of the whole Group.

Established in 1882, the Standard Oil Trust remained top secret from the American public and legislators until 1889. By the end of the century, many more trusts had been established by other protagonists of

American industry in every sector of the economy, representing the most striking sign of the shift of the United States from an agrarian, small-business society to an industrial power. Meanwhile, the use of kerosene was spreading worldwide—probably the U.S.-associated product with the greatest influence on the daily living habits of a large part of the world's population; in the late 1870s and all through the 1880s, kerosene was "the fourth-largest U.S. export in value, and the first among manufactured goods."[15] This triumph of the rising industrial strength of the United States bore the name John D. Rockefeller, who by then had already risen to the status of the richest man in the world.

But his giant creation, which was admired as well as feared and hated, would not remain unchallenged forever. Indeed, the attack on Rockefeller's dominance of the new oil business came from different fronts, and almost at the same time.

While the United States had been virtually the world's only source of crude and refined products for more than twenty years after Drake's lucky strike, competition emerged in Russia by the mid-1880s, posing a major challenge to Standard Trust's overwhelming grip over international markets.

Russian production was concentrated around Baku, in Azerbaijan, and spearheaded by Ludwig and Robert Nobel, brothers of Alfred Nobel, the inventor of dynamite. While Robert was the one who brought the family into the oil business in 1873, it was Ludwig who developed and built up the business, combining the talent of a creative genius and the obsessive attention to detail of a modern manager.[16] He was the first oilman to employ a professional geologist and to improve refining to produce cleaner kerosene. He also was the first to design and commission a tanker to ship oil through the Caspian Sea without the need to first store it in barrels. Under Nobel's leadership, tsarist Russia became the world's second oil producer, attracting new investors and operators in what became a Caucasus version of Western Pennsylvania.

The French branch of the Rothschild family was among the most prominent of the new investors. They entered the Russian oil business by financing the construction of a railroad to transport kerosene from Baku to the Black Sea port of Batum in Georgia, opening a route for Russian oil products to reach world markets. The railroad was completed in 1883. At the same time, the Rothschilds acquired production assets and refineries in Baku, and their company (Caspian and Black Sea Petroleum Company, or Bnito) rapidly moved into second place in the Russian oil market, behind the Nobel family operation.[17]

At the beginning of the 1880s, Russian exports of kerosene began to corrode Standard Oil's control of the European markets. Rockefeller's men reacted by launching an aggressive campaign of price reductions, similar to what Standard had done in the United States to force its competitors out of business.[18] But the Rothschilds did not surrender to the apparently unbeatable might of the Rockefellers. On the contrary, they opted to mount their own offensive by extending their reach into Asia, where Standard's dominance was also overwhelming. The key player in this ambitious adventure was a man destined to enter the Pantheon of the oil industry's fathers: Marcus Samuel, an Englishman of Jewish extraction who later founded Shell.

The son of a merchant who had built a business selling shell boxes, which were very popular in Victorian Great Britain, Samuel expanded his father's operation into an export-import concern with a solid network of buyers and suppliers in the Far East.[19] Leveraging his commercial connections, Samuel took the Rothschilds' Asian ambitions to the extreme by conceiving a world-scale attack against Standard Oil. At the core of his strategic vision was the construction of a brand-new class of oil tanker ships with sophisticated engineering and safety equipment that enabled them to pass through the Suez Canal. This shortened transport routes and slashed costs relative to those of Standard, whose more traditional vessels had to sail around the Cape of Good Hope, at the southern tip of Africa, on their way to Asia.

Despite initial reservations by the Rothschilds and vicious attacks by Standard,[20] Samuel finally succeeded in fulfilling his vision: in 1892, the first oil tanker he had designed sailed from Batum, heading for the Far East via the Suez Canal. Eight more ships of the same type came along in the following two years. Concurrently, Marcus Samuel and his brother Samuel masterminded the construction of onshore terminals and storage tanks in key Asian ports, preparing their target markets to receive increasing quantities of kerosene.

It was an all-out war with Standard Oil, pitting the two companies in self-destructive commercial practices and prompting both to over-invest in infrastructure in an effort to secure a larger market share.

In the process, Samuel emerged as a giant of the industry, a status that was publicly recognized when King Edward made him a Lord. In 1897, at the apex of his success, he reorganized his oil business into a new joint stock company, which—in tribute to his father's original business—he named Shell Transport and Trading Company, or simply Shell.

The combined efforts of the Nobels, Rothschilds, and Samuels spurred Russian oil production and allowed it to briefly outpace that of the United States. In 1900, global oil production had reached nearly 430,000 barrels per day (bpd), with Russia providing around 200,000 bpd and the United States delivering around 165,000 bpd. Five years later, however, the United States had dramatically jumped ahead of Russia, reaching 370,000 bpd, more than twice Caucasian production.[21] (One century later—in 2000—those numbers would appear almost negligible in a world producing and consuming more than 75 million barrels per day!)

Before the nineteenth century expired, another intruder appeared on the once untouched Standard Oil's turf—the Royal Dutch Company. Incorporated in 1890, the company had discovered oil in East Sumatra (then part of the Dutch East Indies, and now part of Indonesia) which it put onstream in 1892. Its dramatic success made the area the third pole of world's oil production by the end of the century, although on a much smaller scale than that of America and Russia.[22] Eventually, Shell joined Royal Dutch in the region, having obtained a concession in Borneo, where it struck oil in 1897.

Now the Asian market was overcrowded. Proximity to consumers enabled local producers to challenge overseas exporters like Standard by offering lower prices, and failure marked repeated attempts at a truce among the participants in the international oil trade. Each sought to maintain its independence, while Rockefeller's men unsuccessfully urged the Rothschilds, the Nobels, the Samuels, and Royal Dutch to become part of Standard Oil, following the model they had developed in the United States face to their domestic competitors.

As this scenario played itself out, Samuel's Shell was the weakest player. It was short of oil of its own, and its deal with the Rothschilds for marketing their Russian petroleum expired in 1900. Shell had invested heavily in refining plants, oil-tanker ships, storage facilities, and pipelines, all of which were increasingly underutilized. The result was simple: Shell was running out of cash because it had little oil to market.

Having refused to negotiate a merger with Standard Oil, Samuel turned to the chief executive of Royal Dutch, Henri Deterding. He hoped to find a formula of association that would preserve some form of autonomy for Shell, and recognize his own role as the leader of the hypothetical combination, something that had proved to be impossible with Standard. But the still young Deterding revealed himself as a tough and unbending leader, characteristics that would eventually make him a dark architect of the oil industry, and he never granted Samuel what he so desperately sought, notably equal treatment. Instead of a 50-50 share in the projected merger, the Dutchman insisted on a 60-40 formula in favor of his company, and told Samuel he could take it or leave it. His pride severely wounded, Samuel had no choice but to surrender.

In 1907 the two companies merged, giving birth to the Royal Dutch Shell Group. In fact, no legal entity bearing that name ever existed, because the merger was somewhat unusual. Royal Dutch and Shell both maintained their formal, independent status and separate listings, while their overall assets were allocated to two sub-holdings. One embraced all production and refining assets and was based in The Hague, while the other was responsible for transport and storage facilities, and had its headquarters in London. Each holding company owned a participation in both sub-holdings, following the 60-40 formula that Deterding had imposed to insure Dutch predominance. Two boards continued to exist, with senior managers sitting on both. The overall activities of this strange construction[23] were coordinated by a "Committee of Managing Directors" that had no legal standing, but enjoyed the support of both boards.

Deterding's autocratic leadership over more than three decades made it possible for the new group to thrive and become one of the biggest oil companies of the twentieth century. Despite the Group's organizational cacophony, determined by the survival of two parallel structures with no real central corporate power, Deterding acted as a one-man band, imposing his overwhelming influence as Rockefeller had done through Standard Trust's informal "executive committee." Remarkably, the weird organization he created survived until 2005, when Royal Dutch-Shell finally decided to evolve into a single company.

Thus, at the turn of the century, Standard Oil was no longer alone in the international arena, where its position weakened day after day. Yet the most damaging attacks on its empire came from within the United States itself.

First, the grip of Rockefeller's creation on the American market relaxed because of the relentless appearance of local competitors, bred by a flurry of new oil discoveries in other parts of the country. If the Standard monopoly had been favored by the concentration of production in Western Pennsylvania, by the 1890s the center of gravity for oil production began to move southwest, prompting a new boom in exploration and production. California led the way in this redeployment of forces when vast amounts of oil were discovered on its territory. To a lesser extent, oil also began to flow from Kansas, Colorado, and other states, but it was particularly in Texas and Oklahoma that a new era of American petroleum began. In 1901, oil was struck in Spindletop, a little hill near the small city of Beaumont, Texas. It happened to be the largest discovery ever in the United States. Oil erupted with such a violence that it formed a huge overflowing fountain, capable of delivering 75,000 barrels per day—a bewildering phenomenon that introduced a new word into oil jargon, *gusher*. Unfortunately, wildcatters assaulted the new El Dorado, torturing it by drilling too many holes too fast, such that the oilfield's internal pressure was quickly exhausted. By the end of 1902, Spindletop was incapable of producing oil.

Nonetheless, this extraordinary Texas find excited the imagination of armies of oil-seekers, who made the American Southwest the new frontier of exploration campaigns and new discoveries. Even more amazing than Spindletop was the discovery of the huge Gleen Pole (1905) field near Tulsa, Oklahoma, which made the state the leader in American oil production up to 1930. Thanks to this and other successes, the United States soon reclaimed the crown of top world crude producer. At the same time, the new boom was accompanied by the establishment of new oil companies that would erode the overwhelmingly dominant position of Standard Oil: noteworthy were the cases of the Texas Oil Company, or Texaco, incorporated in 1902; Gulf Oil Company, which was officially incorporated in 1907 in Texas, under the patronage and ownership of the Mellon family; and California's Union Oil (1893), later Unocal.

More than international and domestic market erosion, however, it was the local and federal governments of the United States that hit a fatal blow to Standard Oil's once absolute power, for reasons that went back to the very roots of Rockefeller's career.

The father of the oil industry was the single most important character of an age that witnessed one of the major social and economic transformations of the United States. In a way, it was Rockefeller who was responsible for the tectonic shift that transformed Jefferson's America—based on the equation between freedom and direct ownership of the means of production and land by every single man—into a global economic power dominated by industrial concentration through the "trust" formula. Rockefeller later proudly described that shift and his own role in it:

This movement [concentration] was the origin of the whole system of economic administration. It has revolutionized the way of doing business all over the world. The time was ripe for it. It had to come, though all we saw at the moment there was the need to save ourselves from wasteful conditions. The day of combination is here to stay. Individualism has gone, never to return.[24]

It was thus inevitable that Rockefeller would become one of the favorite targets of the "trustbusters" once the counteroffensive against monopolies reached critical mass. And indeed, harsh critics did their best to make oil and monopoly synonymous in the mind of American and world public opinion. When Theodore Roosevelt became president of the United States in 1902, the antitrust movement gained its most effective and toughest representative, and John D. Rockefeller his most lethal enemy. By then, the founder of Standard had left the operating leadership of the company (in 1895), retaining only a formal position as its chairman. But Rockefeller had committed a grave mistake in never making his retirement public, partly at the request of his partners.

That secret move—which let him remain the visible symbol of Standard Oil—allowed his opponents to make him the number-one target of a harsh campaign of growing attacks. That campaign reached its climax in 1904, when a destructive portrait of Rockefeller's career was presented to the American public by journalist Ida Tarbell, the daughter of one of the many Pennsylvania oil pioneers whom Rockefeller had driven out of business. Her two-volume *History of Standard Oil*[25] turned "America's most private man into its most public and hated figure," as historian Ron Chernow later wrote.[26] Yet the public perception of his character and personal history did not reflect either the complexity of Rockefeller's personality or the importance of many of his achievements.

Though sometimes confused with his lavish and greed-driven contemporary "robber-barons," Rockefeller was wholly unlike the majority of them. As one of his biographers recalled, he had been "the best employer of his time, instituting hospitalization and retirement pensions,"[27] and was highly regarded by his own employees for his benevolence, kindness, and complete lack of arrogance. At the same time, although he achieved the status of the richest man in the world (with a personal wealth still unchallenged today, unless by Microsoft founder Bill Gates) he was disgusted by the unethical habits and obsession with luxury that characterized the new tycoons of the industrial age. He always lived modestly and far from the limelight.

To his credit, Rockefeller could also claim responsibility for the consumer benefits brought by Standard Oil, which had transferred to its clients at least part of the gains the greater efficiency of its organization had allowed: the price of kerosene in the United States (without taxes), declined from approximately forty-five cents per gallon in 1863 to about six cents by the mid-1890s.[28] Moreover, Rockefeller never speculated on the stock exchange as many of his contemporaries did—particularly, and with great frequency, the robber-barons—and he never made swindling into a business practice. Rather, his overall achievements were mostly based on elusion and encirclement of poor laws.

On the other hand, Rockefeller's role in suppressing competition and winning privileges even by resorting to illegal means, such as bribery, was beyond question.[29] So was his relentlessly imposed policy of "beggaring-my-neighbor," manipulating prices in different regions in order to break his competitors by underselling them. However, what pursued him like an implacable ghost was his original sin: the secret agreements with the railroads. Rockefeller always defended the rationale behind preferential rebates and the discounts he had got from them:

> Who can buy beef the cheapest—the housewife for her family, the steward for a club or hotel, or the commissary for the army? Who is entitled to better rebates from a railroad, those who give it 5,000 barrels a day, or those who give 500 barrels—or 50?[30]

Whatever the logical force of this reasoning, Rockefeller's critics rebuffed it with an argument that would become the key premise behind the modern regulation of utilities, notably the public service nature of railroads, which operated under state charters. This status endowed them with the right of "eminent domain," which entitled railroad owners to expropriate private property to build their routes. In return for this right, railroads had to behave as "common carriers" and provide all clients with uniform conditions to access their services. Consequently, fee discrimination was unacceptable, no matter how great the volume of business that Standard Oil could guarantee to the railroads.[31]

In any case, one of Rockefeller's most serious mistakes was to underestimate the increasing role of the press in the evolving American society, and his failure to understand its crucial importance in shaping the public's perception of reality. For years, secrecy ruled his life, and he viewed press criticism as a temporary phenomenon without lasting significance. As a result, he failed to take any steps to counter the daily attacks on his image, allowing the negatives to stand unchallenged. By the time Rockefeller understood the problem and changed his approach, it was too late, and the harsh portrait of his life and career was engraved in the public mind and that of the oil industry, surviving him by many decades.

After years of attacks, investigations, and trials, in 1911 the U.S. Supreme Court finally ruled for the dismantling of Standard Oil, calling for it to be broken up into more than thirty independent companies.

This decision was a milestone in the history of the antitrust movement and cemented the association in the mind of the global public between the oil industry and everything that was sinister and secretive in modern industrial society. As with the mythical Phoenix, from the ashes of Standard Oil would emerge a host of the eventual protagonists of the oil era—among them Exxon (Standard Oil of New Jersey), Mobil (Standard Oil of New York), Chevron (Standard Oil of California), and Amoco (Standard Oil of Indiana). At the same time, the antitrust ruling did not prevent major oil protagonists from resorting to Rockefeller's anti-competitive practices.

For many decades to follow, monopoly and oligopoly remained the sacred texts by which the oil industry guided its behavior, especially during recurring phases of overproduction—or when seeking to avoid it. Indeed, Rockefeller's curse on free competition in the oil business is active right through our day, its shadow stretching over the latest interpreter of his doctrine—OPEC.

CHAPTER 2

The Age of Gasoline
and Oil Imperialism

Just as Standard Oil was facing its day of reckoning, a profound shift was occurring in the significance of oil in modern life.

By 1900, oil was not just the main source for illumination anymore; at least 200 crude by-products had entered daily use, ranging from lubricants for industrial machinery and petroleum wax for pharmaceuticals and candles, to medicines, solvents, and fuel for stoves and internal combustion engines.

Meanwhile, artificial light had found another powerful source, which was to change the history of the twentieth century. In New York in 1882, Thomas Alva Edison made his first public presentation of his latest invention, a light bulb powered by electricity. The new device soon captured the collective global imagination, even though the need to create a large infrastructure to provide the new tool with power—from the generator to the home—initially limited its diffusion to public lighting or large industrial complexes, without endangering the role of kerosene. (In 1900, only eight million light bulbs lit the United States; in the rest of the world, the total figure was far more negligible.)[1]

While Edison's achievement set the stage for one of the most important revolutions of our time, a flurry of amateur inventors was experimenting with and perfecting the first prototypes of internal combustion engine vehicles, fuelled by diesel or gasoline. In their small assembly shops and garages, those men were paving the way for as revolutionary an invention as Edison's, and one that would dramatically upset the concept of physical distance and change oil's role in contemporary history.

Neither the internal combustion engine nor the automobile had a single inventor. Different models of motors and cars popped up contemporaneously, each somehow taking advantage of improvements and advancements introduced by others.[2] Europe was initially far ahead of the United States in nourishing the new business. It was the birthplace for what is generally credited to be the first commercial version of an internal combustion engine, patented in France in 1860 by Belgian mechanic Etienne Lenoir. Eventually, sweeping innovations were brought about by German manufacturers, starting with Nicolaus Otto's milestone four-cycle engine introduced in 1876 (which first compressed the fuel-air mixture into the working cylinder), and followed by models patented by Karl Benz, Gottlieb Daimler, and Rudolf Diesel.[3] At the same time, the 1901 Mercedes designed by Wilhelm Maybach, inventor of the carburetor and a long-time assistant to Gottlieb Daimler, for the German *Daimler Motoren Gesellschaft,* deserves credit for being the first modern motorcar in all essentials.[4] By the end of the century, dozens of prolific pioneers had delivered original prototypes of four-wheeled vehicles in the industrial countries, proving that the new transportation tool was reliable and sufficiently safe. Nonetheless, it still represented an extravagance limited to a restricted club of very wealthy people.

Merit for removing the automobile from the empyreal grasp of the elite and turning it into a product for mass consumption goes to Henry Ford and his landmark industrial achievement, the Model T (1908).

A farm boy born in a small town near Detroit in 1863, Ford took his first steps in the automotive industry as a machinist, rapidly becoming a self-taught engineer. In 1896, while working for an electric utility company, he built his first quadricycle fed by an internal combustion engine. Eventually, after several ups and downs he established the Ford Motor Company in Detroit in 1903. In the same year, at the age of forty, he set down his basic theory, envisaging his future revolution that would mark forever modern industrial organization and production methods: "The way to make automobiles is to make one automobile like another automobile, to make them all alike, to make them come from the factory just alike—just like one pin is like another pin when it comes from a pin factory." [5] In simpler words, he wanted the standardized production of a single model with the same austere equipment and just one color: black.

Launched in 1908, the four-cylinder Model T was one of the greatest industrial successes ever: between 1908 and 1914, the Ford Motor Company sold 1 million units. Riding the wave of this staggering achievement, Ford introduced the first assembly line in history, in 1914, to speed up vehicle construction. It was the ultimate step toward the age of the automobile, allowing sales to skyrocket to 2 million in 1916 and 10 million in 1924. When the final curtain fell on the Model T in 1927, Ford had sold 15,458,781 of them.[6]

These industrial achievements were underpinned by another key intuition that Ford translated into a consistent strategy: the price of a new car had to be low enough for his workers to afford to buy one. Accordingly, Ford transferred to his consumers the steady cost reductions he obtained through the evolution of his producing processes,[7] incessantly slashing Model T prices from the original $890 in 1908 (when the average American worker's pay ranged between $500–570 per year) to $550 in 1914 and $290 in 1924—then equivalent to about one quarter of the yearly wage of a worker.[8] At the same time, by 1914 he upset most of his narrow-minded industrialist colleagues by increasing his workers' daily salary to five dollars, almost doubling what they had been earning, and reduced the working day to eight hours; in 1921 he cut the work week once again, from six to five days. All of these reforms made him one of the most admired men of his time, such that even Lenin held him in high regard.[9]

At the beginning of the 1920s, Ford accounted for 50 percent of world automobile production and 60 percent of that of the United States. However, in the following years Ford was to be outstripped by another creative genius and revolutionary of modern industry, Alfred Sloan, the head of General Motors Corporation (GM).

Sloan realized that the increasingly wealthy consumers of the "Roaring Twenties" were searching for diversified products, notably different car models made in different colors and with a wide range of accessories and options. It was Sloan who established modern industry's habit of launching new car models each year, mainly through a constant restyling of existing models. And it was Sloan who in 1925 conceived of a new system of industrial organization, based on a multidivisional structure in which each division was an independent business unit, responsible for its own profits and losses, while the corporate headquarters retained responsibility for strategy, central planning, and control. This restructuring turned out to be particularly well suited to Detroit-based GM, which was born in 1910 of a consolidation of several brands, including Pontiac, Buick, Oldsmobile, and Cadillac. Empowered by direct responsibility for their business results, GM's division chiefs supplied the market with the variety of models it hungered for. It was the triumph of Sloan's vision and the demise of Ford, whose inflexible production concept did not fare well in the changing times.

In any case, Ford-GM competition was a blessing both for Detroit, which rose to its long-lasting predominance in world car production, and for the American market. On the wave of both companies' amazing successes, four out of five cars produced in the world in 1927 were made in the United States. In the same year, the country crossed the threshold of mass motorization, with one motor vehicle for every 5.3 people, or nearly 200 cars for every 1,000 people—even today a figure unmatched by most countries in the world. (At that time, in the most mechanized countries in Europe—Great Britain, France, and Germany—there was only one car for every forty-four people.)[10]

Boosted by the rapid spread of cars and other motor vehicles, in 1910 gasoline sales surpassed those of kerosene and other lighting oils in the United States. Symbolically, this development heralded the advent of oil's "age of energy."[11] However, one more transformation would change forever not only the pattern of oil consumption, but also oil's strategic role in power politics: a transformation in the *art of war*.

By the end of the nineteenth century, the new internal combustion engines had been installed on ships and large vessels as well, with consequences that awakened the interest of many military strategists. Oil offered several advantages over coal as naval fuel that could prove to be key in the case of war. First, it had a higher thermal efficiency, which enabled ships to travel faster and cover greater distances while enjoying greater self-sufficiency; for example, naphtha (an oil product) yielded for 50 percent greater mileage than an equivalent quantity of coal. Moreover, with oil a ship could be refueled while underway, whereas the loading of coal required a ship to stop in ports equipped with the necessary facilities. Finally, oil products were far simpler to store and move once on board than coal, and required less space and fewer men, considering that on coal/steam ships, a full three-quarters of the crew was generally devoted to moving coal and controlling related machines.

The revolution caused by the introduction of the internal combustion engine dramatically changed the nature of oil for nations and for mankind, and it took only a few years for "black gold" to rise to its current status as a strategic commodity, vital to the national security of the Great Powers.

The country that best grasped this new reality was Great Britain, at the urging of a young Winston Churchill. As First Lord of the Admiralty, it was Churchill who lobbied for the Royal Navy's conversion from coal to oil in 1911, ultimately winning approval for it in 1913. With that choice, London could hope to preserve its predominance on the seas, particularly against the rise of the German naval force. Yet the shift entailed also a very big problem. While the United Kingdom supplied about half of the coal traded worldwide, and held a virtual monopoly of the hard smokeless coal that had become the maritime fuel of choice,[12] it had neither domestic sources of oil, nor sources in its colonies. Once the country opted for an oil-propelled fleet, the UK's energy self-sufficiency was lost forever, and the search for stable and invulnerable oil sources became a vital necessity for the country. It fell to Churchill again to come up with the solution, notably government control of the Anglo-Persian Oil Company—the progenitor of today's BP—a private enterprise that held a very promising oil concession covering most of the territory of Persia, the future Iran.*

The first venture of its kind in a Middle Eastern country, the Persian oil saga had begun in 1901, when an Irish businessman, William Knox D'Arcy, was granted a sixty-year exclusive oil concession covering the whole extent of the Persian Empire except for five northern provinces.[13] In return for the concession, the Persian monarchy received an up-front payment of 40,000 pounds, the right to a yearly 16 percent cut of net profits, and a royalty of four gold shillings for every tonne of oil sold.[14] D'Arcy's venture was also absolved from paying any kind of tax to the Persian authorities, including income tax. This milestone agreement would become the model for all subsequent oil concessions in the Middle East for five decades. But Persia was not a safe haven for a British company at that time.

Indeed, control of the whole of Central Asia had been the prize behind the almost century-long confrontation between Great Britain and Russia, both countries considering the region key to their security and power. For the UK, Central Asia was the shell around the pearl of its Empire, India; for Russia, it was the soft belly of its own domain, the place whose "open grasslands historically served as the highways of conquest for Mongol invaders,"[15] as well as the door through which Islam could penetrate the heart of the Tsarist Empire. Known to history as the "Great Game" after Rudyard Kipling popularized the expression in the novel *Kim* (1901), that struggle was still under way when D'Arcy's undertaking took shape, making the Russians the more dangerous candidates to replace the British grip on Persian oil.

Furthermore, by 1912 Anglo-Persian/BP found itself in deep financial trouble. Oil production continued at a modest rate, while large capital expenditures were committed to complete the construction of a pipeline

---

* Persia changed its name into Iran in 1935.

network and a refinery in Abadan, on the Persian Gulf. This rapidly exhausted the company's working capital and by 1914 put it on the verge of bankruptcy.[16]

Part of the UK establishment then pushed Churchill to consider what came to seem like an inevitable resolution—the acquisition of a majority stake of the company. Already the chief advocate of a rising British oil lobby, Churchill carefully engineered the final move of a plan that would officially recognize the strategic linkage between oil, national security, and world power for the first time in history. To ease the way for his coup de théâtre, he began catechizing a still skeptical parliament in 1913, arguing:

> If we cannot get oil, we cannot get corn, we cannot get cotton and we cannot get a thousand and one commodities necessary for the preservation of the economic energies of Great Britain.[17]

In that same speech, Churchill introduced the notion that oil sources and commerce had to be directly controlled by the Admiralty, with the target of ensuring Great Britain both an ample diversification of supply and independence from any foreign company. Then in 1914 Churchill launched his ultimate assault, proposing the acquisition of a 51 percent stake in BP for 2.2 million pounds. On June 17, 1914, the British parliament approved Churchill's proposal by 254 votes to 18.[18]

As BP's majority shareholder through the Admiralty, the United Kingdom's government now had the right both to appoint two out of seven board members (including the executive chairman), and to exercise a veto over the others' decisions, particularly in case of politically sensitive issues. Apart from these limits, the company had to run its business according to the financial and industrial strategies typical for any private company, and its bylaws would shelter it from any interference from political forces. This complex architecture rendered BP an oxymoron, notably a state-controlled enterprise with a private soul and mission. But above all, it made it the paradigm of the shift that occurred in the strategic perception of oil.

World War I reinforced that perception. Naphtha, gasoline, and diesel—all petroleum products—emerged as the leading fuels for moving people, armies, airplanes, and naval fleets throughout the world. It soon became clear that both the wealth of modern economies and mechanized war based on mass mobilization could be sustained only with access to ample sources of oil. Thus, after the war the quest for oil became an international phenomenon, also spurred by another powerful factor: the deceptive specter of crude oil scarcity.

The alarm was sounded during World War I in the United States. At that time the country that had been the birthplace of the modern oil industry was still by far the world's largest oil producer, accounting for almost 70 percent of global output, or more than one million barrels in 1919.[19] Yet the findings of a Senate inquiry begun in 1916 suddenly shook its sense of certainty about the future of oil.

In its final report, the Senate stated that most of the American oilfields had already passed their peak production, were in a phase of rapid depletion and were likely to be exhausted within twenty-five years according to the most optimistic view.[20] It was not the first time such alarms had been sounded, but the authority of the institution making the charge gave it rock-solid credibility and assured its worldwide dissemination. A flurry of additional gloomy predictions followed, and in 1919 even the head of the prestigious U.S. Geological Survey delivered his no-exit verdict: American oil would run out in nine years![21]

Everything conspired against optimism. Worldwide consumption of oil products had risen by 50 percent between 1914 and 1918, just when war damages and the Bolshevik Revolution were constraining oil supply from Russia, which was finally curtailed by the revolutionary government's 1919 decision to nationalize the whole industry. America's mounting hysteria about oil was further exacerbated by a remarkable price increase: between 1918 and 1920, it climbed on average from less than two dollars to three dollars per barrel, underscoring the notion that the country's productivity was under stress.

As concern with the supposed end of oil grew, President Calvin Coolidge established the Federal Oil Conservation Board in 1924, a decision he explicitly justified by linking future oil needs with national security, as Churchill had done ten years earlier:

> Developing aircraft indicate that our national defense must be supplemented, if not dominated, by aviation. It is even probable that the supremacy of nations may be determined by the possession of available petroleum and its products.[22]

These concerns had already provoked an early diplomatic reaction by the United States. In 1920, Secretary of State Charles Evans Hughes and Secretary of Commerce (and future president) Herbert Hoover recommended helping American petroleum companies obtain oil concessions abroad before Great Britain and other European powers took most of them.[23]

In effect, by then London was searching to grasp another important potential oil base in the Middle East, Mesopotamia—the future Iraq—as a part of its energy imperial strategy.

Already in 1918, British War Cabinet Secretary Sir Maurice Hankey had written Foreign Secretary Lord Balfour that

> Oil in the next war will occupy the place of coal in the present war.... The only big potential supply that we can get under British control is the Persian and Mesopotamian supply.... The control over these oil supplies becomes a first-class British war aim.[24]

Because the United States was Great Britain's main supplier of oil, fears of American subsoil insolvency reinforced this drive for resource imperialism and added a key ingredient to the British "Arab" policy devised during World War I.

In 1915–1916, London had stimulated the rise of nationalism among Arab populations and convinced them to revolt against the Ottoman Empire, then on the side of Great Britain's enemies in the war. The instrument of that policy was Hussein, the emir of Hejaz, a small kingdom in the Arabian Peninsula. The British felt that Hussein possessed strong personal leadership and sufficient forces to stir up and then direct the Arab upheaval; most importantly, he ruled the Muslim holy towns of Mecca and Medina, and his Hashemite dynasty claimed to be directly descended from the Prophet Mohammed. In return for Hussein's support, London promised him Great Britain's backing in the establishment of a great Arab nation under his own rule, within the boundaries of the Fertile Crescent (MacMahon-Hussein agreement, 1915–1916).

In 1916, Hussein and his sons Faisal and Abdullah launched the Great Arab Revolt, advised by the eventual apologist of their undertakings, the British agent Thomas Lawrence, who passed into history as Lawrence of Arabia. The Hashemites had considerable success and played a significant role in helping the British war effort on the Eastern front. However, at the end of the war Hussein's men abruptly found out that Great Britain's promise was nothing more than a cynical political *escamotage*. Already in 1916, Great Britain and France had secretly agreed (Sykes-Picot agreement) to a partitioning of the territories London had just granted to Hussein as a reward. According to the text of the entente,

Mesopotamia and Palestine would go to the British, while France would get Syria and Lebanon.

The consolidation of this scheme was carried out at the San Remo Conference (1920), where the partition plan was openly blessed also by the newly formed League of Nations. In San Remo, the British oil lobby carved an important clause into the agreement that stipulated that "any company developing oil in Mesopotamia should be under permanent British control."[25] In return for this generous grant, France was given a stake in the Turkish Petroleum Co. (TPC), a company established in 1912 with the mission to obtain an oil concession for Mesopotamia. Dominated by BP (50 percent) and Shell (22.5 percent), TPC was originally participated in by Deutsche Bank (22.5 percent), whose presence mirrored a more general German strategy toward the Ottoman Empire.[26] After the war, German interests in the company were confiscated as a war reparation and given to the French. Five percent of the TPC's stock always remained in the hands of the man who had first outlined the oil potential of Mesopotamia at the end of nineteenth century, and eventually masterminded the compromise between British and German competitors that made possible the birth of TPC; his name was Calouste Gulbenkian, and he may be considered the founding father of the Iraqi oil industry, as well as a major figure in the shaping of the Middle East's petroleum policy.

Now the stage was set for Great Britain to secure its "first-class aim" through direct control of Mesopotamia, but unexpectedly things turned out to be much more difficult than envisaged.

As early as 1920, a revolt started in the southern Iraqi cities of Nasiriya and Falluja as a reaction to the San Remo Agreement. Rapidly and unexpectedly, the revolt spread throughout the whole of Mesopotamia, pushing British forces to launch a harsh and even inhuman repression involving the aerial bombing of cities and villages. The final cost was a staggering 10,000 Iraqi and 400 British casualties.[27] London realized that something more acceptable than overt Arab servitude had to be devised; what was needed was some ornamental façade to disguise its rule at the cheapest cost possible. In fact, the traditional colonial model was revealing its financial unsustainability, putting the British Treasury under considerable stress. In 1920 alone, expenditures on the administration of Mesopotamia reached 32 million pounds; the following year, even though slashed to 24 million pounds, they came to "more than the total of the UK health budget."[28]

At the urging of its newly-appointed Secretary of Colonies, Winston Churchill, the United Kingdom decided to resolve the dilemma of Mesopotamia through a strategy of indirect government. Key to this plan was the establishment of a state built upon the three loose provinces of Kirkuk, Baghdad, and Bashra, ruled by a British-chosen Arab monarch and based on a fundamental law providing for an elective assembly. In order to cement that complex architecture, a British-Iraqi alliance treaty would guarantee London's supervision of any sensitive matter concerning the new state.[29]

Accordingly, in 1921 the British masterminded a referendum in Mesopotamia by resorting to bribe the local tribal chiefs in order to crown Hussein's son Faisal King of Iraq (even if the new name of the state was officially adopted only in 1929). By 1925, they carried out the other points of their agenda, establishing a general assembly resembling a parliament and signing the alliance treaty. To achieve its targets, London did not hesitate to make use of threats and emergency measures against the new hesitating king, who unexpectedly tried to withstand what he considered a complete surrender to the British will. Particularly, it was hard for him to accept a treaty that imposed upon Iraq both an ill-disguised British rule and even its costs. Indeed, while the alliance treaty provided for the King to be assisted and advised by British High Commissioner in Iraq "on all matters affecting the international and financial obligations and interests of His Britannic Majesty,"[30] it also required Iraq to pay half the costs of British engagement in the country. But Faisal could do nothing against the British menace to deprive him of his throne and Iraq of a part of its territory (the region of Kirkuk, claimed by the Turkish), and so Great Britain had its new state built and organized according to its original plan.

Even before the assembly was elected, the treaty ratified, and the constitution promulgated, Great Britain made sure to secure an oil concession in Iraq through the British-controlled TPC. Article five of the concession agreement marked the final accomplishment of such policy, stating that TPC had to remain a British company registered in Great Britain, and its chairman (as chief executive) had to be a British subject.[31]

Thus modern Iraq was born out of the dictates of a foreign government, with oil playing a central role. Yet the imperial drive that led the British oil lobby to shape the destiny of a nation was not unrivaled and actually provoked a major clash with the United States.

Prior to World War I, the American government had consistently refused to become involved in the overseas operations of American oil companies, essentially Standard Oil.[32] This position was abandoned quickly in response to the grim prospect of domestic oil depletion that spread from 1916 on and prompted the country to search for alternative sources worldwide. But when the U.S. companies tried to bid for new concessions abroad, they came up against walls erected by the colonial powers. As a consequence, Hughes's and Hoover's warnings about the risk that American oil interests in the world would remain empty-handed developed into a proactive official U.S. oil diplomacy. Washington proclaimed the "Open Door" doctrine—i.e., free access to all countries of the world for every company, whatever its nationality—and quickly clashed with London over the destiny of Mesopotamia's still-to-be-discovered resources.

After several years of confrontation, in 1928 an agreement was finally reached. BP, Shell, Total, and the American predecessors of Exxon-Mobil combined, became equal partners of Turkish Petroleum Company (later renamed Iraq Petroleum Company), each one with a 23.75 percent stake in the company.[33] The father of the Iraqi oil saga, Calouste Gulbenkian, succeeded in retaining its 5 percent stake, and also convinced his partners to include in the venture's bylaw a clause committing each of them not to initiate without the others' consent any individual oil operation in countries of a large portion of the Middle East, spanning from current Turkey to Saudi Arabia and Bahrain (but excluding Kuwait, Iran, and Egypt). For making clear what the concerned area was, Gulbenkian himself took a map and delimited the area in red ink, thereby leaving to history his anticompetitive device as the "Red Line Agreement."[34]

The case of Iraq marked the beginning of a rush that would place seven major Western oil companies—later to be known as the "Seven Sisters"*—in control of all Middle Eastern petroleum by the early 1930s. By that time, however, some of those companies had already secured the bulk of oil concessions in Latin America, where an oil boom had taken place in the first three decades of the new century, centered in Mexico and Venezuela.

The Mexican oil history began as an appendix to the Texas one, thanks to a world-renowned British engineer and occasional oilman,

---

*Using as a reference their modern name, they were (original name between parentheses): Exxon (Standard Oil of New Jersey), Royal Dutch Shell, BP (Anglo-Persian Oil Company), Mobil (Standard Oil New York), Chevron (Standard Oil Company of California), Texaco (Texas Oil Corporation), Gulf Corporation.

Weetman Pearson. Pearson had conceived and built up the Panama Canal and other engineering marvels of his age, so that Mexico's dictator Porfirio Diaz had asked for his services in order to study and eventually carry out some great undertaking in his own country. But as soon as he entered Mexico from Texas, Pearson was hit by the inhabitants' tales of local oil seepages, exploited since ancient times in plenty of applications. Influenced by the Texan oil euphoria, the British engineer then decided to establish an oil company in 1901, the Mexican Eagle, and plunged in his own quest for "black gold." The concession terms he was granted followed the scheme set up by D'Arcy in Persia and would set the model for all eventual contracts in Mexico: a modest royalty for every tonne of oil produced, a tax on surface occupation, no income tax, and the direct ownership of subsurface findings.

During the first years of his new venture, Pearson went through the same difficulties and financial distress that were quite forcing William Knox D'Arcy to abandon his Persian oil dream in the same period. However, the audacious Pearson was finally helped by the talent of a still young Everett DeGolyer, the later father of modern seismic prospecting. Hired by Pearson as a last chance bet, in 1910 DeGolyer struck the huge oilfield Potrero del Llano 4 in the Tampico area. It was probably one of the largest findings worldwide till then, which rapidly made Mexico the epicenter of a new and successful oil rush. On the eve of World War I, the country was the third largest oil producer in the world—after the United States and Russia—with production topping 100,000 barrels per day. In 1921, it even surpassed chaotic Russia, achieving a peak production of nearly 530,000 bpd. By that time Pearson—then known as Lord Cowdray, having been knighted in 1917—had already left the country and sold Mexican Oil to Shell,[35] fearing the consequences of the overthrow of his protector Porfirio Diaz in 1913 by a revolutionary government. Once again, his intuition would prove correct.

In 1917, a new parliament amended Mexico's constitution to extend government control over oil resources. Afterward, a heated clash over ownership rights of underground mineral resources and taxes to be paid to the central government erupted between the new Mexican leaders and foreign oil companies.[36] With both parties unwilling to accept any compromise, the standoff grew in harsh acrimony and led international companies to shift their sights to another appealing Latin American country, Venezuela, which had the "political advantage" of being ruled by a cruel and corrupt dictator, Juan Vicente Gomez.

Actually, the entire American subcontinent was rife with widespread corruption and satraps, and as one of the Pearson's agents had pointed out clearly in a letter to his chief:

> I have no doubt that you realise that the sort of concession that we are trying to get does not appeal to any government and that it is very difficult to obtain it in a country enjoying a real parliamentary system; it is to my mind only easy in countries of a one man government like Mexico under President Diaz, Venezuela under Gomez or Colombia under Reyes.[37]

Given this context, it was not by chance that oil nationalism first ignited in Latin America, leading to the nationalization of oil resources first in Argentina (1922) and eventually in other countries. Yet in spite of its similarities to other Latin American countries, Venezuela was unique.

As Daniel Yergin wrote, Gomez governed Venezuela as "his own private hacienda," repressing all dissent with terror and brutality and enriching his cronies. His family members sat in key governmental positions, while his brother was his own deputy.[38] Ruling over a barely formed state with no well-established institutions, Gomez could shape as he wished the life of his country and the nascent oil business. In 1912, he forced the country's supreme court to revoke landowners' rights to their subsoil resources, which were redefined as the property of the government—i.e., of Gomez himself. Then he started a dance of oil concessions by selling subsoil resources rights to foreign companies as well as to his cronies and relatives, so that the latter could resell them at a profit. Royal Dutch-Shell won the lion's share of Venezuelan oil and also discovered the first commercial oilfield ever in the country (1914); on the contrary, American companies temporarily backed away from their original interest in the region, and in many cases sold their concessions to Shell itself.

Throughout this first phase of oil development, Venezuela had no oil legislation, and concessions were granted through private negotiations between companies and Gomez's men. The first oil law was introduced in 1920, and was soon rescinded because it did not meet companies' expectations. Finally, in 1922 Gomez gave the green light to another law that had been written by foreign companies' lawyers.[39] The new rules were inspired by the Persian D'Arcy model, but they dramatically improved economic conditions (in terms of royalties and taxes) for private bidders, such that in 1930 they were defined as the "best in the world for companies" in a confidential memo to these companies written by one of Gomez's ministers.[40]

The new legislation lured a second wave of prospectors, bringing the total number of companies involved in Venezuelan oil production to more than thirty by the end of the 1920s, now including Exxon, Gulf, Mobil, Texaco, and Amoco (Standard of Indiana), in addition to other American firms. The effect was dramatic. Venezuela's oil production rose from a modest 19,000 barrels per day in 1919, to 523,000 bpd in 1929 and 779,000 in 1939, making the country the third largest producer in the world after the United States and the Soviet Union, as well as the main oil exporter worldwide.[41] Production growth coincided with its rapid concentration in the hands of just three companies: Exxon, Shell, and Gulf, holding respectively 52, 40, and 8 percent of Venezuelan oil production on the eve of World War II.[42]

This bonanza had its dark side as well. The rapid development of the Venezuelan oil sector discouraged participation in traditional activities such as agriculture and small businesses, and provoked an inflationary spiral that impoverished all who were not benefiting from the oil boom. In the late 1960s, this phenomenon would be given its very own name—*Dutch Disease*—after the discovery of natural gas in the Netherlands brought a sudden infusion of wealth that was concentrated in relatively few hands, driving up all domestic prices and eroding the purchasing power of all those outside the natural gas–based economy. Well before it entered the lexicon, *Dutch Disease* became the common destiny not only of Venezuela and most of the Middle Eastern countries blessed with oil, but also of the majority of countries that polarized their economies by depending on revenue derived from a single resource.[43]

With Persia, Mesopotamia, Mexico, Venezuela, and the East Indies under tight control, all the world's main oil-producing areas outside the United States and Soviet Union were now in the hands of a restricted club of Western companies. There was only one major protagonist of the twentieth-century oil drama that was still missing, the Arabian Peninsula. But it too was about to come onstage.

CHAPTER 3

The Carve-up of Arabia's Oil

At the dawn of the first global struggles for oil, the Arabian Peninsula did not attract much interest and did not trigger any acts of resource imperialism. Odd as it may seem today, there was a simple reason for the lack of interest in the area: no serious person at that time believed the region contained a single drop of petroleum.

It was BP's men who most influenced this conventional wisdom. In 1923, for instance, the General Manager of BP, Sir Arnold Wilson, expressed his judgment about Saudi Arabia in these terms:

> I personally cannot believe that oil will be found in his reign [that of King ibn-Saud, then coinciding with the most oil-rich region in Saudi Arabia]. As far as I know, there are no superficial oil-shows, and the geological formation does not appear to be particularly favourable from what little we know of it; but in any case no company can afford to put down wells into a formation in these parts (however favourable) unless there is some superficial indication of oil.[1]

Given the company's experience in the Persian Gulf area its negative verdict served as a last word on the subject and admitted no reply, the more so since Great Britain had the final word on everything that could take place there.

Starting with Bahrain in 1880 and Kuwait in 1899, the majority of the Arabian Gulf's small sheikdoms had relinquished part of their sovereignty to Great Britain, agreeing, among other things, not to grant foreigners any concession on their soil without British consent. In return, they obtained British military protection and financial support, which were to last until the beginnings of the 1970s.[2] That almost voluntary abdication of power had its own peculiar justification.

Most of the Arabian Peninsula at that time had no fixed boundaries separating its sheikhdoms and tribes. To make matters worse, its deserts were largely inhabited by nomadic peoples unfamiliar with any form of loyalty beyond their own ethnic group or family. For them, moving from one place to another was a rule of life, dictated by millenary habits and daily hardships. No sheikdom could adequately protect its borders from these migratory flows or from the emergence of strong tribal chiefs intent on expanding their power and territory. With the decline of the Ottoman Empire that had ruled over them, the monarchs of these vast regions thus chose Great Britain as an alternative shelter to the vanishing central power of Constantinople in the second half of the nineteenth century.

Having acquired this semi-colonial power, London extended to its Arab protectorates the so-called British Nationality Clause, that required any company operating in any British colony to be registered in Great Britain and managed by British subjects. Moreover, while engaged in its nation-building effort in Iraq, London also drew the lines that mark the current frontiers of the whole region. The man who took on responsibility for that task was Percy Cox, an official who played a special role in shaping today's Middle Eastern boundaries.

After having been the British High Commissioner in Muscat and in Persia, Cox was appointed as Great Britain's top official in Mesopotamia, where he put in place the foundations of the new Iraqi state while masterminding the political architecture of the whole region. In 1922 it fell to him to trace the lasting borders of the Arab states in the Persian Gulf during a meeting of local dignitaries called to discuss the subject. Frustrated and bothered by their failure to reach agreement, Cox picked up a red pencil and drew some lines on a map, which was eventually shown to, and approved by, those in attendance.[3]

Under the forceful guidance of Percy Cox, the entire region remained virtually a private business preserve of Great Britain, despite the clash with the United States on the future of Mesopotamia. Its resolution did not weaken the British grip on the area, thanks to the special role given BP in running the new Iraq Petroleum Company, and the clauses of the "Red Line Agreement" prohibiting companies engaged in Iraq from developing independent initiatives almost anywhere else in the Middle East.

With Great Britain looming large over any significant event in the region, BP dismissing its potential, and a hostile environment making access prohibitive to most Westerners, the Arabian Peninsula seemed doomed to remain on the sidelines of the global quest for oil. But change came in the form of an eccentric possessed by the demon of seemingly foolish undertakings, Frank Holmes, who opened the door to its oil development. A mature mining engineer from New Zealand, Holmes had bought* some oil concessions in Bahrain and Kuwait and in what was to become Saudi Arabia. He had to exercise those rights within a few years, or they would expire. In 1926, strained by financial problems, he sought to sell his rights to Iraq Petroleum Co. through BP, but they rebuffed his offer, reiterating their belief that the Arabian Peninsula did not hold any valuable oil prospects, once again repeating the worst assessment in the history of oil.[4] Having failed to perform any drilling, Holmes lost his rights in Arabia, but succeeded in selling his Bahrain and Kuwait concessions to Gulf, which later resold the Bahraini concession to Chevron. Holmes's Arabian adventure had ended, but the chain reaction he had set off would not be stopped.

Holmes's sales soon provoked another clash between the United States and Great Britain. London invoked the "British Nationality Clause" to deny Gulf and Chevron entry in Kuwait and Bahrain, triggering a strong American diplomatic reaction. Things were less difficult for Chevron, although by no means smooth. The California company found a way around the problem by transferring the Bahrain concession to its own Canadian subsidiary—which was not a British registered company but at least belonged to one of the countries formally ruled by the British monarchy. Bahrain and Chevron signed a formal contract in 1931, and a year later Holmes's intuitions proved farsighted. The Americans found oil in commercial quantities in the small sheikhdom, an unexpected result that put the whole Arabian Peninsula into a new perspective.

Taken by surprise, Great Britain reacted by preventing Gulf from entering Kuwait on its own, and forced it into a joint venture with BP. Established in 1933 as the Kuwait Petroleum Company, the joint venture signed a formal oil concession agreement with the Kuwaiti authorities a year later. In the same period, BP also obtained oil concessions in Qatar and Oman, on behalf of the Iraq Petroleum Company.

---

* Holmes operated as a shareholder and representative of the Eastern and General Syndicate, a British company of which he had contributed to the establishment.

The only area escaping diplomatic quarrels over oil rights was today's Saudi Arabia. Credit for attracting foreign companies to the Kingdom goes to a former British official, who had left his service and his country, converted to Islam, and become a loyal adviser to the Saudi king. His name was John Philby, and beyond his role in Saudi Arabia's oil development, he later gained notoriety as the father of Kim Philby, the senior British Intelligence agent who, in the late 1950s, was unmasked as one of the most important Soviet spies in Great Britain—the same one who inspired John Le Carré's novel *Tinker Tailor Soldier Spy*.

John Philby had long been critical of Colonel Lawrence's passionate support for the Hashemite's dynasty of Hussein and Faisal, and repeatedly tried to convince the top brass in the Foreign Office that Great Britain had to bet on the house of Saud as ruler of the Arabian peninsula. Having lost his battle against Lawrence of Arabia, he had devoted the rest of his life to serve King Abdul Aziz ibn-Saud—the head of the Saud family—which by then ruled over a part of the Arabian Peninsula.

Trying to somehow make money from the radical turn he had taken in his life, Philby soon discovered that all the early business ventures he had plunged into were frustrated, bringing him to quite a difficult financial situation. In his own memoirs, he recalled that he desperately needed money even to sustain his family and pay Cambridge University's fees for his son Kim and tuition for his three daughters' first-class schools.[5] These kinds of worldly worries played a key role in Philby's involvement with oil. Searching for new opportunities to enrich the king and himself, Philby pressed ibn-Saud to open the country to foreign exploration, but the king initially resisted his arguments.

Ibn-Saud had just completed the subjugation of the tribes and emirs of the Arabian Peninsula by the mid-1920s, after a seemingly endless war lasting more than a quarter of a century. In 1925, he had finally defeated his most dangerous rival, the emir Hussein, who had been abandoned by the British government that had once promised him the crown of the whole Arab world as a reward for his active involvement against the Ottoman Empire during World War I.

As the absolute master of the area, ibn-Saud imposed on his people a political-religious system based on a peculiar doctrine of Sunni Islam that his family had embraced in the eighteenth century, *Wahhabism*, which defined every aspect of human life in a restrictive and puritanical interpretation of the Koran, resulting in a strong aversion to foreign habits and lifestyles. The king's religious choice, however, did not depend on his family's traditions alone. It was also suggested by the urgent need to infuse a common cultural element to a scattered and fragmented nomadic population, almost primitive in its habits, shaped by the hardships of daily coexistence with an arid and ungenerous land.[6] As a result, ibn-Saud was reluctant to invite foreign corporations to his newly established domain for fear of destabilizing the delicate cultural equilibrium he himself had established. Moreover, his strongest supporters in the long war to conquer the Arabian Peninsula, the Ikhwan tribes, were the custodians of a radical Islam whose ardent fury sought to eradicate any hint of foreign cultural influence.

Dressed in white robes, with "pointed beards and black antimony past around their eyes,"[7] the Ikhwan had become the subject of several horrible stories that well illustrated their blind destructive determination. According to one source, for example, when they

> first entered Taif and Mecca they smashed all the mirrors they found in the houses, not from lust for destruction but simply because they had never seen such things before. Any visitor to Khurma will see the results of such behaviour—perhaps a fragment of mirror on a wall, somebody's share of the loot—or a window acting as a door because Bedouin do not see the point of windows—or half a door instead of a whole one. Or there may be a quarter or a third of a carpet on the floor, because one big one has been cut up into fair shares. [8]

Some of the Ikhwan's actions "were spectacularly bloody and dramatic,"[9] resembling those of contemporary terrorists because they did not hesitate to butcher men, women, and children during their raids against everything they deemed to be contrary to their values. Increasingly, their fanatical contempt for all things foreign to their culture became a source of discontent toward King ibn-Saud himself, who was held responsible both for promoting innovations like the telegraph, cars, and telephone, and for being too anxious to establish good relations with "impious" countries such as Great Britain.[10] Thus, the constant menace of Ikhwan tribes was an additional reason behind ibn-Saud's very cautious attitude toward any opening to foreign companies.

Yet the king himself had his own worldly needs to take care of, because by any standard he was a very poor monarch. His only sources of income were an annual salary granted by Great Britain[11] and revenues from the pilgrimages of the world's Muslims to the holy city of Mecca, which had come under his control in 1925. On the other side of the ledger, his debts were high and growing, swollen by the need to maintain a court and social consensus among the Arab tribes. In 1928, ibn-Saud also had to finally confront and destroy the Ikhwans, whose radicalism risked undermining the consolidation of his kingdom and the king's very position. This last appendix of his life as a warrior further eroded his finances, and eventually the world economic crisis of 1929 took him to the verge of bankruptcy, as revenues form pilgrimages to Mecca dwindled.

According to Philby's own account,[12] it was ibn-Saud himself who reversed his original stance and asked the former British official to search for oil companies eager to invest in his country, which in 1932 he had named Saudi Arabia. For Philby, that was not an easy task. BP had already dismissed the first Saudi openings by arguing that the country presented no real opportunity to discover oil, and the company's dismissive judgment weighed like a boulder on the future prospects of the kingdom. Eventually, BP took part in negotiations for entering the country, but it was more interested in preventing others from accessing it than in the oil they did not believe was there. In reality, Philby was promoting competition for a goal that no one really wanted to pursue, but eventually a new player materialized: one of the old components of the dismembered Rockefeller empire, Chevron, which had already entered Bahrain.

In July 1933, ibn-Saud signed the royal decree granting the California company a sixty-year oil concession covering the whole eastern portion of the Saudi territory (the *al-Hasa* province), which in succeeding decades would become the richest oil area in the world. Even today, it holds more than 20 percent of the world's proven oil reserves. While the concession model was the same as those used in Persia and Iraq, Chevron was asked to pay in advance 50,000 pounds against future royalties and another 5,000 pounds as rent for the exploratory area.

It took Chevron a few years to realize that ibn-Saud's desperate need for money and the huge investments required for starting operations in such an inhospitable place as the Arabian Desert were too heavy a burden to bear alone. Chevron also lacked adequate market facilities to export Arabian oil, which was already flowing from Bahrain. That is why in 1934 the company relinquished 50 percent of its Saudi venture to a new partner, Texaco, which also acquired half of Chevron's Bahrain operation a year later. In its turn, Chevron acquired 50 percent of Texaco's downstream network east of Suez, which was reorganized in a new jointly owned company, Caltex.

What had begun against the background of concern over a looming oil famine had turned into Western control over all major oil-endowed countries in the Middle East, Saudi Arabia, Iraq, Kuwait, and today's Emirates, with BP already controlling Persia. By then, concern over future oil availability had also captured the attention of many other industrial countries, which moved to shore up their own energy security. European countries in particular started to impose high duties on imported oil products in the 1920s in an effort to stimulate the development of domestic refining industries. At the same time, Churchill's advocacy of oil production controls set a model for other governments.

France was the first European country to decisively imitate the British example. Enticed by the prospect of accessing oil in Mesopotamia, the French government of Raymond Poincaré conceived and promoted the establishment of a national oil company funded by private capital, but whose stock would be granted by the state. For Poincaré, it took a major effort to convince private investors that the uncertain Mesopotamian oil was worth the risk of investing large amounts of money, but he succeeded at last. Thus, in 1924, the Compagnie Française de Pétroles (CFP—the progenitor of today's Total) was established under the protective sway of the French government, which endowed the company with the share it had obtained into the Turkish Petroleum Co. Eager to avoid any foreign interference with its new national champion, the government also acquired a direct 25 percent stake in the CFP in 1928.[13]

Italy also coped with the problem of oil security. In 1927, under Mussolini's fascist regime, the Italian government established the Azienda Generale Italiana Petroli (AGIP—now incorporated into Eni) as a 100 percent state-owned company. Its mission was to develop "a national petroleum policy to secure sources of production both at home and abroad and to encourage domestic refining."[14] To varying degrees, these examples of postwar nationalism were followed in other European countries such as Poland, Romania, the Netherlands, Norway, Sweden, and Denmark.[15] In 1919, the new Soviet government nationalized the entire Russian oil industry and eventually reorganized it into a state concern with a very aggressive operating arm, Russian Oil Products (ROP, 1925).

Germany presented a different situation. Strained by World War I and burdened by heavy war debts and reparations, the country came late to developing an energy strategy. Only after the rise to power of Adolf Hitler in 1933 did it begin to obsessively focus on the search for oil security. In fact, because Germany had no access to petroleum reserves, it opted for independence from it. Since the 1920s, German scientists had perfected two chemical processes (*Bergius* and *Fischer-Tropsch*)[16] that produced synthetic fuels through the reaction of hydrogen with coal ("coal hydrogenation"). Although such products were far more expensive than those derived from oil, Germany embarked on a massive plan to build thirty synthetic fuel plants starting in 1936. This was done at the specific direction of Hitler, under the industrial leadership of the sadly famous chemical group IG Farben.[17] Eventually, this Nazi drive for energy independence would spawn legends and fantastic spy stories such as the movie "The Formula," in which Marlon Brando told the movie's protagonist how Germans had succeeded in finding a low-cost synthetic fuel whose formula had been hidden in a global conspiracy led by U.S. oil multinationals. Naturally, nothing of the sort ever took place. The simple truth was that coal hydrogenation was too costly and inefficient, yielding only modest results that were totally inadequate to solve the Nazis' energy problems.[18]

But, despite all the anxieties surrounding the future availability of crude, and their impact on the shaping of national strategies after World War I, oil once again eluded all dire predictions. Far from running out, as had been so widely predicted, by the end of the 1920s oil once again flooded the world, just as in the glorious and foolish times of Colonel Drake and John D. Rockefeller.

CHAPTER 4

The Oil Glut of the 1930s

Silently but relentlessly, several factors contributed to turn oil market conditions upside-down as the Roaring Twenties came to an end.

To start with, in the first three decades of the new century the oil industry underwent a sweeping technological revolution. At the turn of the century, what was considered the state of the art of the oil industry was poor and rudimentary. Exploration and production techniques were still largely dependent on the "good nose" of the wildcatters, with no contribution from geology or geophysics. Indeed, most wildcatters still believed that oil was contained in huge underground caves or lakes and would occasionally seep to the surface because of some inexplicable act of nature. All of them, moreover, continued to waste oilfield after oilfield by foolish drilling that rapidly exhausted the natural gas pressure, which was responsible for pushing the oil to the surface. The state of refining technology was equally poor, stuck in a primitive mode of oil distillation that consisted of simply heating the crude to higher and higher temperatures to obtain different-quality products at each stage. With this method, less than 50 percent of a barrel of crude could be used to obtain valued-added products such as gasoline, naphtha, kerosene, and gas oil.

The marriage between science and the oil industry was made possible by the praiseworthy work of individual scientists as well as institutions, among them the United States Geological Survey (1908),[1] and the *Oil&Gas Journal,* first published in 1902 (initially with another title) and still today an indispensable tool for all oilmen.[2]

The first major contribution of geology to oil exploration was the Anticline Theory, which revealed how natural gas, oil, and water are trapped together in subsurface porous rock because of their specific gravity; the "trap" containing them forms anticlines, which are upward bulges in rock strata, sorts of underground hills whose peaks can swell the earth's crust leaving peculiar domes that are visible on the surface.[3] Confirmed by a major oil discovery in Oklahoma in 1913, the theory convinced oil companies of the need to establish specific geological departments within their organizations. Over the following decades, many hydrocarbon discoveries were the result of surface mapping of anticlines. (To date, nearly 70 percent of all oil discoveries have occurred in anticline formations.)[4]

The next leap forward in oil geophysics came after World War I, when petroleum geologists first engaged in subsurface analysis, or stratigraphy. Before this, the determination of what lay beneath the ground was largely an act of faith, as surface study was only able to suggest the probability of finding oil in a given place. In contrast, stratigraphy went deep into the secrets of the earth, through the drilling of exploration wells and the careful study of the resulting well logs, core samples, and other data. The new approach was applied after 1916, but only in the early 1920s was it accepted and adopted by major oil companies.

During the Great War, another fundamental discovery took place. As early as 1917, a study published by the U.S. Bureau of Mines had raised the possibility of getting more oil out after the primary recovery phase (which simply exploited the internal pressure created by the natural gas and water contained in the oilfield) by injecting natural gas into the reservoir.[5] This study marked the first intuition regarding so-called secondary recovery of oil. However, because of considerable controversy over production methods, it took more than a decade before secondary recovery methods were accepted and applied.

A fundamental impetus to their entrance into industry practice came in the 1920s, when the issue of conservation of U.S. reserves gained momentum in response to widespread expectations that they would be exhausted. At that time, wild drilling, which had been the rule since the inception of the industry, came under public scrutiny and was thought to be an alarming waste of underground resources. Secondary recovery began to be widely discussed, along with a closely connected principle of correct field development: unitization. The latter involved a core problem of the traditional legal framework that had characterized the infancy of the oil industry worldwide. Both in the United States and in Russia, the so-called rule of capture held that every oilfield was fair game to any wildcatter, all of whom were free to extract whatever they could through their wells. This arrangement took no consideration at all of the unity of an oilfield. In particular, aggressive drilling on disparate parts of a single field rapidly exhausted the internal pressure that forced the oil to the surface, rendering the field unproductive. The correct remedy to this wasteful practice was to develop an oilfield through a unitary approach, i.e., by binding different drillers with a single plan of action and forbidding fragmentation of field exploitation.[6] Like secondary recovery, "unitization" took a long time before it was generally accepted by oilmen and eventually became the norm of the industry worldwide.

The last revolutionary change in exploration and production techniques in those years was "seismic prospecting." The idea was relatively simple: the setting off of small explosive charges on the surface would create energy waves that would bounce off "rock interfaces underground, which allowed the shapes and depths of all kinds of underground structures to be plotted"[7] with geophones or seismographs. The pioneer of this technique was an outstanding figure in the twentieth-century oil industry, Everett DeGolyer (the man who had been the pivotal figure behind Pearson's success in Mexico). He refined a system used by German scientists and later developed a method based on "refraction seismic," the first success of which was the discovery of the giant field of Seminole (Oklahoma) in the 1920s.[8] Since then, seismic has remained a basic component of hydrocarbon exploration, eventually evolving into the contemporary three- and four-dimensional seismic prospecting used today.

While oil exploration and production techniques were experiencing radical innovations, oil refining was also shaken by a major revolution: thermal cracking. Introduced by William Burton in 1913, the new process made it possible to crack heavier oil molecules into lighter ones, which could be then further treated to extract additional volumes of gasoline and other light products. This afforded the industry greater flexibility in "manipulating" crude, instead of simply separating its main components as simple distillation did. Thermal cracking was mainly responsible for the terrific shift in the proportion of gasoline derived from the average barrel of oil in the United States, which passed from around 15 percent in 1900 to 39 percent in 1929.[9] According to some estimates, "it would have taken nearly 268 million additional barrels of crude oil to produce by straight-run distillation the approximate 52 million barrels of cracked gasoline produced entirely by the Burton process between 1913 and 1919."[10]

Together these technological breakthroughs contributed to an overall increase in both the supply of crude oil and the quality of petroleum products.

In addition to the contribution of breakthrough technologies, new waves of oil came onto the market. Soviet oil production rapidly recovered from its wartime and post-revolutionary blackout, rising from a low of 75,000 barrels per day in 1918 to 275,000 bpd in 1929.[11] The Bolshevik government had also forcefully pushed the newly created state oil company to aggressively reenter the international markets with a policy of price discounts aimed at displacing major companies such as Exxon, Shell, and Mobil. At the same time, new large discoveries in the United States revealed how poor a science was the would-be art of predicting natural resource endowments, while the first Middle Eastern production was coming on stream and that of Venezuela was soaring. In all, global world oil production jumped from 1.5 million barrels daily in 1919 to around 4 million in 1929, registering an annual compound growth rate that far exceeded that of consumption.

This dramatic growth in world oil output took place in a landscape of hypercompetition that was not limited simply to a quest for foreign supplies. In fact, both the upsurge in consumption and the emergence of gasoline as the leading product of the industry had prompted a major reshaping of oil companies in keeping with four strategic goals: access to oil resources, vertical integration, size growth, and retail development. At the beginning of this process, many corporations were either purely upstream operators (such as Texaco, Chevron, and Gulf), or gigantic refining concerns (such as Exxon and, to a lesser extent, Amoco and Mobil); others had a mixed profile, such as Shell and BP. As a consequence, companies short on oil reserves began to strive to acquire companies and assets that would enable them to supply their refining and transport networks, while those with ample upstream positions moved to develop their own downstream operations. Furthermore, in the new age of gasoline all of them suddenly realized they had to win over the end consumer in order to secure outlets for their production systems, and this required a completely new orientation toward marketing.

Before the war, gasoline was sold by grocery, hardware, and general stores. The company credited with conceiving of the first dedicated albeit modest drive-in gas station, in 1907, was a small enterprise operating in St. Louis. But with the advent of the Roaring Twenties and the explosion of individual transportation, each company needed to characterize its own product and make it available to potential clients through a vast network of specific and branded selling points. A fierce competition thus ensued to secure the main transportation routes by setting up brand-name, drive-in gasoline stations, each one courting the client with appealing advertisements and small but useful gifts (like maps, for example). By the mid-1920s all oil companies had become highly visible thanks to their widespread networks of stations holding their product brand as a flag. Sometimes, the brand name did not correspond to the name of the mother company but was a simple invention intended to grab consumers' attention; yet over the years, the powerful as well as familiar image it transmitted to millions of clients worldwide convinced oil bosses to adopt it as the company name.

Along with drive-in stations, the transportation revolution in the United States introduced other popular features that would change the landscape of the contemporary age. One of these was the establishment of the first drive-in restaurants, whose progenitor is generally considered Dallas's Royce Hailey's Pig Stand (1921);[12] along with it, the 1920s also saw the birth of the first hotels explicitly created for and devoted to car drivers, the Motor-Hotels—or simply *Motels*—the first example of which was probably established in 1926 in San Luis Obispo, California.[13]

This transformation of the industry into its modern shape involved a vast process of mergers and acquisitions, favored by its growing capital intensity. Searching for new oil sources around the world, developing adequate refining systems, and controlling end markets through transport infrastructure and gas stations all required huge up-front capital expenditures whose returns were usually considerably delayed—as D'Arcy and Pearson had experienced first hand. In the meantime, high fixed costs already incurred obliged small-to-medium operators to sell their products at almost any price to keep the cash flow coming. Yet a prolonged crisis was always a fatal blow for those among them with scant financial resources, which made them easy prey for stronger companies.

Mergers and acquisitions proved a quicker and more profitable way to achieve integration, scale, and market presence than building them step-by-step. The companies that won the survival game emerged as the majors of the twentieth-century oil business, among them Exxon, Shell, and BP at the top, followed by Chevron, Texaco, Gulf, Phillips Petroleum, and few others. Yet even this concentration process did not boost large companies' ability to face a new oil glut. And at the end of the 1920s they finally realized that their self-destructive competitive struggle was once again generating the risk that had been Rockefeller's obsession.

Just before the rising oil tide became a flood, the largest oil companies acted to prevent the situation from spinning out of control by forming the first global oil cartel ever, known as the "As Is" agreement. Negotiated in the Scottish castle of Achnacarry in 1928 by the predecessors of current-day Exxon, BP, and Royal Dutch-Shell, the pact committed each company to freeze the existing status quo among major international companies by fixing their sales on global markets and tying their pro-quota increases closely to consumption growth.[14]

At the same time, a complex system secretly devised by the Achnacarry participants would perversely influence international oil pricing for many years to come. Called the "Gulf-Plus system," it fixed the price of the cheaper oil produced outside the United States at that of oil in the Gulf of Mexico (the main U.S. export point), plus the standard freight charge for shipping oil from the Gulf to its market. All this amounted essentially to the imposition of a phantom freight charge designed to protect the more expensive American oil and keep world prices high.[15] Many other companies endorsed this system, which was applied even by American oil companies selling oil to Allied Naval forces during World War II. What Rockefeller had done by himself was now being implemented by a group of prominent oilmen. However, the extent of the ensuing oil crisis made it impossible for the three giants and their associates to control anything.

The earthquake hit suddenly in the form of the Wall Street crash in 1929, which abruptly deflated United States and world demand for oil; then in 1930 an independent oilman, Dad Joiner, discovered the largest American oilfield ever in the sandy hills of eastern Texas. No name could be more appropriate for the new miracle of nature than "Black Giant," which made Texas the top producing area in the world, to the tune of 900,000 barrels per day in 1931, while total American output stood at 2.4 million bpd and global production had dropped to 3.8 million.[16] For Texas, "Black Giant" heralded the second oil boom portrayed in the James Dean movie "Giant" and a phase of leadership in global production that would last for about thirty years. But for oil companies, it was merely a nightmare.

A huge oil glut submerged the market, and crude prices in the United States plummeted to a few cents per barrel in the summer of 1931, down from around three dollars in 1920, and two dollars in 1925.[17]

The more time passed, the more it seemed the crisis would neither ease nor end. Anarchy ruled in the Texas oilfields, and even the deployment of the National Guard failed to quell the production fury of the oilmen. The other American oil-producing states were prey to chaos as well, starting with Oklahoma, the largest producer after Texas. In 1931–1932, Texas even charged a state entity, the Texas Railroad Commission, with the task of regulating output with a system of mandatory quotas for each producer, but the effort failed. Because of a constant overestimation of the supply effectively needed to cover demand, the Commission's pro-rationing system failed. Moreover, independent producers eluded their quotas by resorting to massive smuggling of crude—soon dubbed "hot oil." After decades of fiercely defending their oil production from any federal interference, the American states had to confront reality. Simply put, the crisis was too big for them to manage, and they had to cede control to Washington.

By then, federal intervention in the economy was the credo of the new Democratic administration of Franklin Delano Roosevelt, elected president in 1932. Oddly enough, oilmen found an unexpected ally in one of the key figures of Roosevelt's "New Deal," Secretary of the Interior Harold Ickes. Long an enemy of oil concerns and always distrustful of what he considered their penchant for intrigues, Ickes was nevertheless convinced that—as he once said—American civilization could not have existed without oil.[18] Paradoxically, it was not the purported scarcity of "black gold" but its ruinous overabundance that was the threat to this civilization, and that risked destroying its irrational producers as well.

Ickes's reaction was to impose a federal system of quotas for each state, which his office would enforce. A part of the framework of Roosevelt's National Industrial Recovery Act (NIRA), the quota system fell apart when the U.S. Supreme Court rejected NIRA in toto in 1935. However, Ickes's campaign for restraining oil production was successful on other fronts. That same year, the U.S. Congress passed a law banning "hot oil" commerce. At the same time, the U.S. Bureau of Mines, part of the Department of the Interior, began to suggest voluntary quotas to each producing state, on the basis of its own forecasts of future demand. Finally, states agreed among themselves to cooperate on exchanging information and tuning their respective plans accordingly. Together with a duty imposed in 1932 on imported oil and refined oil products—intended to keep cheap foreign oil from flooding the American market—the complex regulatory architecture worked. Prices slightly recovered by about 1935, hovering at around one dollar until the end of World War II.[19]

During this period, the Texas Railroad Commission would become the backbone of U.S. oil control, though without any formal power or written mandate. By leveraging Texas's overwhelming leadership in oil production, the Commission succeeded in imposing its influence on all other states. The organization that was born out of chaos and informal agreements turned out to be a very successful and enduring pillar of the oil world. Until 1971, it would perform its duty as the de facto arbiter of American crude by deciding when to switch on and off its taps.[20]

Ironically, the regulatory system envisaged by U.S. authorities in the early 1930s would serve as a stimulus and a model for the future founders of OPEC, a cartel that Western countries deplore today.[21] Also ironic, and lamentably so, is the fact that the collapse of the oil market and the wounds it inflicted on its operators made the notion of scarcity seem unreal, in spite of the long-lasting political consequences it prompted, especially in the Middle East. For the moment, however, the most direct consequence of the new market situation occurred in Mexico. And it was a milestone in the evolution of the oil industry.

After the fall of Porfirio Diaz, the continual quarrel between the new revolutionary government and foreign companies had moved through a spiral of reciprocal recriminations and unresolved claims. The situation worsened as companies shifted their investments towards the much more promising and lower-cost Venezuela, which under the Gomez dictatorial regime offered them red-carpet treatment. Consequently, while the Venezuelan oil sector boomed, Mexican output dived, thereby intensifying the anti-foreign feelings of all Mexicans. The sudden crash of the global oil market deepened the crisis, rendering foreign oil companies even more hard-line in their dealings with the Mexican government and unwilling to make any concessions. The final curtain to that protracted conflict fell as a probe and radical son of the revolution, Lorenzo Cardenas, was elected president of the Republic in 1934.

Having campaigned for an across-the-board revision of all oil agreements, Cardenas forcefully embarked on a national crusade aimed at thoroughly reforming the whole sector and freeing it from corruption and foreign dominance. Companies' local managers realized that it was not possible to delay serious negotiations any further, but their distant and dull-witted headquarters rejected any compromise. This was the case, for example, of Anglo-Dutch Shell, the master of Mexican oil with nearly 65 percent of production. The company's local manager even resigned following his fruitless effort to convince Shell's chairman, the by-then old Henri Deterding, to reach a common understanding with Cardenas.

Sadly, he commented about Deterding's blunt rejection of his advice by underlining that he "was incapable of conceiving of Mexico as anything but a Colonial Government to which you simply dictated orders."[22] But the more time passed without any opening by foreign companies, the more insistent Mexican requests grew, to the point that they became partially unacceptable.

It was in this context that in 1938, in a dramatic and unexpected speech broadcast via radio, President Cardenas announced he had just signed a decree nationalizing the Mexican oil industry. Soon joy broke out throughout the streets of Mexico, as people saw in that decision the real dawn of freedom and national independence; on the other side, efforts by European and American oil companies to enlist their governments' support to block Cardenas' policy failed. In particular, the United States was far more worried that Hitler's Germany might approach Venezuelan leaders to obtain a advantageous oil supply agreement than it was moved by the outcry of its wealthy oilmen. President Roosevelt's decision to leave American companies to their fates was also underpinned by an ethical impulse, which he clearly outlined by saying

the United States would show no sympathy to rich individuals who obtained large land holdings in Mexico for virtually nothing.[23]

The Mexican game was over. On the ashes of evaporating foreign companies, Cardenas established a state oil company, Petróleos Mexicanos or *Pemex*, ushering in a completely new business model for oil-rich nations.

This experiment could not but encounter a wide range of difficulties. The new state entity lacked skilled technicians and competent managers, as well as the capital to self-finance the recovery of mismanaged or declining oilfields. Paradoxically, it was even compelled to reduce workers' salaries with respect to those of foreign companies and to delay implementation of any social programs initially envisaged because of the hardships it encountered early on.

Despite all of this, Cardenas's oil nationalization would represent until our day a powerful symbol of the collective psyche of the Mexican people, considered to be the completion of Mexico's revolution and the country's rise to real autonomy. As an expert pointed out:

The petroleum industry ranks with the presidency, Benito Juarez, and the Virgin of Guadalupe as one of the unifying symbols in a nation riven by differences in language, geography, income, education, social class, and political loyalties.[24]

Yet Cardena's act was also a worrisome warning for the global oil architecture that the great oil companies had shaped over the first four decades of the new century.

CHAPTER 5

Cold War Fears and the
U.S.-Arabian Link

By the eve of World War II, oil had assumed a very important role in modern economies as well as in military strategy. Nonetheless, as an energy source it still lagged far behind coal, which supplied 80 percent of the world's primary energy needs. The United States was the center of gravity for crude production, providing 3.6 million barrels per day, or more than 60 percent of the world output of 5.7 mbd. The whole Middle East was still in its infancy producing about 330,000 barrels per day, less than the Soviet Union and Venezuela, then respectively the second and third largest oil producers in the world.[1] World War II and the Cold War upset this panorama and paved the way for oil's rise to the status of the most vital resource of contemporary history. And once again, fears of oil security and scarcity played a crucial part in shaping this role.

On the many fronts of the war, oil proved to be the winning card in ground attacks and occupation, air campaigns, and naval battles. As part of their strategy, the warring powers devoted themselves to seizing oil-rich areas or denying their enemies access to them. That was the case, among many others, with Hitler's strategy to penetrate the Caucasus to control Baku's oil region, the Japanese takeover of oilfields in Borneo and Sumatra, and the Allied bombing of Rumania's Plotesti refinery complex to halt its supplying of Nazi divisions. By the same token, even brilliant military strategists could do nothing when their troops or vessels ran out of oil—as Germany's General Edwin Rommel was rudely shown in the desert lands of North Africa.

All of this carved into the mindset of postwar strategists the notion that no new war could be won without an ample and secure supply of crude. It was in the United States, however, that this awareness dawned first and most dramatically, amplified during the war by a new wave of *infaust* predictions about the end of domestic crude. The most vocal representative of the new alarmism was none other than Harold Ickes, the Secretary of the Interior who several years before had devoted all his efforts to fighting overproduction.

Already in 1941, Ickes had warned President Roosevelt of the steady decline in the ratio of proven U.S. reserves to production.[2] It was the beginning of a mounting alarm over oil scarcity that reached its climax between 1943 and 1945,[3] after Ickes—who was by then also the Petroleum Administrator for War—made his views public in an article that would soon obtain very wide diffusion. Entitled "We're Running Out of Oil," the article stated:

> If there should be a World War III it would have to be fought with someone else's petroleum because the United States wouldn't have it.[4]

What had happened to completely reverse the situation of oil over-abundance in the United States? In retrospect, the answer was relatively simple.

Investment in exploration and production had been hit hard by the Great Depression and the price collapse of the early 1930s, recovering only from 1937 onwards. In addition, the oil glut and sluggish prices eliminated any incentive to spend on developing new oilfields and implementing new technologies. As a consequence, on the eve of World War II many experts and analysts had begun suggesting that increasing oil reserves in the future would be "more difficult, more limited, and more costly"[5]—a striking resemblance to today's flawed debate about the supposed scarcity of petroleum and the end of cheap oil! In sum, a new bust phase started the pendulum swinging back the other way. It was a dramatic change.

The United States entered World War II quite unprepared to manage the oil supply necessary to meet its unpredictable requirements. Shortages of critical materials such as steel limited the possibilities of increasing new drilling and building much-needed pipelines. Moreover, oil prices at the well had been frozen by the government in 1942 in an effort to limit the cost of the war, but this measure had the side effect of discouraging new investment, which was already endangered by higher steel prices. The fact that the Allied war effort depended almost entirely on American oil resources created an unprecedented drain on them.

Between 1941 and 1945, the United States supplied 6 out of the 7 billion barrels of oil consumed by the Allies (the United States included) for civilian and military purposes.[6]

Echoing the verdict of gloom pronounced by the head of the U.S. Geological Survey in 1919, the head of reserves in Ickes's Petroleum Administration for War declared in 1943

> The law of diminishing returns is becoming operative. As new oil fields are not being formed and the number is ultimately finite, the time will come sooner or later when the supply is exhausted.[7]

The proactive petroleum czar did not hesitate to confront his grim prognosis with radical proposals, which met with strong support from the majority of top officials in the Roosevelt administration.[8] Among these proposals, Ickes stressed the need to nationalize Chevron's and Texaco's concessions in Saudi Arabia in order to assert direct governmental control over them. He also called for federal financing of a pipeline project that would link the Saudi fields to the Mediterranean, the crucial purpose of which was to accelerate the development of Arabian oil.

Despite the support received by the navy and other military branches, these drastic steps were rejected by the U.S. political and business establishment, as well as the oil companies, which considered them as a dangerous sliding toward some form of socialist-like control of private entrepreneurship. However, Ickes's campaign played a crucial role in convincing President Roosevelt to inaugurate a long-lasting, oil-based alliance with Saudi Arabia. It was a U-turn for American foreign policy, the consequences of which are still the subject of heated debate today.

As we have seen, Chevron and Texaco had entered Saudi Arabia in the early 1930s, and in 1938 they had struck oil in commercial quantities in the eastern province of the country. By that time, both companies had already sought political protection from their government because of their fear of a Nazi penetration of the Persian Gulf, as well as of an ever-looming British will to displace American interests in the Saudi Kingdom, a later attempt to conquer what London had lost for its underestimation of the Saudi oil potential. Moreover, King ibn-Saud was continuously pressing them in order to get much more money and advance payments of future royalties, and their refusal to accept the monarch's claims was giving a chance to their competitors. Yet the Chevron and Texaco appeal met with indifference or even hostility.

Washington refused to open any diplomatic channel with the kingdom and only at the beginning of the 1940s did it charge its ambassador in Egypt with representing American interests there. Thus it came as no surprise that in 1941, when pressed by both Chevron/Texaco and the British government, which by then was alarmed by Nazi moves in the area, President Roosevelt openly dismissed the proposal to lend money to Saudi Arabia, observing that the country was "a little far afield" for the United States.[9]

That position was quite an understatement of a more general mood dominating the administration and Congress, which considered the Arabian Kingdom a primitive state, marked by unacceptable habits like slavery, whipping, the cutting off of hands, and decapitation. But as war advanced and oil supply became shorter, Ickes's campaign had its effect, particularly its insistence that Roosevelt secure the huge oil deposits of Saudi Arabia to the control of the United States.[10] Changing his previous position, in 1943 the President informed U.S. Secretary of State Edward Stettinius that Saudi Arabia had become vital to the defense of the United States and authorized the granting of financial support to the kingdom through the Lend-Lease Act. So the first brick in the enduring Saudi-American bridge had been laid, and henceforth its construction went forward quickly.

A further blessing of that alliance came from a technical report prepared by Everett DeGolyer, the brilliant geologist who had masterminded some of the most notable developments in the modern oil industry, including seismic prospecting. After spending a few months in the Persian Gulf on behalf of Ickes's Petroleum Administration for War to assess the extent of the region's oil resources, DeGolyer delivered a verdict that estimated already available reserves at nearly 25 billion barrels, with a high probability of an overall reserve base of 100 billion barrels. Today, we know those figures were far too low. Current estimates put the yet-to-be produced proven reserves of the Persian Gulf area at over 650 billion barrels, without including probable and possible reserves. Yet for the time it was a huge figure, exceeding those for all other known oil regions. And the majority of it was concentrated in Saudi Arabia. This led DeGolyer to conclude:

> The center of gravity of world oil production is shifting from the Gulf-Caribbean area to the Middle East—to the Persian Gulf area, and is likely to continue to shift until it is firmly established in that area.[11]

DeGolyer not only made a fundamental contribution to Washington's recognition of the Middle East's importance with regard to oil. The geologist also suggested two long-term policy goals for ensuring U.S. oil security: first, oil self-sufficiency in the Western hemisphere (the Americas) in order to preserve politically secure resources in case of an international crisis; and second, speeding up Gulf oil development so that it could take the place of American crude on European and other markets.[12]

In 1945, Roosevelt met for the first time with Saudi Arabia's founder and king, Abdul Aziz ibn-Saud, on the USS *Quincy* in the Suez Canal Zone. The meeting reinforced the new relationship born of America's quest for control of foreign oil sources—a quest that in the immediate aftermath of World War II appeared to be justified.

The new global insecurity brought about by the Cold War rendered the United States apparently incapable of serving as the long-term oil supplier for the West. In 1948, the country discovered it had become a net importer of oil for the first time in its history, i.e., that its own consumption needs could not be met by domestic production alone. Given the climate of anxiety about a possible impending clash with the Communist world, this discovery affected American collective psychology far more than it would have in a world open to free trade among free nations. To the mind of U.S. strategists, that purely symbolical passage meant that the country had forever lost its energy independence, and was thus also incapable of maintaining its traditional role as supplier of about 80 percent of Europe's oil needs. This last argument added drama to drama. Washington recognized that if Europe were to avoid becoming easy prey to Soviet designs, it would need to undergo a massive and rapid recovery as well as a leap forward in living standards. But only oil could ensure the industrial and economic reengineering of Europe, as Great Britain had forcefully argued during the 1947 meetings with U.S. officials to discuss the Marshall Plan.[13] For some time, the latter tried to resist British pressures aimed at setting a target of a doubling of oil consumption for Europe as a whole by 1951 with respect to the level of 1939, considering it impossible given the apparent scarcity of oil. But eventually the risk of leaving Europe economically weak convinced the U.S. officials to give up their resistance and to approve the British plan.

In this framework, DeGolyer's suggestions were adopted as guidelines in the shaping of a new U.S. global oil policy. In approving the Marshall Plan, the U.S. Congress asserted the principle of oil self-sufficiency for the two hemispheres, recommending that European energy supplies "to the maximum extent practicable, be made from petroleum sources outside the United States."[14] A direct consequence of this choice had already been foreseen by American strategists: because the Middle East's petroleum reserves were the only ones that could guarantee the future energy needs of Europe and the last-resort needs of America, the region, and first and foremost Saudi Arabia, had to become a focal point of U.S. foreign policy.

In 1948, at the urging of the U.S. State Department, Exxon and Mobil joined Chevron and Texaco in their Arabian oil venture—renamed the Arabian American Oil Company, or Aramco, in 1944. While Mobil retained only a 10 percent share of the company, Exxon took a 30 percent stake, as did both Chevron and Texaco. The final configuration of the most successful venture ever in oil history was thus shaped, with no clear consciousness by the partners of the treasure they were sitting on. For sure, they did not realize that a new oil discovery that very year in Saudi Arabia—the Ghawar field—would prove to be by far the largest petroleum deposit on earth. Mobil would long regret its lack of boldness in entering Aramco, while BP would curse forever those years in the 1920s when it refused to step onto the Arabian Peninsula because of the "high improbability" of finding oil there.

Soon the new American-Saudi partnership moved ahead with construction of the first pipeline linking the kingdom with the Mediterranean Sea, and by the fall of 1949 the Trans-Arabian Pipeline, or Tapline, began transporting Saudi crude to the Lebanese port of Sidon. This important achievement did not placate the ever-growing expectations of King ibn-Saud, which were nurtured by knowledge of a breakthrough contractual formula that oil multinationals had granted to Venezuela in 1948. Known as the "fifty-fifty profit-sharing contract," or simply "fifty-fifty," this formula represented a major revolution in relations between oil companies and producing countries that was to last for nearly twenty-five years.[15] By no means, however, was it a gently won concession.

After the Mexican debacle in 1938, Venezuela had also become a war front for Western oil companies. Nationalist forces started calling for a more equal distribution of the huge profits derived from oil operations, their protests channeled through the voices of two prominent thinkers and leaders: Romulo Betancourt, later Venezuela's president, and Juan Perez Alfonzo, the future "inventor" of OPEC.

Radical democrats in their social inspiration as well as uncompromising and shrewd politicians struggling to improve the well-being of the Venezuelan population, both Betancourt and Alfonzo stood in sharp contrast to the corrupt regime of Vicente Gomez. After the death of

Gomez in 1935 and the beginning of a confused period of political transition, Betancourt and Alfonzo began asserting the need to find a new
equilibrium between the State and foreign oil companies based on an
equal division/sharing of the latter's revenues. In 1945, Betancourt became president of Venezuela and Alfonzo was appointed oil minister.
Soon their campaign to revise existing oil contracts intensified, but
Exxon, Shell, and Gulf strongly opposed the new government's request.
Controlling together around 90 percent of Venezuela's oil production,
they were unwilling to hand over the vast amount of money that the
"fifty-fifty" principle would cost them; moreover, they feared that
any concession would provoke a domino effect in all other producing
countries where they had operations.

Oil multinationals were thus relieved temporarily when a coup in
1947 obliged the advocates of Venezuela's oil rights to leave the stage.
Yet even the new military government had learned their lesson so that
they repeated the very same request to the oil companies. The Mexican
scenario was played out once again. The latter sought help from their
own government, but the Truman administration rebuffed them on the
grounds that America's values were fairness, equality, and anticolo-
nialism, notions hardly to be found in Venezuela. The oilmen had no
choice but to surrender, and in 1948 a new season for the entire oil
industry came into being, shaped by the introduction of the "fifty-fifty"
formula into all of Venezuela's oil contracts.

As soon as King ibn-Saud became acquainted with the new formula,
he asked Aramco partners to apply it to Saudi Arabia as well. Once
again, multinationals tried to resist a concession that might open the
way to an endless submission to producing countries' requests. But this
time they found themselves entrapped by two different and yet con-
verging factors. On the one hand, the United States government was
eager to please the Saudi king in order to cement his pro-American
stance. On the other, at the beginning of 1950 a still quite unknown
independent American oilman had won an oil concession in the Saudi
Neutral Zone* by according the kingdom more than twice the royalty
per barrel paid by Aramco.[16] His name was Paul Getty, and his Arabian

---

* The Neutral Zone was a section of territory between Saudi Arabia and Kuwait
whose existence depended on an unresolved border dispute. From an economic
point of view, sovereignty over its underground resources was divided between the
two countries.

venture was to propel him rapidly into the ranks of the richest men in the world. At that time, however, his company had succeeded only in placing in an embarrassing situation giant corporations whose size completely overshadowed his own.

The solution that satisfied all the parties concerned was finally worked out by the U.S. acting secretary of state for the Middle East, George McGhee, a successful oilman himself, who took care of tuning the priorities of American foreign policy to American oil companies' wishes.[17] McGhee drafted a bill later passed by Congress that made the oil royalties paid to host governments tax deductible for U.S. companies. Thus every dollar paid to oil-producing countries would be one less dollar in the coffers of the American Treasury.[18] In late 1950, therefore, Saudi Arabia was granted the "fifty-fifty" profit-sharing system, which soon was requested by and extended to all other Arab oil producers.

The fiscal *escamotage* devised by McGhee was, in fact, a major foreign policy decision in disguise. Over the following decades, it would cost the U.S. government hundreds of billions of dollars, but there was no other way to accomplish this end. Congress almost certainly would have never approved direct compensation for the oil companies. Consider, for example, that in 1973 nearly 70 percent of net profits of American oil multinationals were made abroad, thus deemed foreign by the American tax system.[19] Twenty-four years later, testifying before a U.S. Senate committee, George McGhee declared that the royalties loophole had been devised in response to a specific recommendation by the U.S. National Security Council, whose target was the consolidation of American control over the most oil-rich countries in the Middle East.[20]

Together with the new profit-sharing formula, another significant innovation occurred in terms of oil pricing, which would remain in place for more than twenty years: the so-called posted price.

Taking advantage of the companies' habit of posting the price of their oil, producing countries asked for and were granted stable "posted prices" as a reference for profit sharing. Those prices became an artificial instrument to cement companies' and countries' interests, a sort of pact that was irrespective of real market conditions. In fact, for several years companies preferred to swallow the loss when real prices declined rather than jeopardize the posted price they had agreed upon with producing countries in order not to destabilize their relations with them.

A few days after the signing of the Saudi-Aramco "fifty-fifty" agreement, President Truman wrote King ibn-Saud a letter affirming:

I wish to renew to Your Majesty the assurances which have been made to you several times in the past, that the United States is interested in the preservation of the integrity of the independence and territorial integrity of Saudi Arabia. No threat to your Kingdom could occur which would not be a matter of immediate concern to the United States.[21]

While America was cementing its long-term Arabian link, it also became engaged in another crucial issue that was to mark its foreign affairs for decades to come: support of the newly established state of Israel. Both diplomatic choices represented surprising as well as contradictory novelties. While the American establishment had long viewed Saudi Arabia as a weird and uncivilized country, the Jewish question had been largely ignored until the advent of the Truman administration.[22]

The leaders of America's postwar foreign policy all strongly opposed the birth of a Jewish state within the boundaries of British-ruled Palestine. Key figures like Secretary of State George Marshall, Secretary of Defense James Forrestal, George Kennan, the head of the State Department's newly created policy planning staff and the architect of the Doctrine of Containment, the Chiefs of Staff, and even the newly created Central Intelligence Agency (CIA) believed that the only American priority in the Middle East should be the special relationship with Saudi Arabia and other Arab oil producers, that required America to reject Jewish demands on Palestine. Actually, the only "ardent champions" of the Jewish cause in the administration were Clark Clifford, Truman's legal counsel, and David K. Niles, special assistant for minority affairs to the President.[23]

King ibn-Saud had also repeatedly warned Chevron and Texaco of his personal opposition to the establishment of a Jewish state in an Arab land, and asked them to urge the American government to avoid becoming involved in the messy issue. Both companies responded by becoming the strongest advocates of the Arab cause, a role they would continue to pursue in the following decades. Thus, by 1947 there was no one in the top ranks of the Truman administration who differed from the view that securing Saudi oil was the most critical imperative of U.S. policy in the Middle East. With one very special exception: President Harry S. Truman himself.

Truman stood up to all his advisers and fought a solitary battle to impose his line about both the approval of the United Nations' plan for the partition of Palestine between Jews and Arabs (November 1947), and the recognition of the newly created state of Israel (May 1948). Outraged by Nazi atrocities against the Jews, Truman had concluded that the United States and the West were morally obligated to make up for their indifference to the Holocaust, as he wrote in his memoirs. At the same time, the president's populist roots had always made him suspicious and hostile toward the oil sector and its protagonists, whom he never loved but grudgingly dealt with for the sake of American national security.

Critics maliciously questioned the president's staunch moral stand, suggesting that his decision to recognize the new state was driven by political opportunism, and particularly by his desire to lock in the Jewish vote in the 1948 presidential elections. A memo by Clark Clifford later became the alleged "smoking-gun" supporting the argument that Truman had used the Jewish question to improve his chances in the coming election, which most pollsters considered very poor. In that memo, Clifford reminded Truman that even though the Jewish vote was important only in New York State, only one presidential candidate since 1876 (Woodrow Wilson in 1916) had won the presidency while losing New York.[24] Yet as David McCullough has pointed out:

> for Truman unquestionably, humanitarian concerns mattered foremost.... When his Secretary of Defense, Forrestal, reminded him of the critical need for Saudi Arabian oil, in the event of war, Truman said he would handle the situation in the light of justice, not oil.[25]

Ethical concerns did not obscure Truman's judgment. Sometimes, his irritation for Jewish propaganda was so high that he once refused to meet the leader of the Zionist movement, Chaim Weizmann. In a letter to Eleanor Roosevelt, he even stated that

> The action of some of our United States Zionists will eventually prejudice everyone against what they are trying to get done. I fear very much that the Jews are like underdogs. When they get on top they are just as intolerant and as cruel as the people were to them when they were underneath.[26]

However, Truman's basic approach never changed, which led to a dramatic clash with George Marshall. During the last White House meeting on the subject of the immediate U.S. recognition of Israel, on May 12,

Marshall vehemently opposed such a move by asserting that, if the president were to take that decision, he would vote against him in the November elections.[27] It took a pause of one day before the Secretary of State called Truman, saying that "while he could not support the position the President wished to take, he would not oppose it publicly."[28] Such a reassurance removed the last obstacle to the de facto recognition of Israel, which the United States announced eleven minutes after its official declaration by the Jews of Palestine, on May 14, 1948.[29]

This set in motion a new political and military process in the Middle East that was to endure until today. As soon as Israel was born, Egypt, Syria, and Jordan attacked the new country in the first of four wars fought over the next twenty-five years. The war also marked the evolution of Arab nationalism to a militant status after years of quiet incubation, and set off deep changes in the environment that had allowed seven Western oil companies to become the absolute masters of Arabian oil.

America's Middle East policy was thus born out of a contradiction, and evolved for decades without resolution. Moreover, the traditionally unstable region became involved in the tensions of the Cold War, fuelling America's obsession with a perceived Soviet design to penetrate the area and control its oil by exploiting the opportunities offered by Arab nationalism. Yet it was not an Arab country, but Iran, to first test the postwar oil order in the Middle East.

CHAPTER 6

The Iran Tragedy and the "Seven Sisters" Cartel

With the spread of the "fifty-fifty" formula, the government-royalty system that William Knox D'Arcy had inaugurated in Persia at the dawn of the Middle Eastern oil saga vanished forever. Yet it was there, where it all had begun, that the new profit-sharing formula triggered the first postwar oil crisis.

After renewing its concession agreement with Persia in 1933, BP had done nothing to comply with its provisions for improving the living standards of local employees, ensuring their technical and managerial education, and promoting them to higher positions within the ranks of its organization. For the Iranian nationalists, the British company's insensitivity to local aspirations and the support it received from its government made Great Britain the symbol of all that was wrong with the recent history of their country, which Reza Shah had renamed Iran in 1935. Then in 1941, an event suddenly changed the climate of Iranian society. Reza Khan was ousted by the British for his pro-Nazi stance and replaced with his twenty-one-year-old son, Mohammed Reza.

The new king appeared to be nothing more than a puppet in the hands of the British government: even in his relatives' judgment, the young man was excessively weak, constantly undermined by doubts, and rather vacuous.[1] Yet the demise of Mohammed Reza's father brought about an easing of the brutal practices commonly used against political opponents, and gave a fresh start to nationalist activism, particularly after the end of World War II. In this context, it was BP's behavior that was to catalyze the new political atmosphere into a radical and widespread revolt.

In 1947, the Abadan refinery workers called the first major strike ever in Iran to protest their inhuman living conditions. In BP's fields, indeed:

> There was no vacation day, no sick leave, no disability compensation. The workers lived in a shanty-town called Kaghzabad, or Paper City, without running water or electricity, let alone such luxuries as icebox or fans. In winter the earth flooded and became a flat, perspiring lake.... Summer was worse.... To the management of AIOC [then the name of BP]... the workers were faceless drones.... In the British section of Abadan there were lawns, rose beds, tennis courts, swimming pools and clubs; in Kaghzabad there was nothing—not a tea shop, not a bath, not a single tree.... The unpaved alleyways were emporiums for rats.[2]

Even Truman's personal envoy to Iran, Averell Harriman, was dismayed by the sight of the hellish barrack town housing Iranian workers, and did not refrain from accusing the British of behaving like nineteenth-century colonialists.[3]

In response, the British organized mobs of paid Arab provocateurs to clash with strikers and thereby give BP's men a pretext to use force to settle the uprising. The final outcome was dozens of dead and hundreds of injured. Yet from that moment on BP's position grew only more and more tenuous. As early as 1947, the Iranian parliament approved a law calling for the renegotiation of the company's concession. Its promoter and the first to sign it was a much beloved and admired figure in Iranian politics, Mohammed Mossadegh.

Mossadegh had risen to wide popularity over three decades of his uncompromising struggle against foreign maneuvering to win influence over Iran, but also by publicly embarking on major campaigns against both domestic corruption and Reza Shah's dictatorial regime. Morally unblemished and uninterested in worldly rewards, since the end of World War I Mossadegh had refused both ministerial positions and other prestigious assignments. He had accepted an appointment as minister only to be expelled shortly thereafter because his plans for modernization and moralization had worried the establishment; another time he had been imprisoned by the Pahlavi's government.

Under the leadership of Mossadegh, Iranian nationalists began to make public the size of BP's gross profits and the meagerness of the company's contributions to the country's treasury, making everyone realize how inequitable the arrangement was for their country.[4] BP dismissed the call for more oil revenue, and when the "fifty-fifty" formula was spreading elsewhere it refused to adopt it in Iran, arguing that the country's oil revenues were already close to 50 percent of the company's net profits. But BP's defense was based entirely on deceitful accounting. Between 1947 and 1950, the British concern paid around 40 percent of its gross profits to its own government as taxes, whereas an average of a modest 20 percent went to the Iranian treasury. Worse, BP continued both to hide its effective net profits (by leaving many of its companies off of its balance sheets) and to forbid Iranians to audit its books, a request made by Iran since the 1920s.

Inexorably tension grew, and by 1949 Iran's claims changed nature. The issue was no longer revising the terms of the concession but getting rid of the predatory presence of a foreign country which for fifty years had masterminded, controlled, and submitted to its interests every aspect of Iranian life. For these reasons, in 1949 a group of members of the *Majlis* (the Iranian parliament) led by Mossadegh proposed to nationalize BP's assets and operations in Iran. The company's chairman and chief executive, William Fraser, then flew to Teheran in order to negotiate a "Supplemental Agreement" to the one signed in 1933, which proved to be only an exercise in public relations. Even the shah's appointed cabinet found BP's proposal unacceptable, but it was forced to sign it on July 1949 under pressure from Mohammed Reza. The new agreement slightly increased Iran's oil revenues, further reduced the area of the concession and again committed BP to improving workers' conditions and ensuring instruction for Iranians engaged in the oil industry. Nonetheless, Fraser did not concede anything close to the "fifty-fifty" formula and rejected even the repeated Iranian request to have BP's books independently audited, making it clear that this was the last word on the issue.[5] In fact, it was not.

In November 1950, the Iranian parliament rejected the "Supplemental Agreement." One month later, the news that Saudi Arabia had been granted the "fifty-fifty" formula ignited Iranian nationalists, leading even the hard-line ultraconservative British ambassador, Francis Shepherd, to urge his government and BP to extend the same terms to Iran. But once again, his interlocutors stood firm against it.[6] Thus in March 1951, parliament approved Mossadegh's proposal to nationalize BP's assets. The same month the shah's appointed prime minister, Ali Razmara, who had tried his best to block the nationalist's mounting upheaval, was murdered outside a mosque. The figure the parliament chose to replace him with was Mossadegh himself. At that point, the young Reza Pahlavi had no choice but to yield to realpolitik and on May 1, 1951, he signed the law providing for the nationalization of the oil industry.

In Great Britain, the Labour government led by Clement Attlee was caught off guard. It labeled the Iranian nationalists "ungratefuls," "paranoiacs," "thieves," and "unreliable," to cite just a few of the epithets wielded by British ministers and diplomats at the time. Labour's obdurate and reactionary attitude was quite paradoxical for a government that made "nationalization" a pillar of its political program and was undertaking several operations of that kind in Great Britain.

In any case, in August 1951 Attlee's cabinet imposed harsh economic sanctions on Iran, asserting that the country's oil nationalization was theft. As a consequence, all those who assisted Iran in producing and exporting oil, as well as those who bought Iranian oil, would be considered accessories to a crime punishable under international law. Western oil companies formed a common front with BP and took part in the "oil blockade," which worked well. Iran's crude production dived from 650,000 barrels per day in 1950 to 20,000 bpd in 1953, while oil export revenues dropped from over 400 million dollars in 1950 to less than 2 million in the period from July 1951 to August 1953.[7] As a result, the country lurched toward economic collapse.

The British government never really considered any option other than showing the Iranians who the boss was, a hard-line position dictated by the hawkish foreign secretary, Herbert Morrison, who also tried to support a covert operation to bring down Mossadegh.[8] This stance became even tougher as soon as Winston Churchill regained the premiership in 1951, following the general elections held in November of that year.

For Churchill, Iran was part of his personal history. It was he who had masterminded the British government's takeover of BP in 1914 to secure the supply of Persian oil; and it was he who had devised most of Britain's oil strategy for the first three decades of the century. Now seventy-seven, the old statesman soon became the most aggressive proponent of a coup d'état against the Iranian government.[9] The British conspiracy to undermine Mossadegh, however, was error-prone and ill-managed and so relatively easy for Mossadegh's men to uncover. In response, Iran closed the British Embassy in Teheran, expelled all British officials, and severed diplomatic relations with Great Britain in November 1952.

Throughout the ineffective first confrontation with Mossadegh, Great Britain was essentially alone. The ideological stance of the U.S. Democratic administration toward Iran and Mossadegh was the opposite of that of its old-time ally. Honoring the legacy of Roosevelt, who had disdainfully refused support to the American oil companies facing the nationalization of Mexican oil in 1938, President Truman felt an ill-concealed distaste for Britain's (and France's) unacceptable colonial claims, and was uneasy about the accountability and honesty of the great multinationals. At the same time, he and his new secretary of state, Dean Acheson, sympathized with emerging national movements in developing countries and therefore rejected and thwarted British proposals to overthrow Mossadegh by means of a covert operation. As proved by the Mexican experience, oil nationalization was not an issue as long as after nationalization oil continued to be available for the "free world" and was not used as a weapon against it. Acheson made this position clear in 1951 by officially stating that the United States recognized "the right of sovereign states to nationalize provided there is a just compensation."[10]

Mossadegh and a large component of the Iranian nationalists looked hopefully to the United States. Moreover, the uncommon figure of the Iranian premier was capturing the sympathy of the American public, with his extravagant pajamas, his even temper, his sincere devotion to his country's well-being, and his penchant for endless discussions. In 1951, Mossadegh was even declared "Man of the Year" by *Time* magazine.

The real worry of Truman and his men was that by exploiting the Iranian Communist Party's support for Mossadegh, should the situation not be resolved by means of an honorable compromise, they might be giving the Soviet Union a pretext for intervening in Iran. For this reason, Truman's diplomats did their best to convince the Iranian leader to reach a settlement, while containing and moderating Britain's desire to oust him. Truman's best international thinkers and diplomats, such as Acheson, Harriman, and McGhee, directly engaged in an exhaustive attempt to achieve a solution, but Britain's firm opposition to recognizing the principle of nationalization made that impossible.

Things changed dramatically after the election of the Republican Dwight D. Eisenhower as president of the United States in 1952. According to the view of the new administration, not only was Iran a traditional target of old Russian and new Soviet ambitions, but it was also the natural corridor to the huge oil reserves of the whole Persian Gulf, which represented a key prize in the struggle for global power. Moreover, after the outbreak of the Korean War in 1950, Iran had become the main supplier of fuel to the American troops engaged in South Korea under the United Nations flag.

Exploiting the new political climate in Washington, Great Britain sent to the United States the chief of its Secret Intelligence Service in Iran,

Christopher Woodhouse, to convince the Central Intelligence Agency (CIA) and the Department of State that the real issue behind the Iranian crisis was Iran's progressive slide into the Soviet orbit. After that visit, the "communist factor" was hyped and twisted—as Woodhouse himself recalled in his memoirs—in order to bring Eisenhower's men over to the British side.[11] That strategy paid off, finding fertile ground in an administration committed to the "rollback" of communism worldwide; in particular, it was well received by two powerful figures who were to become Eisenhower's point men for foreign policy, namely the Dulles brothers: John, the new secretary of state, and Allen, whom Eisenhower appointed director of the CIA.

The Dulles brothers played a decisive role in the plot to overthrow Mossadegh, and it remains to be determined whether they were partly responsible for giving Eisenhower a misleading picture of what was happening in Iran in order to obtain his blessing for the coup. But Mossadegh himself made a tragic mistake with the White House. Desperately searching for financial aid to save Iran from looming starvation, as a tactic he began invoking the specter of an impending Soviet threat to his country and urged Eisenhower to help him to keep Iran out of the communist camp.[12] Given the obsession of the new U.S. government with Moscow's expansionist aims, Mossadegh's argument had the opposite effect of that he anticipated and probably helped trigger the American decision to launch a coup d'etat, which was endorsed by President Eisenhower in the fall of 1953.[13]

The plan was given the code name *Operation Ajax* and its execution was entrusted to the CIA, which took advantage of the work already done on the subject by British intelligence. The CIA's responsibility in getting rid of the democratic government of Iran has been well documented in the memoirs of the main actors of that event and by the growing availability of declassified official documents. But in 2000 our knowledge was enriched by the publication by the *New York Times* of a detailed internal CIA reconstruction of the whole operation, the accuracy of which was confirmed by the Agency (the newspaper published a summary of the document, while the entire memo has been made available on the *NYT* website).[14]

According to the document, the first blueprint of the plan was drafted by an American and British senior intelligence officers in Cyprus. With minimal changes, it reflected the actual stages of *Operation Ajax*. Its main guidelines are worth reading:

- "Through a variety of means, covert agents would manipulate public opinion and turn as many Iranians as possible against Mossadegh. This effort, for which $150,000 was budgeted, would "create, extend and enhance public hostility and distrust and fear of Mossadegh and his government." It would portray Mossadegh as corrupt, pro-communist, hostile to Islam, and bent on destroying the morale and the readiness of the armed forces.
- While Iranian agents spread these lies, thugs would be paid to launch "staged attacks" on religious leaders and make it appear that they were ordered by Mossadegh or his supporters.
- Meanwhile, General Zahedi would persuade and bribe as many of his fellow officers as possible to stand ready for military action as necessary to carry out the coup. He was to be given $60,000, later increased to $135,000, to "win additional friends" and "influence key people."
- A similar effort, for which $11,000 per week was budgeted, would be launched to suborn members of the Majlis [the Iranian parliament].
- On the morning of the "coup day," thousands of paid demonstrators would stage a massive anti-government rally. The well-prepared Majlis would respond with a "quasi-legal" vote to dismiss Mossadegh. If he resisted, army units under Zahedi's control would arrest him and his key supporters.[15]

The effective puppeteer who organized and directed the whole operation and ensured its final success was a thirty-seven-year-old secret agent, Kermit Roosevelt, who was the head of the Middle East Department of the CIA. Using the U.S. Embassy in Teheran as headquarters for all his plots (a detail that would turn out to be very important twenty-seven years later, when Iranian fundamentalists stormed the U.S. embassy in order to avoid another countercoup), Roosevelt failed in the first coup attempt on August 16, 1953, because of the unexpected reaction of Mossadegh's loyalists. Having secretly blessed Roosevelt's coup, the shah soon left Iran for Iraq, from where he finally flew to Rome. But Roosevelt did not give up.

Although he had been ordered to stop the whole operation, Roosevelt organized a second coup, which succeeded on August 19, only three days after the first attempt. Mossadegh surrendered and was arrested. A few days later, Mohammed Reza Pahlavi came back to Iran and began a repressive campaign to eliminate his enemies. Mossadegh received relatively gentle treatment, given his worldwide fame and popular support. He was sentenced to three years in prison followed by house arrest for life, and died alone behind a curtain of silence in 1969. In contrast, many of his supporters were sentenced to death. A dictatorial regime then replaced the only democratic and—paradoxically—Western-oriented experience Iran would ever know.

It was a tragic mistake for American foreign policy. Obsessed by its reading of world events through the Soviet lens, it tarnished its credibility as an anticolonialist country defending the rights of those struggling for freedom and independence. But the most immediate consequence of the Iran crisis was for British interests.

Great Britain had ignited this fire in the name of oil, with its staunch refusal to agree to any of Mossadegh's demands. But despite the apparently favorable ending, London did not get what it wanted. In effect, the United States could no longer trust its ally and disapproved of its colonialist approach to world affairs, fearing that its absolute control of Iranian oil could be a permanent source of instability for the country. As a consequence, the U.S. Department of State promoted the establishment of an "International Consortium" to produce, refine, and market Iranian oil in 1954. BP, which formerly controlled all of the country's oil resources, received only a 40 percent stake in the new entity. The rest was divided up among American companies Exxon, Mobil, Texaco, Gulf, and Chevron (which together received 40 percent, in equal parts), Shell (14 percent), and Total-CFP (6 percent).[16] The state company established by Mossadegh to run Iranian petroleum, the National Iranian Oil Company (NIOC), outlived its creator and remained the owner of the country's oil reserves. Above all else, it was the image crisis that BP suffered after the events in Iran that obliged the company to change its name from the Anglo-Iranian Oil Company (Anglo-Persian until 1935) to British Petroleum, or BP.

Historians have proved that the communist threat in the Middle East in the early 1950s was nonsense. It was even incompatible with the local nationalistic movements formed in rejection of colonialism. Yet moved by this fear, the Truman administration established a set of guidelines for U.S. postwar oil policy that provided the frame of reference for future Middle East policy. Classified as National Security Council 138/1,[17] the document laid out a basic premise for government action by recognizing oil as a vital source of power in the postwar world, and its free supply as a prerequisite for the survival of Western economies. Based on this premise, it set out several key lines of action:

> Since Venezuela and the Middle East are the only sources from which the free world's import requirements for petroleum can be supplied, these sources are necessary to continue the present economic and military efforts of the free world. It therefore follows that nothing can be allowed to interfere substantially with the availability of oil from these sources to the free world.[18]

An important corollary of this position factored in the critical role played by the Western multinationals, which were deemed to be the only actors capable of "maintaining and expanding the production of those areas to meet the rising demand of petroleum of the free-world." As a consequence, the document stated that:

> American and British oil companies thus play a vital role in supplying one of the free world's most essential commodities. The maintenance of, and avoiding harmful interference with, an activity so crucial to the well-being and security of the United States and the rest of the free world must be a major objective of the United States Government policy.[19]

The philosophy of the NSC document reflected oil's rapid ascent to a key factor in global power after World War II, and projected the consequences of this development for long-term American politics. Sadly, the first victim of the association between the new quest for oil and the life-or-death struggle against the Soviet Union was the first attempt at democracy by a Middle Eastern country.

Although the conclusion of the Iranian drama seemed to reward American oil multinationals, the latter had been reluctant to comply with the solution envisaged by the U.S. State Department because of a sword of Damocles hanging over their heads: a new antitrust inquiry had begun an investigation of them for suspected suppression of competition and price fixing. These allegations were as potentially devastating as those that had destroyed Rockefeller's Standard Oil in 1911. And their entrance into Iran could only worsen the picture.

The problem had started with a flurry of inquires undertaken by various U.S. agencies into oil companies' activities. The issue was always the same. Oil prices of the largest companies were excessively high in certain markets and inexplicably lower in others; in addition, some buyers were overcharged while others received significant discounts. All of this called attention to the opaque pricing mechanisms of the oil giants.

Public outcry exploded in 1947 when it surfaced that Chevron and Texaco had sold oil to the U.S. Navy at $1.23 per barrel, well above the $0.95 it charged to French buyers and $1.00 to Uruguay.[20] Even worse, a U.S. Senate committee discovered that the overcharging of the Navy occurred even during World War II, a shocking betrayal of U.S. national interests. The final report of the committee stated:

> The oil companies have shown a singular lack of good faith, an avaricious desire for enormous profits, while at the same time they constantly sought the cloaks of United States protection and financial assistance to preserve their vast concessions.[21]

Soon the organization that oversaw the administration of the Marshall Plan[22] began its own investigation and found that it, too, had fallen victim to similar practices that had been carried out by multinationals operating in the Middle East. This revelation obliged the latter to shut down the "Gulf-Plus" system agreed to in the Achnacarry Castle in Scotland in 1928.

Catching wind of the growing evidence and mounting public outrage at the large oil concerns, the U.S. Federal Trade Commission (FTC) initiated its own inquiry in 1949 and released its final report during the Iranian crisis, in November 1951. Entitled "The International Petroleum Cartel," it was the most damning portrait yet of the secret inner workings of the largest oil corporations.

After examining oil agreements dating back to 1928, the Commission concluded that domestic and international oil trade, as well as U.S. and world production, had been artificially restrained and manipulated by the largest multinationals in an attempt to fix oil prices. Moreover, the oil giants had divided up among themselves the world's most important markets and producing areas with a view to suppressing competition.[23] The report also pointed out that in 1949 seven major oil corporations controlled 82 percent of world crude reserves, 80 percent of world production, and 76 percent of world refining capacity, excluding the United States and the Communist Bloc.[24] The names of the companies were already familiar to the public: Exxon (Standard Oil New Jersey), Texaco (Texas Oil Company), Chevron (Standard Oil California),

Mobil (Standard Oil New York-Vacuum Oil), Gulf Oil, Royal Dutch Shell, and BP (Anglo-Iranian Oil Company).*

For the first time ever, the immense and silent power of the Seven Sisters was revealed, threatening to bring down the entire architecture of U.S. postwar oil policy. Indeed, the FTC formally requested that the attorney general initiate a criminal investigation of them. It was one of those vital paradoxes of an affluent democracy that the FTC, an organ of the Department of Justice, could openly clash with the very administration that it was a part of. In any case, President Truman gave the green light to the inquiry in June of 1952, and ordered it released in August. Both decisions, however, were reached in the midst of a storm of doubts and worries.

Any public airing of the report's findings would have been a spectacular blow to America's credibility as a pillar of fairness and equal justice, offering the Soviet propaganda an unexpected bounty of material for attacks against Western capitalism. Worse yet, after conducting their own supplemental inquiry, senior officials in the Department of Justice decided to initiate criminal proceedings against oil companies.

The more public outcry mounted, the more the Seven Sisters struggled to regain an impossible purity. It was easy, they argued, to criticize their current windfall profits without taking into account the enormous risks, investment, and environmental hostility they faced at the beginning of their bold ventures in many producing countries. High risk, high reward, was their reply. Without the prospect of great returns, no one would be mad enough to explore and develop places as hostile and dangerous as the deserts of the Arabian Peninsula. Furthermore, returns were a long time coming in such a capital-intensive industry, where the immediate spending of incredible amounts of money was a necessity, partly because of the need for constructing facilities and providing services for their workers in the middle of nowhere.

However, this reasoning provided meager defense against the allegations of overcharging the U.S. Navy or the portrayal of the Iranian oil crisis as the struggle of a rapacious Goliath against an innocent David. And given the general climate, there was no way the defendants could escape a guilty verdict, which would have shattered their control of Middle Eastern oil. For this reason, on January 12, 1953, Truman finally decided to resolve the matter through civil litigation rather than criminal proceedings.

---

* In parentheses, the name of the companies at the time of the FTC's report.

This would be Truman's last decision as president of the United States. Just a few days later, Eisenhower took office. It was the tortuous culmination of a protracted internal debate; it was also the logical consequence of NSC 138/1. The Eisenhower administration endorsed the content of NSC 138/1 and also ratified Truman's decision about the case against the Seven Sisters. This was instrumental to letting them enter the Iranian Consortium without drawing fresh allegations of oligopolistic control of the world markets. Accordingly, in January 1954, the U.S. National Security Council stressed to the U.S. Attorney General, who was responsible for antitrust actions, that

> the security interests of the United States require the United States petroleum companies to participate in an international consortium to contract with the Government of Iran, within the area of former AIOC [BP], for the production and refining of petroleum and its purchases by them, in order to permit the reactivation of the said industry, and to provide therefore to the friendly Government of Iran substantial revenues which will protect the interests of the western world in the petroleum resources of the Middle East.[25]

National interests had to prevail, and they did. The attorney general complied with the NSC's order and gave his waiver for American oil companies' entry into the Iranian Consortium.

Thus, in the short time between the end of World War II and the overthrow of Mossadegh, a new oil order had been forged. But events would show that the directives set out by the American government were far from clear. On the contrary, they created a misleading framework that often brought the interests of the U.S. government and multinationals into open conflict. Sometimes, as Daniel Yergin wrote, "Washington would champion the companies and their expansion in order to promote America's political and economic interests, protect its strategic objective and enhance the nation's well-being."[26] But in many cases, the U.S. government supported, or at least did not oppose, assaults against oil majors by antitrust authorities, independent oil companies, and other domestic players. At the same time, while oil multinationals were naturally inclined to take a pro-Arab stance, postwar U.S. foreign policy remained obsessed by the possibility of Soviet penetration of the Arab world and always defended Israel as the only stronghold of democracy and anticommunism in the Middle East. This impulse, albeit unevenly acted upon by U.S. administrations, clearly represented the major source of destabilization for Western oil interests in the region. What's more, it was a sad irony that the fears of oil scarcity that had shaped U.S. postwar foreign policy once again proved totally unjustified, and that, contrary to the gloomy prognosis of Ickes, a new era of overabundance became the Seven Sisters' next major problem.

CHAPTER 7

The Golden Age of Oil
and Its Limits

In the twenty-five years between the launching of the Marshall Plan in 1948 and the first oil shock in 1973, world oil consumption grew more than six-fold, ushering in the Golden Age of Oil. Whereas North American* consumption "only" tripled (given its higher initial level of demand), in the other industrial countries it increased eleven-fold, which meant an average compound growth rate of more than 11 percent per year, or a doubling of oil consumption every six and a half years. Thus, while the world consumed 9.3 million barrels per day in 1948, by 1973 its daily requirements had risen to 56 million barrels.[1]

Oil's success in fuelling modern economic development brought about the fastest process of energy source substitution in the history of mankind, whose victim was coal. As late as 1950, the chief energy source of the first industrial revolution still reigned over all rivals, supplying about 65 percent of world energy needs. But by the mid-1960s, oil had supplanted coal as energy king.[2]

The most striking effect of the triumph of oil, and its major source of consumption, was mass motorization. Between 1950 and 1973, the number of automobiles and other passenger vehicles worldwide climbed from 53 million to nearly 250 million. These figures continued to soar year after year even after the 1973 oil shock, so that by 1980 the number of passenger vehicles in the world had surpassed 440 million (148 million in the United States).[3] By the same token, car manufacturing shifted from being a primarily American activity to a feature of Europe and even Japan.

---

* United States and Canada.

This shift started in Europe, where local carmakers realized that they could exploit consumers' growing appetite for motion by producing cars cheap enough for everyone's pocket, much like Ford's original idea for the Model T. So with more than four decades' delay, they plunged into mass production of small-to-medium "people's cars," which hit the target: German Volkswagen's "Beetle," British Motor Corporation's "Mini," Italian Fiat's "500" and "600," French Renault's "4CV," and Citroen's "2CV" became the popular symbols of Europe's rebirth from the ashes of World War II. All over Europe, car production jumped from just over 1 million units in 1950 to more than 11 million in 1972, with West Germany taking over production leadership from Great Britain in 1956.[4]

Even more amazing was the case of Japan. At the end of War World II, the country's automotive industry was very small, fragmented, and mainly concentrated on the manufacture of trucks and other commercial vehicles. But after the Korean War (1950–1953) three factors spurred its rapid transformation into a world giant: the facilities and plants created by U.S. investment for wartime purposes (Japan had become a base for assembling and supplying American military needs), a strong domestic economic recovery, and a protectionist policy that included 100 percent tariffs on imported vehicles.[5] With these supports, Japan—which still in 1960 produced 165,000 units per year—by 1971 overtook West Germany as the world's second largest car producer, and by 1973 it was able to deliver a staggering 4.5 million passenger vehicles yearly.[6]

Along with cars, all other forms of transportation developed at an astonishing rate. Civil aviation, for example, increased by eight-fold in terms of miles flown, making even airplanes vehicles for mass transportation. Oil also challenged coal as a source of heating and power generation, prompting a radical shift in the way industrial plants, cities, and houses were fed by energy, and thus contributing to cleaning up the skylines of most urban agglomerates from the smoking chimneys that were the most familiar heritage of the first industrial revolution. Weird as it may appear today, this revolution for those times was an environmentally friendly one that greatly improved people's standard of life.

The postwar years also witnessed another striking oil and natural gas based revolution, that of petrochemicals. As we have seen, petrochemistry took its first steps with the advent of thermal cracking in 1913. The period between the two wars represented the incubation era for far-reaching scientific research into the possibility of manipulating coal and gaseous hydrocarbon molecules to obtain cheap and flexible feedstocks for synthesizing new materials. Centered in Germany, Great Britain, and the United States, these efforts led to the discovery of several processes and products whose commercial development occurred in the 1950s. Because of the postwar economic boom, massive volumes of cheap raw materials and intermediates were required, and the petrochemical industry responded to the new challenge by supplying an incredible flurry of products. Probably the most visible aspect of this revolution can be summed up in one word: plastic.

The spread of modern plastics changed human daily life by replacing all of the most common materials used since the early days of civilization. In so doing, plastics forever altered the setting of cities, homes, and the very relation between human beings and objects, making the latter much cheaper and thus affordable by everyone. From a technical point of view, plastic materials had existed since the mid-nineteenth century. In fact, according to a scientific definition, plastics are solid synthetic materials that may take the place of other materials by miming them without having their composition. The first successful commercial product with these features was celluloid, a solid solution of nitrocellulose and camphor patented in 1870. The second milestone in plastics was bakelite, whose commercial success was far more long-lived and extensive. Invented and marketed in 1907 by the genial Belgian chemist Leo H. Baekland, bakelite heralded the era of thermoplastics, synthetic resinous materials solidified through heating, which gave them their final form. In the following decades, several common objects (pens, telephones, etc.) were manufactured in bakelite—and are now sought after by antique dealers around the world.

Yet what we are accustomed to calling plastic today is considerably more flexible and adaptable than the first plastics, and much cheaper: a thermoplastic polymeric material that is a mixture of one or usually more polymers and different chemical additives that change the polymer's composition and features making them solid but, at the same time, flexible enough to be adapted to many forms and uses. Among the first and most famous of these plastics derived from oil and natural gas were PVC* (1927), polystyrene (1935), and polyethylene (1940), which began to be commercially available by the end of the forties. In 1954 polypropylene was patented, the material that before long most household objects would be made of.

---

* PVC stands for polyvinyl chloride.

This extraordinary boom could not have occurred without cheap and abundant oil. Defying all predictions of its impending exhaustion, crude once again flooded the world right through the early 1970s, with production keeping well ahead of the impressive growth in demand. Indeed, overproduction was the hallmark of the Golden Age of Oil.

Even the United States, the most "tapped" region in the world, registered almost a doubling of oil production from 1948 to 1970.[7] But the superstar of the era was to be Persian Gulf petroleum, with its unrivalled low cost. Under the aegis of the Seven Sisters, the combined production of Saudi Arabia, Iran, Kuwait, Iraq, and the Arab Emirates skyrocketed from 1.7 million barrels per day in 1950 to 13.3 mbd in 1970 and to 20.5 mbd in 1973.[8] In the same period, average production costs in the Middle East declined from about 20 cents per barrel in 1948 to around 11 cents in 1970,[9] versus more than one dollar in Venezuela or nearly $1.30 in Texas (1970).

Global proven oil reserves jumped from nearly 70 billion barrels in 1948 to 667 billion barrels in 1973, extending their life-index from 20.5 to 32.7 years.* More than half of this quantity, or 355 billion barrels, was concentrated in the Middle East, which in 1948 was estimated to hold a mere 28.5 billions barrels.[10]

Surprisingly, these figures were deliberately kept low by the Seven Sisters' secret policy of underestimating Middle Eastern production and reserves. In effect, large multinationals conspired to limit exploration and production investments in the area so as not to "carry vases to Samos," as the ancient Greeks would have said—i.e., not to pour more oil onto an already flooded market. They were also careful to issue conservative estimates of the oil resources of Middle Eastern countries in an effort to restrain the latter's eagerness for more production and revenue. Such deception was possible only because of the Sisters' absolute dominance of Persian Gulf oil concessions through a system of cross-holdings. Since the 1920s and the 1930s, as we have seen, those concessions covered virtually the entire region, for periods of between sixty and ninety-nine years, granting Western companies quasi-extraterritorial rights and leaving host countries no possibility of taking part in oil-related decisions.

Leveraging on these favorable conditions, oil giants assigned each producing country an oil production quota that could change only if

---

* Reserves life-index expresses the ratio between proven oil reserves and current production.

consumption increased or events created a need for revision. Called the "Aggregate Programmed Quantity" (APQ), this system of prorationing represented a quite usual response to the never-ending struggle between the oil industry and overproduction. The system worked for about twenty years before Senate investigations revealed it to the world. Other than the oil companies, only the U.S. Departments of State and Justice had been previously informed of its existence.[11]

While oil production and reserves were being manipulated in the Middle East, in the United States the Texas Railroad Commission was presiding over a general effort by the American producing states to conserve subsoil resources and maintain an adequate margin of unused output capacity. In this case, the mission of the Texan entity was to both prevent a repetition of the ruinous and wasteful competition that had shattered the industry in the 1930s, and to guarantee the country the capacity to boost output in case of a sudden crisis. A consequence of the U.S. postwar oil strategy, this last issue became particularly sensitive after the Suez crisis of 1956, when the U.S. government secretly required the commission to ensure there was a production cushion sufficient to compensate for the temporary disruption of international oil supplies.[12]

As a result of the artificial limitation of production in the Middle East and the United States, from the early 1950s to the end of the 1960s the unused world oil production capacity—otherwise called "spare capacity"—remained substantial. Even today it is difficult to say how large it was. The director of the U.S. Office of Oil and Gas estimated that, in 1960, it was equal to more than 42 percent of the world's actual production capacity, excluding the Communist Bloc.[13] In 1961, the Chase Manhattan Bank placed it at 50 percent, also excluding the communist countries.[14] According to various sources, in the United States alone spare capacity in this period hovered at around 3 million barrels per day. Whatever the exact figure, even the forced withholding from the market of such a massive amount of crude oil did not succeed in preventing prices from dropping further and further.

Cheap oil was the rule right up to the eve of first oil shock in 1973. A barrel of Saudi "Arabian Light," which would become the crude of reference for the international markets, was officially priced at around two dollars in 1950, dropping to about $1.80 in the 1960s, and bottoming out in 1970 at about $1.21[15] (Free on Board, or Fob price). Even in nominal terms, these values were much lower than those registered on the eve of World War I or during the "Roaring Twenties." In real terms, they were astonishingly lower.

It was the longest and most stable period of cheap oil since the birth of the market. The abundance of such a low-cost, highly efficient, and versatile source of energy was a key factor in the most spectacular economic leap forward ever experienced by the world. Indeed, the postwar economic miracle of the industrial countries would have been unthinkable without the supply of cheap oil.

The age of oil's rapid entry into daily life coincided with the apogee of the Seven Sisters, which had by then achieved control of almost 80 percent of world reserves, production, and refining capacity outside of the United States, Canada, and the Communist Bloc. Including those areas, their share of global oil production remained at around 50 percent for the entire period under consideration.[16] Nonetheless, even the power of the major oil multinationals had its limits. New actors and new forces challenged their supremacy, relentlessly chipping away at it until they finally eclipsed it altogether.

The first of the new actors to appear was the Soviet Union, whose oil industry speedily recovered from the hardships it had faced throughout its tempestuous history. After being essentially shut down by the Russian civil war, it recovered during the 1920s but then stalled through the end of World War II. It was the discovery and rapid development of the huge Volga-Urals oil deposits in the early 1950s that ushered in the new era of Soviet oil. Between 1948 and 1973, daily Soviet oil production increased nearly fourteen-fold, from 616,000 to 8.5 million barrels.[17] Reinvigorated by this stunning accomplishment, Moscow promptly set about breaking into European markets by offering 20 or 30 percent discounts on its oil with respect to Seven Sisters prices, and bartering crude for industrial machinery, steel pipe, synthetic rubber, and many other goods. Washington warned its allies to beware of the Machiavellian strategy behind this generosity, notably the Soviet desire to draw European governments into a relation of growing energy dependence. But the alarm went unheeded. Countries such as Italy, Austria, Germany, France, and Sweden could not resist the siren song of Russian oil, and became its main buyers.[18]

This honeymoon with Europe convinced Soviet strategists to embark on an ambitious and far-reaching plan to create a vast transportation network to supply the new markets. The blueprint envisaged two major systems: the "Friendship Pipeline" would link the Volga-Urals area to Poland, East Germany, Czechoslovakia, and Hungary; a second pipeline would connect the same area with Stalingrad and the Black Sea ports of Novorossiysk and Tupsa. Other internal pipelines were built to connect the most important cities and industrial complexes, so that by the end of the 1960s the foundation of the contemporary Russian oil transportation network had been laid.

Together with the ever-increasing oil flows from the Middle East, the USSR exports drove down international prices, plaguing the profitability not only of the Seven Sisters but especially of another powerful category of oilmen: the so-called independent U.S. producers.

Globally overshadowed by the giant Seven Sisters, the independents were nonetheless the protagonists of oil exploration and production in the United States. Consisting of hundreds of minuscule-to-large firms, they were crucial to preventing the country from growing increasingly dependent on foreign oil sources. Politically, they were much more influential at home than were the multinationals, which were seen more as virtual super-states than as American companies, whereas the independents were very visible and active in local and national politics. They employed local people, directly affected their communities, and contributed heavily to political campaigns. With considerable wealth, the self-referent owners of these companies were often people of extremes who saw life in black and white and had little use for restraint or diplomacy in their relations with the rest of the world. They defended their interests bluntly using whatever means were effective, by themselves and through their common lobbyist organization, the Independent Petroleum Association of America (IPAA).

By the mid-1950s, oil overproduction and falling prices had put the independents and the Seven Sisters in sharp conflict. Large multinationals were importing growing volumes of cheap Middle Eastern and Latin American oil into the United States, displacing the high-cost domestic oil of their smaller competitors. All voluntary attempts to limit imports had little effect, and neither official plans nor the continuous effort of the Texas Railroad Commission succeeded in improving the situation. A direct consequence of this competition was a remarkable reorientation of U.S. companies' investments toward low-cost producing areas that could guarantee them higher profits. For the first time ever, in 1956 the American oil industry as a whole invested more abroad than domestically.

However, international expansion was not in the cards for the majority of small- and medium-size companies, whose world was limited to a few American states. Being substantially deprived of access to the Seven Sisters' foreign treasuries, the independents launched a lobbying campaign to stop foreign oil from flooding the country, and they finally got what they wanted. In 1959, President Eisenhower approved the Mandatory Oil Import Program, which capped oil imports at 13 percent of domestic consumption.[19]

What was a victory for the independents was a betrayal for the multinationals. The United States, which consumed nearly 40 percent of the world's oil, was the only major market that could possibly absorb the growing glut of Middle Eastern production. Already prorationed by the Seven Sisters, the latter could not be cut back any further except at the risk of provoking a major backlash by producing countries. Complicating matters further, the majors were already selling oil at a discount on the international markets, while in the artificial limbo of their pricing regime with producing countries, official posted prices remained unaffected, thus placing the entire burden of losses from low prices on the Sisters. In this framework, Eisenhower's decision forced them to make a momentous resolution.

In September 1959, Exxon's board unilaterally reduced its posted prices by about 5 percent, bringing the price of benchmark Arabian Light to $1.91 per barrel. The other oil multinationals followed suit, opting for reductions in the same range. In August 1960, posted prices were lowered by an additional 10 percent. The effect was as sudden and jarring as an earthquake. The Seven Sisters' move vaporized millions of dollars of producing countries' expected revenue and provoked a wave of rage among them. The most lucid interpreter of their revolt was Juan Pablo Perez Alfonzo, the man who had fathered the "fifty-fifty" profit-sharing formula in Venezuela shortly before being ousted by a military coup.

Reentering the political arena in the 1950s as the generals withdrew, Alfonzo became the country's Minister of Mines and Hydrocarbons and, once again, a thoughtful and dangerous enemy of oil multinationals. Alfonzo had a complex agenda. He wanted a sort of general agreement among producers to stabilize oil prices and production over the long term, one that would allow producing countries to accurately plan their future income and not fall prey to destructive hypercompetition. He also dreamed of a major and active role for all developing countries in the decision-making process as well as in the management of oil activities on their territory, and he hoped to establish a common front of major producers capable of collectively negotiating prices, production, taxation, and overall oil policy with the Seven Sisters.

But as a nationalist devoted to Venezuela's economic and social well-being, Alfonzo was also preoccupied by the rising tide of Middle Eastern oil, which was far cheaper than Venezuela's and could thus displace it on the international markets. In 1959, the country was struggling to remain the second largest producer in the world, after the United States. It produced more than 2.8 million barrels per day—less than it had in 1957, and the Soviet Union was on the verge of surpassing it (this would occur 1960).[20] But while Moscow saw Europe as its potential market, the booming oil production of the Middle East threatened Venezuela's most vital outlet, the United States.

With his double-sided strategy in mind, Alfonzo found an ally in the young director of the newly established Saudi Directorate of Oil and Mining Affairs, Abdullah Tariki. Even more radical than his Venezuelan counterpart, Tariki had spent his life in the oil business, beginning with his studies of chemistry and geology at the University of Texas on a scholarship from Texaco. Eventually he worked for Texaco in Aramco, before being appointed director of the Saudi oil sector in 1955. But despite his American training, job, and even his American wife, Tariki had become an admirer of Nasser and a fervent Arab nationalist.[21] The first meetings between Alfonzo and Tariki occurred before Eisenhower's decision to cut oil imports and the subsequent downward revision of "posted prices" by the Seven Sisters; yet these events acted as a catalyst for the two men's plans, leading to an unexpected agreement among producers that would have been considered mere wishful thinking only a few years earlier.

In September 1960, representatives from Venezuela, Saudi Arabia, Iraq, Iran, and Kuwait met in Baghdad and established the Organization of Petroleum Exporting Countries (OPEC) as an instrument for collective bargaining and self-defense.[22]

In its first years the new entity was all thunder with no storm. It soon surfaced that the members were far apart in their positions and claims, riven by jealousies and eager to win the best conditions for themselves to the detriment of the group as a whole. Rivalry prevailed over concerted effort, particularly in the case of Iran and Saudi Arabia. Both countries repeatedly clashed, each competing to produce more oil and attain the special status of the West's most important ally in the Persian Gulf. Iran had just surpassed the symbolic benchmark of 1 million barrels per day, and Saudi Arabian production stood at 1.3 million barrels, but the other countries of the newly established organization were no less eager to improve their positions. In 1960, Kuwait was the largest Middle Eastern producer, at more than 1.6 million barrels per day, having benefited from an impressive output escalation effort that had begun in 1950, when it produced only 340,000 barrels per day; meanwhile Iraq reached the 1-million-barrel mark.[23] None of them was willing to tighten the spigot of its source of wealth to benefit the others.

The struggle for more production was the real issue for OPEC, further exacerbated by the deep political antagonism among its members, which in turn fed the basic indiscipline that would become its norm. Above all, in a world where the consumer had the upper hand, the producers—both companies and countries—had to surrender to the hard reality that they were prey to the market's iron laws.

The inflationary tendency of the oil market was further aggravated in the 1960s, because of major new crude discoveries in Africa, particularly in Nigeria, Algeria, Egypt, and Libya. The rise of African oil proven reserves was impressive, from virtually nothing in 1948 to around 110 billion barrels on the eve of the first oil shock in 1972,[24] with a combined production of nearly 5 million barrels per day in the same year.

Another major blow to oil prices was the evisceration of independent American oil companies that successfully rushed to exploit opportunities outside of the United States—a phenomenon whose epicenter was Libya.

Thanks to a law aimed at avoiding the concentration of oil assets in the hands of few operators,[25] as well as the excellent quality of its crude (with one of the lowest sulfur contents of oil anywhere) and its closeness to the European markets, Libya became a sort of new Promised Land for many companies whose rewards were extraordinary by any historical standard. In 1955 the country did not produce oil at all and had no tapped wells; in 1970 it achieved an output of 3.3 million barrels per day, which made it the sixth largest producer in the world.[26]

More than OPEC and Africa, however, it was Iraq that was the agent of the most striking evolution in producers' politics during the 1960s. In September 1958, the country created by cynical British diplomacy in the wake of World War I was rocked by a dramatic coup d'etat. The British-invented monarchy was swept away, its tragic and horrible fate decided by the angry triumphant mobs of Iraqis. The latter literally got rid of the royal family and Prime Minister Nuri Said, the regime's hated strong-man. King Faisal II (the son of the first Iraqi King) was beheaded; his young son, the Crown Prince Abdullah, was shot, his body torn into pieces, and then his poor remains brought in procession through the streets of Baghdad; Nuri Said, who had managed to escape disguised with a woman's black cloak, was identified and lynched by the mob.

For the Seven Sisters and Total's oil concessions in the country, the revolution was more than a warning. For many years, even pro-British

Iraq had contested the Iraq Petroleum Co.'s policy, which, it argued, kept to a bare minimum investment in the development of Iraqi oil resources.[27] No one could know better than the Western companies that this charge was true. The Iraqi government did not even know about the Sisters' secret plan for prorationing Persian Gulf countries' oil production; even worse, it did not know that this policy was not one of equal sacrifice: politically stronger or more attractive countries such as Iran, Kuwait, or Saudi Arabia were penalized far less than Iraq, whose government was deemed a minor player. Consequently, as of 1960 the IPC had developed only eight out of the thirty-five oilfields that it had discovered, while 70 percent of Iraqi production came from the Kirkuk field alone.[28] Meanwhile, an official inquiry by the former government had also revealed that the company was using accounting tricks to reduce its payments to the government.[29]

After three long years of useless negotiations, during which the IPC conceded nothing, doomsday came in 1961. Through Public Law 80, Iraq took back from the Western company all areas that it had not yet developed, an amazing 99.5 percent of its original oil concession.[30] Protests by the IPC's parent companies failed to achieve any results, also because the Iraqi military government had not formally nationalized the country's oil resources but simply revoked a right granted to the company on the ground that it had not honored its own commitments.

Coming after the birth of OPEC, the de facto nationalization of Iraqi oil was a further sign that something was cracking in the Middle East's oil architecture under the impetus of the appealing pan-Arab vision of Egyptian leader Gamal Abdel Nasser. But before it grew to preoccupying levels, another breach in the postwar petroleum order was opened by a country that was apparently beyond suspicion in this regard, Italy, and by a man who, at the beginning of the 1950s, could be called anything but an oilman: Enrico Mattei.

In 1946 Mattei had been named vice president of Agip, with a mandate for liquidating the state oil company created by the fascist regime in 1926. Yet while a heated debate raged in Italy over the future of its oil industry and legislation related to hydrocarbon extraction, between 1947 and 1949 Agip made a series of important discoveries of natural gas in the Po Valley in the north of the country. Mattei immediately took advantage of the situation to oppose liquidation of the company and, to the contrary, proposed to set up a full-fledged state entity responsible for the full cycle of production, refining, and marketing of hydrocarbons; the company would also be entrusted with the security of the national oil policy and the exclusive control of the Po Valley as a monopoly. In February 1953, the Italian parliament passed a law establishing the entity Mattei had in mind, which was called Ente Nazionale Idrocarburi (National Hydrocarbon Entity), or Eni.[31]

The birth of the new state agency was the first area of conflict between Mattei and the Seven Sisters, particularly the U.S. companies. Opposing the substantial nationalization of the Italian oil sector, the latter embarked on a multipronged campaign against the plan that included attacks in the press, pressure on the government and individual politicians, as well as requests by the U.S. State Department that the Italian executive exert pressure to block it.[32] This first scuffle with the oil multinationals would have specific consequences for Mattei, particularly at the time when he decided to internationalize Eni in an effort to procure the company its own sources of crude outside Italy, which lacked significant oil deposits.

All available evidence supports the contention that when Mattei decided to focus on internationalization he had little familiarity with global problems and a rather vague grasp of the oil industry. This is not necessarily a strike against him. Like all pioneers, he was driven by a sense of mission, an extraordinary enthusiasm for great challenges, and an opportunistic perception of the weak points of his adversaries; he also had an indomitable desire to make Italy energy–self-sufficient, providing a basis for the industrial resurgence and modernization of the country. Mattei's most important achievement in those years, and his major contribution to Italy's growth, was the methanization of the country, a bold and monumental project for an era in which in most of the world natural gas was burned off at the well and accounted for less that 1 percent of Europe's total energy sources.[33] Thanks to Mattei's Eni, at the end of the 1950s, Italy had the most extensive natural gas distribution network in the world after the United States and Soviet Union,[34] making Eni one of the major players in the Italian economic miracle. But going worldwide was another game altogether for a small Italian company.

After an initial brief and fruitless foray into Somalia in 1953, Eni went into Egypt, where it signed its first important oil agreement, acquiring in 1955 a minority stake in the International Egyptian Oil Company (IEOC), the state oil company created by Nasser. Yet through some paradox of history, it was the Iran "normalized" by Reza Pahlavi that provided Mattei with the opportunity to decisively test the strength of the international oil system.

In March 1957, Eni and the National Iranian Oil Company signed a contract creating a joint company (SIRIP) that would introduce a new oil formula. The Iranian government would receive 50 percent of the gross profits of the company in the form of a tax, and then share in the fifty-fifty division of the company's net profits: as a consequence, 75 percent of profits would go to the Iranians. Moreover, the contract stipulated that ENI would cover the initial costs of exploration; only after oil was discovered would NIOC pay half of these exploration costs, assuming joint administration and management of SIRIP. Eni's Iranian contract thus contained two potentially revolutionary features: first, it superseded the "fifty-fifty" formula that the major multinationals had reluctantly accepted only few years before; second was the inclusion of local representatives on the board of the company, with equal power and in equal number as the Western company, a move that was far ahead of its time. The essential assumption in relationships between major companies and producer countries, in fact, had always been the extraterritorial character of the concessions, in other words, the exclusion of the national government from any decision regarding the administration or management of oil production.

In a single blow, Mattei seemed to have mined the foundations on which the system of the Seven Sisters rested. His action, moreover, seemed to be the product of an organic strategy to penetrate the whole Middle East. On March 25, 1957, just nine days after signing the contract with Iran, Mattei signed an analogous contract with Libya. The reaction was immediate. The U.S. State Department, the American embassies in Rome and in Teheran, and the top brass of the oil multinationals all moved in unison in an attempt to block approval of the agreement by the Iranian and Italian parliaments, fearing that its echo would be felt throughout the entire region, triggering requests for the revision of existing contracts.[35] On his own, Mattei did not hesitate to publicly proclaim the coming of a new era:

> The time has come, for all of us, to sit down at a table and talk. The old "fifty-fifty" formula is obsolete and we must find a new one. If the major companies want to fight me, let them do so. They can't say I didn't warn them.

On September 1957, the office that coordinated U.S. intelligence agencies submitted to the National Security Council a report on Mattei, stressing that the Italian oilman constituted a global threat to U.S. policy whose interdiction had to be a top priority of the government.[36] But in spite of the alarmist tone of this report, Secretary of State Dulles stated that he was not preoccupied by the actions of Mattei and that there was "nothing sacred" about the "fifty-fifty" formula. Similarly Eisenhower rejected every request for intervention against Mattei, holding that what he had done fell within the scope of free market competition. Both, moreover, raised doubts about the ability of Mattei to upset the world oil order given Eni's insignificant size.[37]

The British government also analyzed and weighed the Mattei case, somehow fearing the destabilizing effects Eni might have in the Middle East. But after requesting Shell's and BP's opinion of the Italian company and its leader, it issued a comprehensive position paper that echoed Eisenhower's remark. Consonant with the assessment of Shell and BP, the paper labeled Mattei's ENI a "paper tiger," too small and financially precarious to really pose a threat to the interests of the major Western multinationals in the Middle East. To reinforce their argument, Shell and BP had also added a table to their report showing that, in 1961, ENI produced almost 35,000 barrels per day of oil, in contrast to Exxon's 2 mbd, while all the other "Sisters" produced over 1 mbd. Even a small U.S. company, Skelly Oil, was producing 60,000 bpd, and abroad ENI had fewer than thirty insignificant productive wells. According to the oil majors, all Mattei wanted was a place in the sun, i.e., a stake in the Iranian Consortium, or something similar. But he had nothing to offer in exchange.[38]

In effect, the logic of the postwar oil system did not offer much room for a new company lacking both massive means and above all outlets on the international market. Oil overproduction, the Seven Sisters' global control of oil refining and marketing, and the pricing system then in place all provided powerful reasons for real complicity between the major multinationals and the governments of the producing countries. Only the former were actually capable of guaranteeing the latter remuneration insulated from drops in the market. Moreover, Mattei's ENI would not have been capable of assuring the marketing of discovered oil, because it lacked the huge downstream capacity of the largest multinationals. Nonetheless, Mattei dealt a serious blow to the postwar oil system. As early as the second half of the 1950s, it was ENI that signed the most important contracts (exceeded, however, by Germany and Austria) for oil purchases from the Soviet Union, attracting yet again the attention of the American authorities. Furthermore, apart from openly supporting Nasser, Mattei financed various nationalist movements in Arab countries and North Africa fighting the colonial powers, most importantly the National Liberation Front of Algeria. When in October 1962 Mattei suddenly died in a suspicious airplane accident—which only in 2004 was proved to be the result of a bomb—it was natural for Italians, by now conditioned by his propaganda against the Seven Sisters and the colonial powers, to suspect the participation of the United States or some other hostile country in his death.

In reality, the new Kennedy administration had decided to seek an accommodation with Eni. This turnaround was dictated by the evolution of Italian politics rather than by oil matters: having by now agreed to support the entrance of the Socialist Party into the Italian governing coalition, the Kennedy administration wanted to ensure that a similar development would occur only under the complete control of its own players and instruments. In this framework, Mattei had to become an "ally," being consistently considered by the United States to be the most powerful figure in Italy.[39] A few months before his death, the "father" of ENI adapted to this possibility in an act of realism. Largely because of its immense outlay of funds on fruitless international oil exploration, ENI was going through a period of considerable financial difficulty and the Italian Central Bank had even barred it from issuing new bonds (1960).

Thus from 1961 on, Mattei began to have confidential meetings with top emissaries from the Kennedy administration, like Averell Harriman, George McGhee, and George Ball. In the course of these meetings, the United States delineated the political path to the agreement it desired.[40] Things progressed as hoped. After the meetings, Secretary of State Dean Rusk personally invited the number two of Exxon, Howard Page, to open negotiations with the Italian company, suggesting that his company offer ENI oil at an attractive price.[41] A few days before Mattei's airplane exploded in mid-air,[42] a memorandum of understanding between the two companies was ready, and on that basis Exxon would provide ENI with oil at a discounted price for five years. It was Mattei's successors who would sign the agreement.

The disappearance of the father of ENI deprived the oil industry of a personality resembling those of the great founders of the oil industry. Yet his death did not allow the oil multinationals to breathe even one sigh of relief.

CHAPTER 8

Oil and the Explosion
of Arab Nationalism

While the global oil market was weakening, growing nationalism throughout the Arab world began to destabilize the postwar petroleum order. Centered in Egypt, this phenomenon had its catalyst in the new leader of the country, Gamal Abdel Nasser, who also succeeded in pointing out to the Arab countries how they could use oil as a weapon to free themselves from domination by the West.

Nasser had been one of a group of young officials who in 1952 staged a coup that put an end to the Egyptian monarchy and set in motion a process of profound renewal of the country's political system. In the two years that followed the coup, he kept a low profile while the older General Neguib ran the newborn republic. Only in 1954, at the age of 36, did Nasser assume direct power, pushing Neguib aside and naming himself president of the Republic.[1]

The political plan of the young Egyptian statesman soon showed an evocative power that shook the ideological panorama of the Arab world. Even though he was essentially an Egyptian nationalist, Nasser portrayed the vision of a unified Arab nation, a society without internal borders inspired by socialist principles that would allow the Arab peoples to fully recover their freedom, dignity, and values.[2]

To achieve his political goals, Nasser did not hesitate to install a single-party political system and to violently eliminate his rivals, even those from the Muslim Brotherhood and the Egyptian Marxist movements, who considered his doctrines a travesty of their ideas. With absolute control of Egypt, by 1954 he was able to concentrate on spreading his vision beyond its borders, generating propaganda that inflamed the popular psychology of other Arab countries and furnishing them with an ideology and sense of mission that had been lacking until that time. From 1954 on, insurrections shook Tunisia, Morocco, and particularly Algeria, countries under French domination. In Algeria, in November 1954, a revolt began that would grow into a prolonged and painful civil war and one of the darkest chapters of French history. By the same token, Nasser's political vision and opportunism led him to employ the principle of oil control as a tool of political blackmail.

In his "Philosophy of a Revolution," the Egyptian leader defined oil as one of three fundamental pillars of Arab power, as well as the most efficient weapon in making the rights of the Arab nation prevail over those of the Western powers.[3] This predicament was supported by an unexpected and unconscious ally, the U.S. Federal Trade Commission, that through the publication of the results of its inquiry into the "International Petroleum Cartel" furnished many Arab intellectuals and Nasser himself with a clear sense of how much power many countries of the region had unwittingly placed in the hands of an exclusive club of Western multinationals. Moreover, precisely in those years, the oil production of these same countries was growing at a dizzying rate, while others, though they lacked oil deposits, were benefiting from guaranteed income from the oil pipelines that crossed their land (e.g., Syria and Lebanon) or from the passage of oil tankers through their territory, as in the case of Egypt and the Suez Canal. Nasser thus began to wield the weapon of oil, proclaiming the right of Arab governments to sovereignty over their natural resources against the control of foreign powers, and calling for the withdrawal of international oil companies and the constitution of local consortiums to produce, transport, and refine crude oil. The consequences were felt immediately.

From 1955 onward, oil sector workers in various Persian Gulf countries called a series of strikes in the name of Arab nationalism. After several riots, in June 1956 an official visit of the new Saudi King Saud to Aramco headquarters sparked an uprising against Aramco itself. The king ordered a severe crackdown by the military, forbidding from then on any unionization or political activity by the oil workers. Saud had succeeded to the throne of Saudi Arabia after his father ibn-Saud's death in 1952; soon after the rise of Nasser to power, he had tried to establish good relations with the Egyptian leader, apparently reciprocated by the latter. The king even invited Nasser to Saudi Arabia in September 1956, a visit that turned into the "first massive political demonstration in Saudi history"—clearly in support of the Egyptian leader who had also sponsored some of the unrest against Aramco and the very Saudi monarchy.[4] It was not long before the action of a personality in revolt generated an international crisis of major proportions.

When in July 1956 the United States officially informed Nasser of its unwillingness to finance the construction of the Aswan Dam, Nasser's most ambitious project, the Egyptian leader's reaction was dramatic. Within a few days he announced the nationalization of the Suez Canal Company, with the goal of using the income from the canal's tolls to finance construction of the dam. At the same time, Nasser froze traffic in the Gulf of Aqaba on the Red Sea, which was vital to Israel's oil supply. This news reverberated throughout the world and aroused deep concern among the major Western governments: just three years after the defenestration of Mossadegh, a new threat was casting its shadow over the interests of the European powers in yet another Middle Eastern country.

The Canal Company was in fact one of the world's great Franco-British jewels and an important source of income for both countries; but it was also a geopolitical garrison of extraordinary importance. The majority of international trade to and from the Persian Gulf passed through the canal, but more than anything else it was the primary route for transporting Middle Eastern crude to European markets. In 1956, the Middle East produced 3.5 million barrels of oil per day, exporting about 90 percent of it; around 1.3 million barrels passed through the Suez Canal every day, supplying more than half of European petroleum needs.

With Nasser's popularity among the Arab masses reaching spectacular heights, France and Great Britain decided to hit back hard. Considering the Egyptian president a mortal threat to their interests throughout the Middle East, the governments of the two countries worked out a secret military plan to take back control of the canal and drastically prune Nasser's power. Israel was also involved in the plan, worried as it was by the unscrupulousness of the Egyptian leader and firm in its refusal to passively accept the closing of the Gulf of Aqaba.

The essential outlines of the plan were rather simple. Israel would militarily occupy the entire Sinai Peninsula, thus getting close to Suez, with the goal of reestablishing freedom of trade in the Gulf of Aqaba. Meanwhile, after delivering a useless ultimatum to Egypt and Israel, Great Britain and France would intervene militarily on the pretext of creating a security buffer between the two countries and protecting the canal. On paper, it seemed simple, but it soon proved to be a political and diplomatic fiasco. The three aggressors had not factored in the reaction of the international community and particularly that of the United States, which had no intention of getting involved in actions and policies that seemed to have been dictated by a desire to perpetrate neocolonial control over an independent country. For these reasons, on November 1 the country presented a resolution to the UN (approved by the General Assembly), demanding the immediate cessation of hostilities and the restoration of the preexisting situation—i.e., the withdrawal of Israel, France, and Great Britain from the positions they had occupied.

Washington's net condemnation of the action was motivated not only by a real discomfort with the methods and antihistorical claims of the French and British but also by the explosion of another international crisis. Just as these events were unfolding in Egypt, Soviet tanks were entering Budapest to crush Hungary's uprising against Moscow's rule, arousing the ire and condemnation of much of the world. In blaming Soviet imperialism, the United States was bound to consider the incursions into Egypt in the same light.

Israel, France, and Great Britain found themselves internationally isolated and even threatened with intervention by the USSR. With no alternative, on November 6 they were thus obliged to call off their operation and accepted the conditions imposed by the UN.

For Nasser it was a triumph.[5] In the eyes of the Arab world, the Egyptian leader had challenged Israel and the colonial powers and won, thus demonstrating the possibility of deliverance for the Arabs. Word of Nasser's victory spread throughout the Middle East through *Radio Cairo* and *The Voice of Arabs*, entirely new tools of mass propaganda for an Arab leader that permitted the Egyptian leader to speak directly to the Arab people, bypassing their governments and presenting himself as a reference point for the entire Arab universe.

The growth of domestic pro-Nasser movements led Jordan and Lebanon to request U.S. military intervention in 1957. In February 1958, Syria and Egypt brought into being the United Arab Republic (UAR), which Nasser intended to serve as the embryo of a united Arab nation. Notwithstanding the velleity of the political plan (which was soon to fail as a result of rivalry between the two countries and Nasser's desire for hegemony) the Seven Sisters and the governments of the principal Western countries were duly alarmed. Syria in fact represented the juncture of the two largest Middle Eastern oil pipelines, one Iraqi, one Saudi. During the 1956 Suez Canal crisis, pro-Nasser Syrians had blown up the oil pipeline carrying oil from Iraq to the Mediterranean in support of the Egyptian leader's cause. With the Suez Canal now tightly controlled by Nasser, the UAR was in a position to control the entire flow of Middle

Eastern oil to the Mediterranean. Then came the bloody coup in Iraq in July 1958, and the eventual consolidation of power by the pan-Arab, lay, socialist-leaning Baath Party in both Iraq and Syria. In 1964, the Palestine Liberation Organization (PLO) was formed in Alexandria, Egypt, under the auspices of the Arab League, but in reality at the behest of Nasser himself, who intended to take full advantage of the Palestinian question with a view to advancing the larger project of Arab unification.

As a result of this propagation, in the early 1960s Nasser's influence over the Arab masses seemed unstoppable, as did his capacity to destabilize the precarious equilibrium left in the wake of the decolonization of the Middle East. Washington began to study ways to reduce European dependence on Middle Eastern oil, fearing that sooner or later tensions in the region could disrupt the constant flow of oil to the allies. Still, while the world oil market was drifting steadily toward ever greater overproduction, Nasser's successes were obscuring unresolved problems that would soon erupt to plague him. The most important of these was the reality of the Arab world itself.

Secular divisions, elites set against other elites, deeply rooted local features, made it impossible to call for the dissolution of all borders and the overcoming of local identities. Beyond the requisite speeches about solidarity and agreement, each Arab government eyed the others with diffidence and rivalry. The regional and international interests of the various countries of the Arab League were irreconcilable, suspended between socialist or neutralist positions and pro-Western stances, between oil wealth and the relative or absolute lack of raw materials, between circumstances and history that might legitimize the ambitions of local leadership, and the absence of the minimum requirements for power.

Riding the wave of success, Nasser did not stop even when these problems began to erupt, continuing to spread Egypt's energies and resources too thin on too many fronts. Of particular importance was his attempt to set up a friendly regime in Yemen, backing a coup that put in power pro-Egyptian elements in 1963. The Zaydite Imam (the Zaydites are a Shiite sect) who ruled the country retreated into the mountains of the North with his followers, unleashing a guerrilla resistance that before long grew into a civil war. Thus Egypt found itself dragged into an unplanned military escalation that by 1968 saw 70,000 Egyptian troops engaged in Yemen fighting an anomalous war, more like what the United States was attempting in Vietnam than a conventional conflict, albeit on a much smaller scale. That conflict was destined to consume lives and funds without offering any concrete possibility of success,[6] and it had also the negative consequence of damaging the already tense relations between Egypt and Saudi Arabia.

The friendship nurtured by the weak King Saud toward Nasser had already been a major factor behind the intense power struggle between the king himself and his brother Feisal, each one representing two faces of the Saudi society and body politic. In 1962, Feisal had ousted Saud with a de facto coup; also it took two more years before Feisal formally assumed the crown of Saudi Arabia. With Faisal in command, the Saudi stance toward Nasser changed, and his Yemenite adventure made it precipitate.

The Egyptian *rais* had toppled a traditional dynasty in an Arab country and replaced it with a socialist lay regime. Moreover, this happened on the Saudi border, threatening from close up the established order of the principal monarchy of the Persian Gulf. Worse, resembling that of contemporary Arab or Islamic radicals, the Egyptian propaganda portrayed the Saudi monarchy as a corrupt, lavish, and impious servant of the American interests:

> O! my brother Arab! American imperialism stresses that it should remain in Saudi Arabia, and this is the price of protecting the Saudi throne from destruction. Thus Saud's throne will remain under the protection of American battleships and of the atomic base in Dhahran [which did not exist]. But Saud should know that by this he is speeding his end. Imperialism will never save thrones from the will of the people. The people endure, but they do not forgive. The free men in Saudi Arabia do not fear any power on earth and they shall soon strike. [America], which is controlled by Zionism, wants to liquidate the Palestine problem with the help of Saud, the criminal king and the protector of the Muslim countries.[7]

Given the state of the art, Riad could not but move rapidly on two fronts. To start, it backed the anti-Egyptian resistance in Yemen, which it furnished with arms and funds; second, it sought to counterbalance the influence of Nasser in the Arab world with its own influence through an instrument that would have enormous consequences on the political evolution not only of the Middle East but the entire Muslim world.

In 1962, Saudi Arabia masterminded the creation of the World Islamic League, using its own prestige as both protector of the holy sites of Islam (Mecca and Medina, the two cradles of Islam), and representative of a "pure" fundamentalist form of Islam. From its conception, the League attempted to promote and spread Islamic culture by supporting associations and movements and through the construction of mosques and cultural centers throughout the Islamic world or wherever a Muslim minority was to be found. At the same time, the League was charged by its founders with organizing and administering economic aid programs for the poorest communities and families, using both state and private donations.[8] In 1969, then, Saudi Arabia pushed for the creation of the World Islamic Conference, which in 1971 was endowed with a permanent secretariat. The United States approved of and supported this strategy as antisocialist and antinationalist, considering it to be a conservative response intended to guarantee the existing order. Almost forty years later, after the tragedy of September 11, 2001, the United States would rail against this form of Islam spread by the official Saudi channels, forgetting the reasons they supported it for one decade after another.

In any case, the juxtaposition of Nasser's lay vision and the Saudis' religious one upset the balance of elements on which Nasser had grounded his power in Egypt. The persecution and silencing of the Egyptian opposition had hit the Muslim Brotherhood particularly hard, forcing many of its members to seek shelter in other Arab countries. Saudi Arabia treated these refugees with prodigious generosity, which incited further retaliation by Nasser. In 1966, at the peak of the Yemen civil war, Nasser authorized the execution of the leader of the Brotherhood, Sayyid Qutb, then in prison in Egypt. At the time this event passed largely unremarked, but the works and thought of Qutb, most of which were published outside of Egypt with Saudi financing, were to provide the essential component of the theoretical kit of the fundamentalist Islamic movements for decades to come.[9]

While the errors in judgment and rivalry of Arab countries undermined Nasser's political project, his popularity was severely tested by a new conflict with Israel.

In part, the Egyptian leader had provoked the spiraling of events that would set off the new conflict, but he was caught off guard by Israel's decision to aggressively address the growing threat of war by initiating war. On 5 June 1967, Israeli troops and planes attacked Syria, Jordan, and Egypt, and in a mere six days—thus the name the Six-Day War— they completely destroyed the Arab resistance and took positions that Israel would never give back, despite subsequent UN resolutions. Among those positions was the Arab part of Jerusalem (which made it possible for Israel to impose total control over the entire city and proclaim it the "eternal capital of the state of Israel"), the West Bank, the Gaza Strip, and the Golan Heights.

In the wake of the total rout of the military apparatus that was fighting Israel, the Arab world responded with a solidarity it had never known, and oil would play a big part in this response. On June 6, Arab oil producers imposed an oil embargo on Europe, Japan, and the United States. This decision represented a completely new fact in postwar international politics. Following the lines suggested by Nasser himself, that experiment was the fruit of the alignment of Saudi Arabia and the other moderate monarchies of the Gulf with the anti-Western policies of the pan-Arab front, but it was also a matter of political necessity: in the tumult created by the Israeli offensive, the Saudis and other major oil producers would risk isolating themselves dangerously in the Arab world were they to abstain from participating in the collective effort to contain Israel's success.

Everything indicated that the petro-retaliation against the industrialized countries guilty of supporting Israel would produce significant effects, because the Middle East supplied about 80 percent of European and Japanese oil needs. Nonetheless, the halting of Arab oil exports did not have the desired effect. The United States reacted by increasing domestic production, eventually to full capacity, to make up for the shortages in Europe and Japan; additional quantities of crude were guaranteed by Iran and Venezuela, while the system of world oil shipping was reorganized to allow the new flows of exports to reach their destinations. Thus, just a few weeks after their decision to wield oil as a weapon, the Arab countries grew convinced that conditions were not right to use it efficiently and reversed course. The embargo against the United States and United Kingdom, seen as the most direct supporters of Israel, lasted a few more weeks, but the effects of this embargo were even feebler than those of the other one.

The defeat in the Six-Day War put a crimp in Nasser's ascendancy but not in his message of revolt. Socialist-leaning Arab nationalism had spread to many countries of the Middle East, adapting itself according to the organization of the state and the Arab masses in the postcolonial era. But in Egypt, as in the rest of the Middle East, its most immediate and visible consequence was the militarization of civil society, which drained resources from the satisfaction of basic needs that the new governing Arab elites should have addressed: better living conditions for the people, more possibilities for social and cultural growth, modernization, and free individual and group expression.

By 1969, Egyptian military spending was devouring 19 percent of the country's GDP, true to the pattern in other Middle Eastern states. This tendency was reinforced by the conviction that maintaining leadership in the symbolic war against Israel was the prerequisite for maintaining leadership in the Arab world. Even after 1967, Nasser did not seek nonmilitary avenues and initiated a war of attrition against Israel using border skirmishes and guerrilla actions. Egypt grew increasingly dependent on Soviet arms, financing, and advisers to the extent that by the early 1970s it seemed to have been drawn into Moscow's orbit. Meanwhile many of the processes that Nasser had catalyzed had grown independent and uncontrollable, particularly in Palestine.

In 1969 Yasser Arafat took control of the PLO after having been kept out with his organization, al-Fatah, because he disagreed with the way in which the PLO had been conceived. Al-Fatah had always refused to consider the independence of Palestine as part of a broader goal of reconstituting the integrity of the united Arab nation proposed by Nasser and the pan-Arab movement. As a consequence, once Arafat took charge, the exclusive goal of the PLO became the fight for the liberation of all of Palestine, including the sections ceded to Israel by the United Nations decision of 1947. Also in 1969, a coup in Libya brought to power a group of young officials of pan-Arab, socialist inspiration lead by Colonel Muhammar al-Qaddafi.

But by now Nasser's time was up. The Egyptian leader barely lived to see in 1970 the massacre of Arabs by Arabs perpetrated by Jordanian troops against the Palestinians who had taken refuge in Jordan under the aegis of the PLO, an event that entered history as "Black September." A few days later Nasser, who had mediated in an attempt to effect a reconciliation between Jordan's King Hussein and Arafat, had a heart attack and died. Retrospectively, Nasser had also been a victim of oil overproduction. His attempt to use oil as a weapon had crashed against the continuous flow of oil from everywhere and the American cushion of unused oil production capacity. It fell on the initially ardent pro-Nasser Arab leader Qaddafi to be much more effective in playing the oil card as a destabilizing tool. Like Nasser, Qaddafi had come to power in Libya through a coup of "young officals" in 1969.

Libya's oil bonanza in the late 1960s nurtured Qaddafi's drive to impose on the Western oil companies his own agenda, which was made up of two primary goals: raising the price of high-quality Libyan crude and increasing the government's share of oil profits. In pursuing those targets, the Libyan leader was helped by both the peculiar fragmentation of oil properties on his country's territory, and the overnight fortune built up in Libya by a small American independent company, Occidental Petroleum, owned by American millionaire Armand Hammer.[10]

Under Hammer's autocratic reign, Occidental had won two concessions in Libya in 1965. One was in an area abandoned by Mobil, which had drilled without discovering anything. In an ironic twist of fate, it was there that Hammer's company discovered a huge oilfield (initial estimates indicated nearly 3 billion barrels of recoverable reserves, which eventually proved far greater), which transformed it overnight into an international major, with a daily production of 800,000 barrels per day by 1969. But this terrific success was built on sand. Libya had become Occidental's only source of reserves and production, which made it vulnerable to any request by the Libyan government, and Qaddafi soon realized this. On January 1970, the Arab leader called for a revision of the "fifty-fifty" formula, and after the oil companies' initial refusal, he imposed upon them progressive production cuts.[11] While the other companies stood firm, Occidental had no choice but to accept the terms after seeing its production cut by half.

In September 1970, Hammer signed a new agreement with the Libyan government providing for a 55–45 oil profits split in favor of the Libyans, and a 30 percent increase in the country's posted price for oil.[12] The breach opened by Hammer's company weakened the positions of the others, which had refused it any help, and it was only few weeks before they too acceded to Qaddafi's request.

However, while Libya's success was the decisive spark that set off an unprecedented chain reaction in the postwar oil order, it was the preceding twenty years of overproduction and low prices that had created the conditions for the shocking conclusion of the Golden Age of Oil. Moved by the almost inexorable law of opposites, the pendulum was about to swing the other way.

CHAPTER 9

The First Oil Shock

A ll of a sudden, in 1971 oil overproduction and the decline of prices ended. And, all of a sudden, it surfaced that twenty or so years of foolish consumption habits and excess availability of crude had relegated to a no-man's land all security issues that the United States had tried to address in the aftermath of World War II. However, the spiral of events leading to the first oil shock was by no means triggered by the physical scarcity of oil. Rather, it was a sort of perfect storm of diverse circumstances and events that converged to bring on the crisis, or at least generate the collective psychology that would sustain it.

Even today, it is difficult to say to what extent the shock was the self-fulfillment of the prevalent gloom regarding the future. In retrospect, the period of the early 1970s appears like an ancient Greek drama, whose actors, incapable of escaping their fate, precipitated it with their irrational actions.

At the beginning of the decade it became clear that the postwar exploit of oil, based on low prices and an ever-growing supply, depended more and more on the Middle East, the only area in the world capable of continuously expanding production to satisfy global consumption. This was partially due to the huge shift begun in the mid-1950s in exploration and production investment from more politically stable United States and Venezuela to Middle Eastern and North African countries, where production costs were lower. In the United States, for instance, drilling activity fell dramatically from 1955 to 1971: the number of drilling rigs at the end of this period was "a little more than one third the level of the mid-1950s."[1] In broader terms, the shift in investment and decline in drilling were a general phenomenon provoked by the steady erosion of

upstream margins. Because of competitive pressures and overproduction, throughout the 1960s upstream exploration and production capital
spending dropped to a mere 25 percent of total investment by the oil
industry, which found it far more convenient to allocate its money in
its refining and petrochemical operations.[2] In the West, all of this had a
sharp effect.

In an attempt to meet the ever-climbing demand for oil, in 1971 the
de facto arbiter of U.S. crude production, the Texas Railroad Commission, approved putting onstream all available production capacity of
the country. This unprecedented decision was the first clear sign that
something was wrong with the available supply, and indeed it produced
only a useless palliative. By the end of 1970, production in the lower
forty-eight states (i.e., without Alaska and Hawaii) had peaked at more
than 11 million barrels per day, and subsequently entered a phase of
long-term decline. While useless in addressing the real market needs, the
TRC decision eliminated the sole disposable spare production capacity in
the Western world, depriving industrial countries of the security cushion
they had relied upon since the 1950s. It was a remarkable change in the
history of oil, one that apparently made the West vulnerable to the use of
the "oil weapon" by the Arab countries. But this would be a simplistic
and incorrect way to deal with the origins of the first oil shock, whose
deflagration was the result of many more causes.

A peculiar role in contributing to the degeneration of the oil crisis was
played by the economic and regulatory policy introduced in the United
States by the Nixon administration. In August 1971, the president announced the unlinking of the dollar from the gold standard in a dramatic
move to remedy the overvaluation of the currency. At the same time, he
implemented a vast system of price controls intended to relieve the other
plague of the U.S. economy—rampant inflation—which was exacerbated by the financing of the Vietnam War without raising taxes.[3] Accordingly, price ceilings were imposed on oil at the wellhead and burner
tip, which made them artificially low at a time when domestic demand
was rising and production was at maximum capacity. The result was
two-fold: on the one hand, Nixon's measures provided an incentive to
increase domestic oil consumption; on the other hand, it discouraged
domestic exploration and development in favor of imports.[4]

By this time, the notion of the possibility of an energy crisis had already
begun to circulate in the United States, particularly after the terrible
winter of 1969–1970—the coldest in thirty years—when many utilities
supplying electricity had been forced to interrupt service because of oil

and natural gas shortages.[5] The most prominent figure warning of a looming storm was the State Department's Middle East expert, James Akins.[6] A sober and highly erudite Quaker with "uncompromising principles,"[7] Akins repeatedly warned that the world would risk running out of oil in a few years if there was not an immediate shift to other sources of energy; in his view, the latter would be made possible by higher oil prices, which would render investment in oil alternatives attractive.[8]

Actually, this line of reasoning was wrong in strict economic terms. There were abundant oil resources in the world that could yield double-digit profits if developed even with oil prices below two dollars per barrel,* but that simply were not developed because of the oil companies' persistent fear of overflowing world markets. More than economics, however, Akins grasped the nature of the Arabs' mounting expectations and changing psychology which, right or wrong, were driving the world toward a different oil order.

After Qaddafi's success with Western oil companies, the major producing countries realized that the wind had shifted in their favor, though it was also fanning rivalries among them. In November 1970, Iran's Reza Pahlavi angrily demanded a profit-sharing formula and a posted price in line with those obtained by Libya. In December 1970, OPEC approved a resolution requesting both the application of the 45–55 profit-sharing formula to all its members and separate negotiations with Persian Gulf and Mediterranean/African countries to review price levels. Soon thereafter Libya proclaimed that it would demand higher prices and taxes on its oil should the Gulf countries obtain the same conditions it had won in September.[9] Oil companies watched this charade in dismay, and yet had no alternative but to accept negotiations with producing countries, which were held in Teheran and eventually in Tripoli in the first months of 1971.[10] It was the beginning of an endless retreat in the face of the new strength of producing countries.

Indeed, while the Teheran and Tripoli agreements gave all producer countries a 55 percent share in oil profits and a higher posted price, the new terms did not placate them. Ignited mainly by the Libyan-Iranian rivalry, new calls for additional price and contractual revisions became the norm of oil producers' behavior in the face of any agreement with oil companies, generating a leapfrogging process that left the latter at the

---

* As observed in the previous chapter, total production costs (or technical costs, notably capital plus operating costs) in Saudi Arabia were 11 cents per barrel of oil.

mercy of the former. In this case as well, the U.S. policy aggravated the situation.

In his 1972 State of the Union address, President Nixon had formulated the so-called Nixon Doctrine, the key goal of which was preventing direct U.S. involvement in future conflicts that were "peripheral to the central interests of the great powers." A direct consequence of this strategy was the empowering of those allied countries that could withstand regional threats without relying on U.S. troops, a new approach to the containment of Soviet expansion that needed regional "policemen" capable of acting as surrogates for an American presence on the ground. As far as the Persian Gulf was concerned, the U.S. strategists had chosen Reza Pahlavi's Iran to perform the role of pro-Western security "pillar," and they were thus favorable to indirectly finance the country's rise to the status of regional superpower through oil price increases. Given this overall picture, the Seven Sisters and their governments were vulnerable as never before. And like a flooding river swells as it rushes toward the sea, this awareness incited the producer countries to play not only for higher prices but also for two recurrently evoked objectives, namely participation or, worse, nationalization of the oil concessions.

In general, proponents of nationalization within OPEC wanted to have complete control of foreign companies' assets, strategies, and operations on their land. To the contrary, supporters of participation had a more cooperative approach to the issue. As explained in 1969 by the father of the concept participation, Saudi Oil Minister Zaki Yamani:

> For our part, we don't want the majors to lose their power and be forced to abandon their roles as a buffer element between the consumers and the producers. We want the present set up to continue as long as possible and at all costs to avoid any disastrous clash of interests which would shake the foundations of the whole oil business. That is why we are calling for participation.[11]

The architect of OPEC's ascent in the 1970s as well as a harsh critic of its eventual follies, Yamani was to become one of the most popular characters in the history of oil.[12] Fond of long-term goals, he saw participation as the only way to align the interests of producers and marketers, thereby safeguarding an orderly development of supply and prices; for this reason he also remarked that participation had to remain "indissoluble, like a Catholic marriage."[13] But while discussions on nationalization and participation were taking place, militant states acted.

After Algeria expropriated 51 percent of all foreign companies' hydrocarbon holdings between 1968 and 1971, it was once again Libya that set the pace for the takeover of Western holdings. In 1971 the Colonel ordered the complete nationalization of BP's Libyan assets. Eventually, Libya applied the same scheme as Algeria, taking over 51 percent of all foreign oil ventures. In 1972, Iraq followed suit by completing the expropriation of Iraq Petroleum Company assets that it had initiated in 1961.

While all Seven Sisters' walls were crumbling down, inflation and the devaluation of the dollar blew another ill wind upon them. Day after day, the downward spiral of the American currency significantly eroded the revenues of crude producers, the dollar being the reference for all oil transactions worldwide; in addition, because their economies were almost exclusively oil-based, they had to rely for the bulk of their purchases on Western imports, which were made more expensive by inflation. This vicious circle gave an additional incentive to oil producers to raise the price of crude and curb in their favor contractual conditions in order to counter the erosion of their wealth. The irrationality of the oil market only made things worse.

In an attempt to secure supplies for their refineries, independent companies engaged in a frantic struggle to buy up every available barrel, nurturing the emergence of a spot market where oil was sold to the highest bidder, outside of the system of posted prices and long-term commitments that regulated the vast majority of oil contracts. Prices of these "spot" transactions reached six dollars per barrel or more for a single cargo, when posted prices were two or three dollars. Even if these transactions amounted to less than 3 percent of all oil traded on the international markets, they had a huge impact on the sellers' psychology.

Romans said *res tantum valet quantum vendi potest*—a good is worth as much as it can be sold for. For the producers, the emerging spot market proved that consumers were so oil-crazy that they were ready to pay well above official prices, which they consequently raised in a rush to catch up with spot values.

The final blow to market sclerosis was dealt once again by the Nixon administration's changes to its oil policy in 1973. With shortages occurring periodically as a result of the odd price controls imposed in 1971, Nixon's men reacted by devising an oil regulatory system that took complexity and contradiction to the extremes. In April 1973, the U.S. government decided to cancel the oil import quotas fixed by President Eisenhower in 1959. Because Middle Eastern oil was so much cheaper than that from the United States, quota elimination provoked an immediate surge in domestic demand, providing an additional reason for prices to soar. Quotas were replaced by a more complicated system based on tariffs, then by petroleum price freezing. In August 1973, price controls were partially removed according to highly complicated rules;[14] eventually, the government launched an "entitlement system" in order to guarantee importers or domestic refiners access to cheaper oil.[15]

Bad regulation always seeks to remedy the damage it has caused by imposing new rules; in most cases this only makes matters worse. Indeed, the regulatory militancy of the Nixon administration made the United States and the world more vulnerable to the eventual oil shock. For some energy experts, like Edward Morse, it even "made the Arab oil weapon usable."[16]

Between December 1970 and September 1973, official oil prices* jumped from \$1.21 to \$2.90 per barrel, while spot values topped \$5.00. These figures, however, do not convey the extraordinary fragmentation and volatility that rendered the market an unreliable source of information. By mid-August 1973, as reported by *Petroleum Intelligence Weekly*, prices had actually become "imprintable" because of schizophrenic differentials among the various qualities of crude, spot transactions, discounts, and so on.[17] Surprisingly enough, demand also continued its upward rush, shooting from 46 million barrels per day in 1970 to around 58 million barrels in September 1973, the bulk of that increase concentrated in industrial countries, with the United States at the top of the list. Consumer countries' thirst for oil seemed to be unaffected by rising prices, convincing many that oil demand was impervious to price concerns.

The "perfect storm" was already in motion, with all its components raising expectations of a major energy crisis. Yet this was only one part of the story. Actually, one cannot appreciate the collective psychology that shaped the crisis without taking into account the political circumstances in which it unfolded.

After the dramatic "Black September" in 1970, the Palestine Liberation Organization began to scale up its head-on confrontation with Israel and its supporters by launching a terror campaign designed to capture the world's attention and focus it on the unresolved Palestinian issue. Among the airplane hijackings, the bombings, the killings of

---

* Using Arabian Light quality as a reference.

random civilians and prominent figures, the most spectacular act of this strategy occurred in August 1972 during the Munich Olympic Games. Before an astonished world, PLO operatives kidnapped several members of the Israeli Olympic team. The operation ended in tragedy when German special forces attacked, placing in grim new focus the political and military radicalism that gripped the whole Arab world.

Spurred by the Arabs' humiliating defeat in the Six-Day War, anti-Israeli and anti-Western feelings had exploded throughout the Middle East. In part, this was the result of Arab leaders' wielding of the issue of Israel and political maximalism to deflect attention from their own poor performance in securing their people an over-promised brighter future. But it was also an expression of frustration at a world perceived, rightly or wrongly, as dominated by the *arcana imperii* of an apparently unbreakable American-Israeli alliance. In this climate, Arab governments' support for the PLO's claims and strategy made likely an imminent clash, the dimensions of which were difficult to grasp. And indeed it was this collision between the Arab-Israeli issue and that of oil that ultimately set off the energy crisis many had feared.

In retrospect, the new clash of Arabs and Israelis and its implications for the world oil market were also the consequence of underestimation by the U.S. government of what was brewing in the Middle East. Neither Nixon nor Kissinger believed either that Arab militant propaganda could set off a war or that Arabs could use oil as a weapon to force the West to rein in its support for Israel. Unlike Akins, whom he had no love for, Kissinger had always viewed the Middle East through the lens of the American-Soviet confrontation and thought that the problems that were igniting the Arab political arena could not be resolved except by eliminating Soviet influence in the entire area. He did not judge Arab governments capable of wielding the oil weapon against the West, and was skeptical even of their ability to successfully implement a political strategy—whatever its goal—without Moscow's support. Finally, Kissinger completely misread—by his own admission—the personality of the new Egyptian leader, Anwar el-Sadat, who would prove to be the decisive factor behind the outbreak of a new Arab-Israeli war and the first oil shock.

Sadat had inherited from Nasser a wrecked country, overextended in military spending and patrolled by Soviet advisers and forces. All of the promised achievements of Nasserism had fallen short of expectations, with little or no improvement in the condition of the Egyptian population. In this context, Sadat had decided on a profound de-Nasserization of his country's domestic and foreign policy, which he strenuously pursued. To preempt a possible reaction against this ambitious plan, he needed new friends, and he found them in the Muslim Brotherhood. Harshly persecuted by Nasser, the father of modern Islamic movements began a new life with Sadat: its members were freed from prison, and the government began to support their activities, allowing them to become a significant force in Egyptian universities as well as in the country's social and cultural landscape. In a short time, Sadat acquired the necessary strength to expel Soviet forces and to put the brake on the leftist movements, in this case using the same violent instruments as Nasser. But Sadat's departure from his predecessor's messianic vision risked isolating him from the rest of the Arab world and made him vulnerable domestically, so that only by appearing to the Arabs as a staunch anti-Israeli warrior could he survive. Moreover, Israeli forces still held parts of Egyptian land in the Sinai Peninsula that it had occupied during the Six-Day War, and this outrageous humiliation left him no option other than forcing Israel to leave. Although he was not a pan-Arabist, it thus fell to Sadat the decision to launch a new war against Israel to break the impasse left by the previous war and force the rival country itself to come to terms.

In preparing his war strategy the Egyptian president focused on oil, which he perceived as a key factor in his chances of success. In particular, he was convinced that only if Arab oil producers had used oil as a weapon could he have forced the United States to stay out of the impending war and to refrain from supporting Israel. According to most sources available, Sadat informed King Faisal of his plans in May 1973, after having already obtained a Syrian commitment to share the military burden of the attack against Israel. During his meeting with King Faisal, he also asked for the Saudis' commitment to use oil as a weapon against any Western country eager to help Israel, and reportedly he received the king's blessing.[18] It was a Copernican revolution in Saudi Arabia's stance on the issue, and a decisive factor in the degeneration of the crisis.

Since the Six-Day War, the kingdom had consistently maintained a complete separation between oil policy and the Israeli issue, but now things changed radically. After a long competition with Iran, Saudi Arabia had emerged as the Middle East's major oil producer, with an output that by mid-1973 had surpassed 8 million barrels per day, up from less than 5.5 mbd in 1970. Saudi Arabia was soon to become the second largest producer in the world, the largest oil exporter globally, and—above all—it appeared to be quite a virgin territory, with huge possibilities for enhancing its production capacity in a relatively short time frame. Proof of this situation was given by Aramco's approval of a plan (in spring of 1972) to boost Saudi production to 13.4 million barrels per day in 1976 and to 20 mbd in 1983.[19] As a consequence, while the Kingdom's importance to the world's oil supply rendered it crucial to Sadat's war strategy, it also made it extremely difficult for the Saudis to invoke a position of neutrality. Simply put, if Saudi Arabia refused support for Egypt, it would be held responsible for any failure in the upcoming war: already under accusation by Arab radical forces for its moderate stance toward Israel and its long-term friendship with the United States, the Saudi monarchy could not afford further isolation within the Arab world. Finally, King Faisal was personally a staunch anti-Zionist, and seemed quite obsessed with a sort of conspiracy theory by which Israel and the Soviet Union were plotting together to subvert the political order of the Middle East.[20]

American companies knew quite well the Saudi king's position on the disturbing effects provoked by the American-Israeli alliance and had never hesitated to side with Arab claims. But the influence of the apparently mighty oil multinationals over leadership in Washington was negligible. In May 1973, Aramco's top management met King Faisal in Geneva and received a clear message. The Saudis, the king explained, risk "becoming more isolated in the Arab world, and they cannot permit this to happen, and therefore American interests in the area must be removed." Unless the course of American foreign policy changed immediately, there was no escape for U.S. oil companies: "You will lose everything," the king finally warned.[21]

What followed was a frantic public relations campaign by the oil majors, whose main target was the Nixon administration. Aramco's men reported the king's message to Washington and tried to reinforce it with their own "on-the-field" analysis of the dangers they and the United States were facing in Saudi Arabia and the whole Middle East. Once again, they got no result. Frustrated, they reported to Aramco's boss, Frank Jungers, that the U.S. top officials they had met with had showed a general disbelief that the Saudis would act on their threats, and observed, "some believe that His Majesty is calling wolf where no wolf exists except in his imagination."[22]

These months of blindness to the real extent of Arab unrest represented a substantial rupture in the U.S.-Saudi relationship established more than twenty years earlier. In 1953, the United States had officially recognized that the operations of the large Western oil companies in the world were instruments of American foreign policy. But for a long time,

Washington forgot the spirit and the letter of NSC Resolution 138/1, letting companies face the Middle East's conflicts alone, and often actually causing their problems, particularly as far as their position on Israel was concerned.

Thus when the fourth Arab-Israeli conflict in twenty-five years broke out during the Jewish festivity of Yom Kippur (October 1973), it finally precipitated a crisis in the precarious oil situation.

OPEC suspended a meeting in Vienna scheduled to negotiate higher prices with Western companies. For a few days, all stood still. Then, on October 13, Israeli premier Golda Meir wrote President Nixon that her country was on the verge of collapse and faced a severe shortage of arms, making a surprising victory of the Arab forces seem at hand. The United States then authorized a secret night airlift to supply Israel with weapons, and to respond to Soviet entanglement in the conflict. But the covert game was discovered: high winds forced U.S. airplanes to land in Israel in full daylight. The reaction was immediate, and produced two different outcomes.

On October 16, an OPEC delegation from six Persian Gulf countries* met in Kuwait City and decided on a unilateral price increase of benchmark crude Arabian Light from $2.90 to $5.11 per barrel. Then, on October 17, members of the Organization of Arab Petroleum Exporting Countries (OAPEC, a parallel OPEC made up of solely Arab producers)[23] announced an immediate oil production cut of 5 percent, to be followed by additional cuts of the same size for each month Israel failed to withdraw from the territories it occupied in 1967. They also declared that "Arab-friendly states" would not be affected by that decision, which meant producers pointed to a "selective embargo" directed against Israel's supporters. In fact, a parallel but secret target set by OAPEC was to completely shut down oil supplies to the United States, the Netherlands, South Africa, and Portugal in stages; other countries would face a partial supply shutdown depending on their position on Israel; friendly countries would be exempted from any cuts. Iraq left the OAPEC meeting after its proposal for a more severe resolution was rejected by other members; Iran and other producing countries refused to take part in the embargo, and in fact increased production.[24]

---

* The countries were: Saudi Arabia, Iran, Iraq, Kuwait, United Arab Emirates, Qatar.

It is said that the perception of reality is itself reality, in spite of the facts. But behind the perceptions, the reality of this crisis was not as ominous as the reactions it provoked. One thing is certain: the effective shortage of oil created by the OAPEC decision was relatively small, for several reasons. As professor Morris Adelman has pointed out:

> From October to December, total output lost was about 340 million barrels, which was less than the inventory built-up earlier in the year. Considering as well additional output from other parts of the world, there was never any shortfall in supply. It was not loss of supply, but fear of possible loss that drove up the price.[25]

Adelman's position was quite extreme, but not far from the truth. Based on figures available today, total Arab production in September 1973 had reached 19.4 million barrels per day; in November, when the cutbacks were the most severe, it dropped to 15.4 million, which meant a loss of 4 million bpd. By that time, production and export increases by other countries had added 900,000 bpd to the picture, leaving the effective shortfall 3.1 million bpd at its apex—around 5.5 percent of world consumption, or 10 percent of oil traded internationally. Such an amount could be largely compensated for by drawing on existing inventories.[26]

Actually, various observers at the time doubted the true extent of the crisis. In the United States, for example, there were heating oil stockpiles higher than a year earlier, while there were many signs that refiners could access oil from different sources, even though oil companies refused to supply information when asked to explain and divulge statistics on runs of crude. Noting the contradictions in market perception, the *New York Times* wrote that it was a "dramatic paradox," remarking that crude oil flowed "in huge quantities," but information about it had been cut "to a murky trickle."[27]

Also the so-called selective embargo was largely a myth. The oil market was—and is—like a sea, drawing its water from many rivers, each with its own tributaries, and whatever the course it follows, all water will ultimately find its way into that sea. Accordingly, oil buyers unaffected by Arab cuts could resell their crude, or the products derived from it, to whomever they wanted, as long as they took care not to do so openly, which could hurt their suppliers' dictates. And indeed, Western oil companies did their best to spread the burden of the apparent oil shortfalls among all countries, adopting a policy dubbed "equal misery."[28]

However, at the time no one had access to reliable figures on the impact of cutbacks and their geographical distribution. The issue was further complicated by the relative shortages of specific grades of crude that were essential to a significant proportion of the world's refineries. Thus, although the size of the Arab oil embargo was not great, ignorance and confusion greatly amplified its effects, feeding the panic worldwide.

Between October and December prices went wild, reaching absurd levels at auctions for single cargoes of crude: in mid-December, for instance, Iran obtained a price of $17 per barrel in an auction for 450,000 barrels.[29] And there were cases with even higher prices. The companies and countries that were most severely hit by the cuts went looking for crude wherever it could be found, offering prices that a month before would have seemed insane. Finally, in December, OPEC decided to raise the official posted price of benchmark crude—Arabian Light—to $11.65 per barrel, which meant a four-fold increase in less than four months; even more shocking, oil prices had skyrocketed by almost a factor of ten since 1970.[30]

By early 1974, oil cutbacks were silently ended, with no formal announcement, without satisfaction of any of the conditions Arab countries had imposed, and despite the fact that Israel had finally won the Yom Kippur War. Yet the earthquake they provoked had already shattered postwar economic certainties.

It is hard to find in history a comparable revolution in the price of a strategic resource. And given oil's centrality to industrial economies, this revolution helped bring the curtain down on the most extraordinary period of development ever registered by the advanced countries, opening the door to a severe stagflation that hit the non-oil-rich developing countries as well. At the same time, in the winter of 1973–1974, the endless lines of cars at undersupplied gas stations in the United States, and the various programs intended to limit the use of cars, central heating, and lighting in Europe and Japan shaped the collective psychology of the people of the industrial countries, threatening their already precarious belief in an ever-better future that now appeared to be at the mercy of a group of countries they knew almost nothing about.

This situation offered the prophets of doom an ideal backdrop to stage their grim plays. The list is too long to remember them all. Ideally, it would start with the Club of Rome's report "The Limits to Growth" (1972),[31] issued one year before the first oil shock, which envisaged the advent of an era of oil scarcity by the mid-1990s. The report's basic assumptions, of course, were wrong. It assessed the world's remaining oil reserves at 550 billion barrels and projected an average yearly compound growth rate of demand of 4 percent for twenty years. In contrast, over that period demand increased at less than half that rate, while between 1972 and 2004 the world produced more than 700 billion barrels of oil—leaving proven reserves that still today exceed 1 trillion barrels.[32] However, the report had the ethical merit of focusing the West's attention on the risks entailed in its foolish consumption habits, particularly as far as the environment was concerned.

In 1973, some months before the embargo, the first prophet of the crisis, James Akins, published a seminal article in *Foreign Affairs*, "The Oil Crisis: This Time the Wolf Is Here," in which he set out his pessimistic view that the inexorable exhaustion of oil would strike humanity within the next few decades.[33] There soon followed innumerable books and articles, as well as studies and computer models performed by the top universities, institutions, and companies—all substantiating the inevitability of the destiny that would deprive mankind of its most important source of energy in a few years. Animated by a range of motivations and goals, both liberals and conservatives throughout the world lined up to present their most dire scenarios.

At that time everything in the Western world seemed to corroborate the most pessimistic visions. The United States, in particular, was entangled in the Watergate scandal, which exploded in the middle of the "selective embargo" depriving Nixon of the political strength necessary to deal with the Middle Eastern turmoil. The U.S. president tried to figure out something that could placate the worries of the public opinion. On November 7, he broadcast a message to the nation calling for "energy independence," using a highly emphatic tone:

> Let us set as our national goal, in the spirit of Apollo, with the determination of the Manhattan Project, that by the end of this decade we will have developed the potential to meet our own energy needs without depending on any foreign energy source.

The goal Nixon indicated, however, was a public relations exercise, rather than a serious response. The very presidential staff had made clear to Nixon that it was simply impossible to achieve that target, both from a technical and from an economic point of view.[34] Moreover, American consumers were so addicted to oil that the prospect of their breaking free of it was sheer nonsense; on the other hand, for the government to impose a real cure would have been political suicide. Unequivocally confirming the seriousness of this addiction, U.S. oil consumption rebounded vigorously by the end of 1974 despite high prices, recovering its previous rate of growth and finally peaking at a staggering 18.5 million barrels per day in 1978—entailing the country's highest per capita consumption ever;[35] demand rebounded in other industrial countries as well, but it did so far more slowly than at any other time since World War II. Nonetheless, the comforting myth of energy independence would prove to be a useful political slogan throughout the world, a way to publicly address the issue while doing nothing about it. More than twenty years later, in January 2006, another President of the United States, George Bush, Jr., would resort again to that myth in his State of the Union address in the midst of a new oil crisis.

A concrete consequence of the first oil shock was instead the establishment in 1974 of the International Energy Agency (IEA), a governmental association of industrial oil-importing countries based in Paris. Promoted by U.S. Secretary of State Henry Kissinger as an instrument to counter OPEC, and even to break it, in its early years the Agency had to content itself with being a technical forum that gathered data on energy supply and demand, generated studies and scenarios, and proposed policies and measures for its members to adopt. The Arab embargo had so strained relations between Americans and Europeans that it was impossible to find common ground for a more clear-cut and cohesive action. Europe, in particular, tried to inaugurate a new policy of Euro-Arab dialogue aimed at differentiating itself from the United States vis-à-vis the Arab countries, something that most observers considered an early step of an inevitable day of reckoning for the Atlantic Alliance.

The large oil multinationals also had their day of reckoning. Inhabitants of the no-man's-land between their own countries and the major oil-producing nations, they found themselves under attack from both sides. In the United States, their operations came under intense scrutiny from the press, analysts, and finally the Senate. At the same time, anti-oil-industry sentiment was running high in the country, prompted by public outrage at high oil prices and the widespread suspicion that giant oil companies had secretly plotted the first oil shock in alliance with the Arab countries—a suspicion seemingly corroborated by their apparent compliance with the Arabs' "selective embargo" and the windfall profits they made from it. After an initial inquiry conducted in 1973, a new, major investigation of oil companies was begun in 1974 to examine the connections between the "Multinational Oil Corporations and United

States Foreign Policy" since World War II.[36] Motivating the investigation was the implicit charge that large corporations' policies and their influence on the government had made the United States a hostage to Arab oil.

Contrary to most people's expectations and the leading investigators' negative predisposition against oil companies, however, the final report by the Senate presented another story: what emerged "was a more intricate and fascinating tale of the interplay of government and companies, with a gaping void of abdication and evasion in the middle."[37] Having backed the Aramco agreement and forged the plot of the Iranian Consortium, the U.S. government had substantially delegated its oil policy to the Seven Sisters for reasons of pragmatism and political expediency. This was partly because of the government's faith in the excellent industrial, financial, and logistical abilities of the multinationals, and partly to sidestep the extremely delicate issue of Palestine with Arab countries while the government maintained its support for Israel. Oil overproduction in the 1950s and 1960s had eased scrutiny of U.S. oil policy. In fact, sometimes the U.S. government left oil companies alone at crucial moments, concerned more about restraining them at others. Most important, the White House had continued to nurture its alliance with Israel while oil companies did their best in Washington to support Arab demands. It was inevitable that these contradictions between the two faces of America's Middle East policy would erupt and so come to light.

The U.S. Senate investigation also showed how oil multinationals had restrained production worldwide to keep from flooding the market, and how they had tried to suppress competition; yet it dismissed the notion of a vast Seven Sisters–led conspiracy behind the events leading up to the oil shock. As a consequence, the latter emerged unscathed from the hearings, albeit with a tarnished image that would endure for a long time.

Beyond any myth, the most concrete and destructive attack on oil multinationals came from the producing countries, and it struck what had been the core of their power since the 1920s and 1930s: the major Middle East oil concessions.

After those in Iraq and Libya, the first concession to fall was that granted in 1934 to the Kuwait Oil Company, established as a BP-Gulf joint venture: between 1974 and 1975, the Kuwaiti government expropriated the entire company. Then it was the turn of Venezuela's oil concessions, which had been concentrated mainly in the hands of Exxon and Shell since the 1930s. In 1971, the country had adopted a "law of reversion," which provided that all concessions would revert to the state once they expired, beginning in 1983. However, the evolution of a new oil order accelerated this process, bringing Venezuela's new leading party, Acion Democratica, to call for an immediate nationalization of all foreign assets. Exxon, Shell, Gulf, and the other companies barely resisted, as they were already resigned to this eventuality and eager to preserve preferential treatment for the future marketing of Venezuelan crude. The nationalization was effective as of January 1, 1976, and gave birth to a state oil company, Petroleos de Venezuela, or PDVSA.

The final curtain on the age of the Seven Sisters came down with the end of the most lucrative and important concession of all, that of Aramco in Saudi Arabia, which alone represented more that a quarter of the proven oil reserves at that time. In June 1974, 60 percent of the company was acquired by the Saudi government; in December of the same year, Riyadh informed Exxon, Chevron, Texaco, and Mobil—which still held 40 percent of Aramco—that it intended to completely nationalize the company. In the midst of the surging Arab turmoil, the four American companies could not but capitulate.[38] Once the agreement was worked out, the Saudis allowed Aramco's foreign management to proceed with operations, delaying completion of the company's nationalization until 1981 so as not to damage it.

In many other Middle Eastern and African countries a new contractual formula linking oil companies and producing countries began to gain widespread acceptance, replacing the previously prevalent "concession" formula. Called "Production Sharing Agreement," or PSA, it provided that foreign firms did not own the underground reserves, but only have rights to contractually defined shares of current and future production of the fields they operated. One component of this was called "cost oil," and was used to cover the company's costs. Another component, called "profit oil," guaranteed the company a profit over the full term of its contract with the producing country. The "cost oil" components were calculated on the basis of crude oil price assumptions, and varied as prices change, providing for preestablished cash flows. Thus, when the price of crude increased significantly, the company would see its share of cost oil drop, and vice versa. The total of current and future production that a company expected to have over the term of a contract—twenty years, for instance—was referred to as its "equity reserves." Naturally, the company bore all the risk and costs incurred in projects that failed to come up with either oil or gas.

All these changes symbolized the end of the Sevens Sisters' era and the emergence of a new one, that of OPEC. What did not change at all was the fundamental oligopolistic command of the oil market. It simply passed from a group of actors—which had inherited it from John D. Rockefeller—to another.

CHAPTER 10

The Second Oil Shock

The age of OPEC was shorter and more dramatic than the reign of the Seven Sisters. At the peak of their influence, OPEC countries controlled about 55 percent of world crude production (1973–1974), maintaining over 50 percent control throughout the 1970s. Never, though, did they attain the pervasive presence on the global markets which had been the hallmark of the Sisters as a consequence of their integration in all segments of the oil industry, from exploration and production to refining and marketing. Most OPEC countries, in fact, were poor in midstream and downstream activities, such that Western oil corporations maintained a prominent role in securing producing countries outlets for their oil.

OPEC briefly assumed the role of price-setter previously performed by the Sisters, by simply deciding the official price of crude during specific meetings, and eventually applying it to contracts, as the Organization had first done in October 1973. However, it was far less successful at this than the Sisters had been. Unlike the latter, the new cartel lacked both cohesion and internal discipline, which sapped its ability to manage prices in a balanced and realistic way.

Facing a new consumption paradigm in a world marked by a more modest growth in oil demand, OPEC countries continued nonetheless to forecast a strong upsurge in oil consumption, thereby inducing the cartel to support crude prices. In a short time, this policy left OPEC with too much oil without a market, such that by the spring of 1975 it had lowered production to 35 percent below total capacity.[1] It was only a contradiction that saved OPEC from a collapse of prices. Though there was a surplus of oil, consuming countries were gripped by the fear that they had entered an age of crude scarcity. Once again, doomsayers were hard at work fanning the flames of hopelessness and pessimism. In April 1977, the Central Intelligence Agency (CIA) delivered a highly influential report stating that the growth of world oil demand would soon outpace production because of constraints on OPEC potential and the impending peak of Soviet production. By the 1980s, the report argued, oil would be scarce and very expensive.[2] Though it was only one of several bleak visions of the future, because of the mythological status of the CIA's supposed insight into world events, it was taken as definitive proof that what many feared was true. In October of that year, the U.S. Secretary of Energy, James Schlesinger, warned in the *New York Times* of a "major economic and political crisis in the mid-1980s as the world's oil wells start to run dry and a physical scramble for energy develops."[3]

A compendium of gloomy predictions was produced by the fledgling International Energy Agency, itself a source of pessimism: an Agency's study envisaged no escape from the world's absolute dependence on OPEC by 1985. (As we will see later, in 1985 OPEC would be forced to reduce production to half the level in 1979, due to a glut on the market.)[4] Above all, the apparent omnipotence of Arab countries in setting the pace of OPEC policies seemed to provide a sinister glimpse of their inner will to hold the West in check indefinitely, using oil as a virus capable of destroying the circulatory system of advanced economies.

Driven by reactions to this type of forecasts and analyses, by irrational anxieties, and by OPEC's behavior as an assertive cartel, oil prices thus continued to climb gradually after the first shock, until they leveled off at around $13 in 1978. By then, their departure from market reality had already seduced most OPEC countries, silently nurturing one of the main factors behind the second oil shock. Eager to exploit the bullish market, producers began selling oil outside of the traditional long-term contracts that had been the rule for decades, and turned their appetites on the so-called "spot market."

A marginal spot market for oil had existed for years but in an extremely limited fashion. Centered mainly in Rotterdam, at the beginning of the 1970s it comprised only 1–2 percent of the international oil trade, consisting of sporadic shipments of oil transported primarily by independent commercial operators who bought it from producer countries, expecting to sell it for more to independent refiners or those experiencing temporary shortages. Even if limited in size, by the eve of the first shock the spot market had already exerted a significant distorting effect on oil prices because of the opportunity it introduced for speculation. True, if an independent refiner needed a single cargo of oil at a certain date, the spot market was his last resort, provided he was able to pay whatever was requested by the seller. Based on a free negotiation of occasional deliveries between two parties (the buyer and the seller), spot transactions could diverge dramatically from the official prices set in long-term contracts, and herein lay its potential for destabilization. Indeed, in a frantic and bullish market, daily spot transactions tended to move strongly upwards, signaling to the broader market that there was room for an increase in official prices as well.

In 1978, the spot market still accounted for a meager 3–4 percent of the total oil market, but its scope was broadening thanks to many OPEC countries' increasing use of it to sell more oil than they were entitled to through their long-term contracts. At the same time, spot prices rose as well above those officially set by OPEC, once again for a quite simple reason: market psychology.

Fostered by catastrophic visions of future oil shortages, numerous oil companies around the world did not hesitate to bid for oil whatever the price, particularly independent companies with less access to the global oil network that scrambled to satisfy their short-term needs. In addition, almost every actor on the stage bought oil to stockpile in preparation for the future shortage everyone was predicting.

The limited size of spot transactions contributed most to their terrific impact on the market as a whole. It was understandable that small and independent operators would willingly pay prices well above the official ones set by OPEC for small volumes of oil. But OPEC producers took this peculiar phenomenon, confined to a margin of the overall market, as a strong signal that their official prices were generally too low. A Kafkaesque liturgy then began to unfold.

In the period before each OPEC pricing meeting, spot market quotations rose, a sign that the operators expected the organization was going to increase prices. The OPEC "hawks" (principally Iran, Iraq, and Libya in those days), meanwhile, seeing the spot quotations rise, called for additional increases, convinced that the market would be able to bear them. As a consequence, the expectations of the operators became self-fulfilling, or at least created a state of constant tension that left the producer countries with plenty of room for maneuvering.

Within OPEC, only Saudi oil minister Yamani consistently tried to resist the deceptive logic of this vicious circle, calling for long-term strategies that would bring stability to the market. Worried by the interdependence of consumers and producers, and fearing the inevitable reverberations of a contraction of crude demand in the industrial countries due to high prices, Yamani proposed the adoption of a policy based on limited, periodic price increases in order to provide certainty and tranquility to operators. But his effort was unsuccessful, and partly ambiguous. Already in 1978, OPEC's Long Term Price Policy Committee had fixed a long-term price target "to a level just below the cost of producing synthetic liquid fuels," supposed to be near 60 dollars per barrel.[5]

Yamani himself presided over that committee, but he was probably unable to temper its conclusions because of the aggressive stance of other OPEC members, particularly Iran. Indeed, it was the most pro-Western head of state in the Persian Gulf, Shah Reza Pahlavi, who came out as the staunchest supporter of the alignment between the price of oil and that of synthetic fuel.

However, everyone in the cartel but Yamani was convinced that oil prices could defy the law of gravity as long as there was no economic alternative to petroleum. And OPEC was not alone in this schizophrenic departure from reality.

In March 1978, a Rockefeller Foundation study involving several leading energy experts affirmed that the world was "heading toward a chronic tightness, or even severe shortage, of oil supply."[6]

Thus that wild beast of any market, distorted psychology, had already set the stage for the second oil shock, which lacked only a catalyst to be put in motion. But instead of a single spark, it was a storm of major events that switched it on and kept it going for about two years, lifting oil prices to their highest peaks ever.

It all started with the Iranian Islamic revolution, which began in the last months of 1978 as a broad popular revolt against the regime of Mohammed Reza Pahlavi and provided the epilogue to the never-successful fifty-year marriage between the Pahlavi dynasty and the Iranian population. Paradoxically, the misleading windfall oil revenues generated by the first shock precipitated an outcome that was perhaps already inscribed in the destiny of Iran.

The sudden wealth of the 1970s had strengthened the shah's ambitions to speed ahead with Iran's transformation into a military power and a modern, Westernized country. Supported in his goals by the "regional pillar" policy embraced by the White House, Reza Pahlavi spared no expense in stockpiling armaments, particularly after he received the green light from the U.S. administration in 1972 to buy whatever conventional weapons he wanted.[7] At the same time, he extended the reach of state control into every aspect of Iran's social, political, and economic life, while eradicating the practice and cultural influence of traditional Iranian values, including religious ones. But what Mustafa Kemal had imposed in the 1920s in Turkey—an overnight change of a nation's culture and destiny—did not succeed for the Iranian monarch. Instead of setting in motion an orderly development, Reza Pahlavi's policy generated broad economic inequality and increased poverty and marginalization in large parts of Iranian society, and provoked a massive displacement of peoples from rural areas to the major cities, like Teheran (which in those years would absorb around 15 percent of the total Iranian population). This process of eradication was carried out with no preparation whatsoever, in the midst of a speculative bubble in the real estate sector that rendered significant segments of the population incapable of even paying rent.[8]

Thus, while the oil wealth flowed to a few fortunate enclaves within Iranian society, the majority saw its relative position decline, as the terrible secret police of the shah—the *Savak*—held in check dissidents and banned any form of protest, usually resorting to torture and killing. In the second half of the 1970s, violent riots began to occur, leading to a violent campaign of repression by the shah's forces. The world looked on in alarm at the outcry of the Iranian population, while Reza Pahlavi found himself more and more isolated internationally. Even the United States under the newly elected Democrat Jimmy Carter distanced itself from him, as Carter made global human rights a central theme of his foreign policy. The rapidly convoluting situation finally found its political catalyst in the charismatic and yet enigmatic figure of Ayatollah Ruhollah Khomeini.

Exiled in Turkey and then in Iraq since the 1960s after serving time in prison for his implacable criticism of the Pahlavi regime, Khomeini was an irreconcilable enemy of Iran's Westernization. Though hampered by the difficulties of exile, he never relented in his attacks on Iran's moral decay and corruption in vibrant sermons, recordings of which he had one of his daughters smuggle into Iran so his message could reach the people. Year after year, his stubborn opposition to the shah and his impeccable moral credentials propelled his growing ascendancy with the Iranian people, which had been by that time deprived of the leadership of Mossadegh, who had died in 1969.[9]

Khomeini's theocratic conception of political and social order left no room for compromise with other anti-Pahlavi movements. Even a character like the widely beloved Mossadegh was incompatible with his vision because of the latter's interpretation of Iran's redemption through the prism of nationalism, reformism, and democracy—all categories that Khomeini held belonged to the Western system of thought. But as rigid as he was in his private views, the ayatollah was also an extraordinarily pragmatic politician who knew that he also had to win to his side the secularist forces of Iranian society if he wanted to change the course of his country's future.

When his campaign against the Pahlavi monarchy ramped up after the mysterious murder of his son in the late 1977, the shah pressured the Iraqi government to expel the ayatollah, and it complied. In October 1978, Khomeini left Iraq for Kuwait, but he was refused access and had to accept a sojourn in Paris as a last resort. In a way, this was a fortuitous accident. While the growing rebellion in Iran was turning the country into a battlefield, with deaths and mass arrests daily, in France Khomeini's popularity was peaking thanks in part to the quite obsessive attention the foreign press was devoting to him. As a consequence, he had a far more powerful pulpit from which launch his incitements to revolt against Reza Pahlavi, along with his message of unity of all the forces hostile to his regime. In this manner, as a prominent scholar pointed out, "he was careful to sidestep in public the nature of an Islamic state," whereas he spoke about a "progressive Islam" in which "even a woman could become president."[10] His political shrewdness and apparent moderation brought all of the opposition parties together around him, partly because the hard-line instruments used by *Savak* even against moderate opponents left the latter no alternative but to join the only voice capable of standing up to the regime.

Meanwhile events exploded in Iran after a bloody massacre in Teheran on September 8, a day remembered as "Black Friday." In November the government imposed martial law, but more bloody episodes followed. In December a massive demonstration of around 17 million people took place all across Iran calling for Khomeini's return and the shah's departure, while the oil workers called for major strikes, which paralyzed Iranian oil production. Secretly affected by cancer, Reza Pahlavi refused to use force against the population to reestablish order, an option most of his loyalists pushed him to embrace.[11] It was the final curtain for the regime. In January 1979 the shah left Iran, never to return. As the Pahlavi regime collapsed, the oil world went into panic.

Before the revolution, Iran's production had peaked at slightly more than 5.5 million barrels of crude per day,* or slightly less than 10 percent of global production at the time—making it the fourth largest producer after the United States, USSR, and Saudi Arabia; more important, it was the second largest oil exporter (after Saudi Arabia), with 80 percent of its production destined for world markets. By January 1979, however, Iranian output had dived to 40,000 barrels per day, rebounding in April to over 4 million bpd. Taking into account additional production from other sources of around 2 million barrels, the net loss during the height of the crisis corresponded to 5.5 percent of oil demand, which was then peaking at 64 million barrels per day.[12] But despite the brevity and marginal size of the oil shortage—particularly in the face of the previous huge inventories buildup by most companies in the world—those first months of 1979 set off a new escalation in OPEC official prices that would not stop until 1981, helped along by subsequent events that deepened the perception of a no-way-out situation.

To begin with, in March 1979 the near meltdown of the Three Mile Island nuclear reactor in Pennsylvania cast doubts on the reliability of nuclear energy as an alternative to oil, on which many industrial countries had focused their hopes of becoming less dependent on "black gold." That same month, the proclamation of a theocratic government in Iran led by Ayatollah Khomeini heralded the radicalization of religious militancy and anti-Western feelings in Iran and all over the Middle East. In proclaiming the birth of an Islamic Republic in Iran, in fact, Khomeini appealed to Muslims to rise up against the West, which was deemed responsible for the erosion of traditional customs and values and the loss of cultural and political identity in the Islamic world.

Even if Iran's Shiia Islamism was foreign to most of the Arab world, and Iranians were ethnically foreign to Arabs, the success of Khomeini's revolution sent a powerful message to the Arab masses, signaling the possibility of using religion as an instrument to forge a political program as well as build a new kind of state in the Middle East. Consequently, tensions and upheavals grew outside of Iran as well, reaching a breaking point when students and militants of the revolution occupied the American embassy in Teheran in November 1979—the same place from which Kim Roosevelt had direct the coup against Mossadegh in 1953— holding more than fifty U.S. employees hostage for 444 days, an

---

* Average production in September 1978.

operation that provoked a major clash with the United States. A few weeks later, Saudi Arabia was shaken by an unprecedented attack on the foundations of the Kingdom when more than five hundred armed radical Islamists assaulted the Great Mosque in Mecca.

This striking challenge to the Saudi monarchy did not come from pro-Khomeini Muslim groups but largely from the heirs of those Ikhwan tribes that had first supported and then fiercely struggled against the founder of Saudi Arabia, Abdul Aziz ibn-Saud, because of the latter's departure from a fanatical interpretation of Wahhabism. In fact, the leader of the rebels, the self-proclaimed *Mahdi* Saif al-Othiba, was grandson of an Ikhwan leader.[13] What appeared as the onset of a new revolution in the Persian Gulf ended in carnage, with 177 rebels and 127 Saudi soldiers killed during a siege that lasted two weeks. The rebels were finally defeated, but the episode demonstrated the seminal influence of the Iranian revolution on radical religious groups. A new Islamic upheaval in Saudi Arabia's oil-rich eastern region followed shortly thereafter, in December 1978, and was cracked down on. This time, the rebels were part of the Saudi Shiite minority.

On December 27, then, the Persian Gulf situation reached its climax when Soviet Union invaded Afghanistan, a country sharing a one-thousand-kilometer border with Iran.

The Soviet move aimed both at supporting the Afghan Communist Party, then in power, and at stopping the upheaval of Islamic forces (the so-called *mujaheddin*) that had allied with local tribes, which risked igniting the Muslim populations spread throughout the southern extremes of the Soviet Union. This risk was also a consequence of Khomeini's revolution in Iran, which did not affect Western interests alone. Yet for U.S. analysts and policymakers, the invasion of Afghanistan resounded like a fearsome alarm bell, making them see the possibility that Moscow's move was only a first step in a new "Great Game" played out over Central Asia, whose ultimate target was the Persian Gulf and its huge oil reserves.

Displaced by the events, President Jimmy Carter reacted by proclaiming in January 1980 what would pass into history as the "Carter Doctrine":

Any attempt by outside forces to gain control of the Persian Gulf region will be regarded as an assault on the vital interests of the United States, and such an assault will be repelled by any means necessary, including military force.[14]

That pronouncement made clear and public what had been the *fil rouge* of American foreign policy in the Gulf since Truman's 1950 letter to the king of Saudi Arabia, followed soon after by National Security Council resolution 138/1. Following the presentation of the Carter Doctrine, the United States started a lengthy military build-up in the Persian Gulf, beginning with the creation of a Rapid Deployment Joint Task Force, which could be dispatched to the Middle East. That was the premise for President Reagan's decision to establish a Central Command (CENT-COM, 1983), with the task of defending the American interests in East Africa, the Middle East, and Central Asia.

However, Carter's credibility had already been undermined by his administration's inability to resolve the hostage crisis in Iran, the general upheaval in the Middle East, the energy crisis, and the apparent penetration of Soviet Union into the Arab world and sub-equatorial Africa. Everything played dramatically against him. In April 1980, a military rescue operation conceived by Washington to free the hostages ended in tragedy. Eight helicopters and six Hercules C-130 transport planes sent by the United States in a desert strip 275 miles southwest Teheran never fulfilled their ill-prepared mission; environmental conditions put three helicopters out of order, while one of them crashed against a C-130, provoking the destruction of both. Eight American soldiers died, the other quitted the stage by leaving everything on the field, including weapons and secret documents.

All of these events seemed to illustrate the West's startling impotence before a world that was rebelling against it, in a void of international leadership that seemed to condemn it to exist at the mercy of external forces. In this situation, the oil market too was subject to the law of the strongest. And OPEC was trapped in the illusion of being able to perform that role, irrespective of the ongoing world economic crisis and the policies already implemented by most of the industrial countries to reduce oil consumption.

When in September 1980 war broke out between Iran and Iraq, depriving the world market of around 3 million barrels of combined exports from the two countries, oil prices registered their last, foolish jump. By that time, the spot market had grown dramatically to around 10 percent of the international oil trade, mostly because of producing countries' policy of selling more and more of their crude directly or through intermediaries at the higher spot prices. Excited by the upward spiral of spot prices, OPEC soon calculated their higher values into its own official prices. In late 1980, the Arabian Light spot price reached its historic peak at $42 per barrel for single lots; in that same year, the average price of gasoline in the United States reached its record level at $1.42 per gallon.

True, OPEC was not alone in being galvanized by the rise of spot prices. Even the British National Oil Company (BNOC—a creation of the British government entitled to acquire 51 percent of oil produced in the British North Sea) by 1979 had begun to add a surcharge to its oil prices in response to spot values and continued to do so throughout the ultra pro-free-market administration of Margaret Thatcher, who became prime minister in 1979.

As a prominent expert of OPEC affairs remarked, "if BNOC, and by implication the British government, were behaving like OPEC, who could expect OPEC to put an end to the oil price spiral?"[15] In short, fear turned into panic.

Research and planning offices of the oil companies, government experts, and independent analysts all agreed that prospects were hopeless, estimating that prices would rise to between sixty and one hundred dollars per barrel by the 1990s, and that regardless there would not be enough oil to meet the needs of all.[16] Even the queen of the once-powerful Seven Sisters, Exxon, went along with the general mood. In its 1979 annual report, the American giant underlined that "the balance between world energy demand and available supplies" would be "precarious" in the following years, and that "virtually all new petroleum reserves" would be "expensive to find and develop."[17] BP endorsed the same view in a study published in 1979, where it predicted a world oil production peak outside the Soviet bloc in 1985, followed by a rapid fall of oil supply that, in 2000, would have been 25 percent lower than in 1979.[18]

Among the crowd of catastrophists, it was once again the CIA that summed up in a comprehensive verdict the gloomy widespread expectations for the future:

We believe that world oil production is probably at or near its peak. Simply put, the expected decline in oil production is the result of a rapid exhaustion of accessible deposits of conventional crude oil. Politically, the cardinal issue is how vicious the struggle for energy supply will become.[19]

Echoing that assessment, President Carter declared that oil wells "were drying up all over the world."[20]

But contrary to the CIA's analysis and the dark visions of other grim prophets, no life-or-death struggle for oil was necessary. Once again, the oil market had blinded the judgement even of those who were most acquainted with its unpredictable nature. And, once again, it was to remind them how the science of forecasting its behavior could make also the questionable ancient art of predicting the future from animals' intestines seem respectable.

CHAPTER 11

The Countershock

One of the few oil actors who was not blinded by the irrational spiking of prices was Saudi oil minister Yamani. Since 1979, he had thought that world oil consumption was slowing, and that consequently the steady upward pressure on prices had to be artificial. In fact, that year marked the peak of global demand, at around 64 million barrels per day. But after coolly observing the relenting pace of world economic activity, Yamani grew increasingly suspicious that a significant part of that demand was inflated by the buildup of oil inventories by private companies, a sort of defensive (or speculative) hedge against further price increases. In his view, once everyone realized that there was no shortage in sight, this phenomenon would lead to a price collapse. "There will be a glut in the market," Yamani commented in private during OPEC's Caracas meeting of December 1979. "It's coming."[1]

Yamani was almost right. But as is always the case during speculative bubbles, no one can precisely predict when and why they will vanish, just as no one can anticipate their coming. Simply put, as long as most oil operators were convinced that oil scarcity was the rule for the future, no force on earth would be able to bring prices back in line with reality. Nonetheless, inner market forces were silently at work, as Yamani had somehow understood.

From 1980 onward, world oil demand abruptly declined, bottoming out in 1983 at 58 million barrels per day—a drop of 6 million barrels from the peak of 1979.[2] In retrospect, it was inevitable. For several years, industrial countries had struggled to preserve their consumption habits, accustomed as they were to the unrestricted use of oil. It was a textbook case of inertial behavior, which economists call the "ratchet effect": when the purchasing power of an individual decreases, he will try to maintain his standard of living as long as he can by ceasing to save, or by borrowing, before he is forced to give up his consumption habits. This is what finally happened in the oil market, as the sky-rocketing prices of the late 1970s forced western countries to heed the iron laws of economics. And contrary to most people's analysis, oil had proved to be price-sensitive, like any other good.

The fall of demand was also brought on by the energy conservation policies that most countries had promoted since the mid-1970s. Of particular importance was the legislation passed in December 1975 by the Ford administration in the United States, as part of the Energy Policy and Conservation Act. Even if highly criticized at the time, it nonetheless introduced structural reforms that served as milestones for the country's energy and environmental policy. Among other things, the law provided for the doubling in ten years of the average efficiency of passenger vehicles, raising it from 13 miles per gallon to 27.5 mpg—a limit that is still in effect today. Moreover, it established a Strategic Petroleum Reserve (SPR) of about 600 million barrels of oil under governmental control. Along with conservation policies, the industrial world promoted a shift to energy sources other than oil, such as natural gas and nuclear, which developed rapidly after the first oil shock. Indeed, the 1970s and the early 1980s were a golden age for nuclear energy in particular, which was a relatively new source of electric power.

The first experimental power reactors had been built in the early 1950s in the United States and the Soviet Union, but the first commercial nuclear plant—the *Yankee Rowe*—began operating in the United States in 1960, with an installed capacity of 250 megawatts (MW). By 1980, there were already 243 nuclear plants in the world, with a total capacity of about 140,000 MW; the bulk of them (162) had been built in the 1970s. Despite the Three Mile Island incident in 1979 (see previous chapter), the rush for nuclear energy continued into the 1980s with the construction of another 176 plants globally, which brought the world's total nuclear capacity to 325,000 MW, or 325 gigawatts (GW) by 1990. By that time, however, the catastrophic accident at the nuclear reactor in Chernobyl, in the Soviet Republic of Ukraine (1986) had provoked a general rethinking of nuclear energy, after its security and environmental risks had proved to be so huge and uncontrollable.

What made matters worse for oil, however, was that while its consumption plummeted, the market was hit by a flood of oil from new producing areas.

Actually, the high prices of the 1970s made very profitable various high-profile investments in areas and fields whose development would have been otherwise delayed by their higher costs and greater technical difficulty. This was the case with Alaska, Mexico, and, especially, the North Sea (British and Norwegian). In a few years, these areas—outside of OPEC's dominion—saw an escalation of production that no one had anticipated. Each of the three had a different story.

The crude of Alaska remained closed in the vast fields discovered at Prudhoe Bay in the second half of the 1960s because of opposition by ecological groups to the construction of an 800-mile pipeline that would carry the oil to the port of Valdez, California. It was not until 1974 that the U.S. Congress managed to put together an agreement and win approval for this project, which was completed and went into operation in 1977. By 1983, Alaska was producing almost 2 million barrels per day, or slightly less than 20 percent of total U.S. output, which was 1 million barrels below its 1970 peak regardless of the new Alaskan production.[3]

Despite its long tradition of production, Mexico stayed largely outside the world oil system after the nationalization of its oil industry, and was actually on the verge of becoming a net importer of oil in the early 1970s. Two events forestalled this eventuality: the discovery in 1974 of large oil deposits in the south of the country, and a departure from the country's long-standing oil conservation policy. The latter change encouraged a flood of major international loans, which in turn made possible a significant increase in production, from 500,000 barrels per day in 1972 to almost 2.7 millions bpd in 1983.[4]

The obstacles to developing the huge fields in the North Sea were of another sort. The area had been the subject of hydrocarbon research since the second half of the 1950s, but what attracted the attention of many operators later was the major discovery in 1959 of a natural gas field off of the Netherlands, in Groeningen, by Shell and Exxon. The similarity of the undersea geological structures of Groeningen and certain parts of the North Sea enticed many companies to aggressively explore the latter. But it was not until 1969, with the discovery of the supergiant oilfield of Ekofisk in the Norwegian part of the North Sea, that the boom of the area began. Between 1970 and 1971, other huge oilfields were discovered, including the British Forties and Brent (the latter was to lend its name to the crude benchmark that is used today). The first shipment of crude from the North Sea reached a British refinery in 1975, and from then on production increased exponentially, reaching

3 million bpd in 1983. The North Sea became not only a prime world oil frontier, but also a veritable technological school that allowed the oil industry to carry out a true revolution in its technology, in response to the extreme environmental conditions of the area.

In addition to that of Alaska, Mexico, and the North Sea, USSR production rose as well to a record level of 12 million bpd, up 40 percent from 1973, which made it the top oil producer in the world. As a consequence, its exports also grew by an additional 1 million bpd. As a whole, huge volumes of new, non-OPEC oil production reversed on the market whereas demand had already plummeted by 6 million barrels per day. While all this happened, Western oil companies started refraining from irrationally buying oil and drew it from their vast stockpiles, which helped to further ease demand.

The glut envisaged by Yamani had thus arrived. Yet OPEC was completely incapable of dealing with the new situation, prisoner as it was to its faith in the inelasticity of oil demand regardless of price. In October 1981, at a meeting in Geneva, the Organization reached a useless agreement on price reunification, setting the new benchmark at $34 per barrel. This was the meager result of endless negotiations between "hawks" and "doves," while the market was already sagging and everyone was overproducing. By 1982, the glut was mounting and spot prices were falling. Symbolically, it was the announcement by the British National Oil Corporation of a $4-per-barrel price reduction for March that signaled the world was changing.[5]

In an attempt to stem the swelling flood, in March 1982 OPEC transformed itself in a formal cartel by imposing a production quota on all its members and a total production ceiling of 17.5 mbd, almost half the level of 1979.* In order to strengthen the new system, Saudi Arabia decided to act as a "swing producer," adjusting its production to compensate for the void left by other members of the cartel.[6] But the new strategy proved unmanageable. OPEC, in effect, found that it was impossible to reconcile the different economic and political interests of all its members, a problem that had dogged the American oil industry as well in several phases of its history.

Soon after OPEC's decision to impose the quota system, Iran announced that it would maximize production, whatever the price it could sell it for. Khomeini had no intention of using oil as a weapon, and he

---

* In 1979, OPEC countries produced 31 million barrels per day.

never did. Partly revenge against the Arab oil countries that were supporting Iraq in the Iran-Iraq War, the Iranian policy boosted the country's production to nearly 3 million by the end of 1982 from 1.1 million bpd registered in March of the same year, exceeding by three-fold the quota assigned to Iran by OPEC.[7] Nigeria also appeared incapable of complying with high OPEC prices. In the Atlantic Basin, its light crude was by now competing with growing volumes of the North Sea crudes, which were cheaper but of the same quality of the Nigerian ones. The African country, thus, was obliged to lower its oil price and scramble for recovering market. Other producing countries within the Organization only formally respected their quotas, while selling oil "under the table" on the spot market.

A breathtaking succession of events in March 1983 marked a watershed transition for the industry.

To start with, British National Oil Corp. slashed its price by another five dollars per barrel, making Nigerian and many other OPEC crudes uncompetitive. OPEC was forced to do what it had deemed impossible since the early 1970s: for the first time ever, it reduced its official price by five dollars as well, setting it at $29 per barrel.[8] As with the Seven Sisters more than ten years before, it was the beginning of an unstoppable retreat, this time before the impetuous advance of overproduction.

More than anything else, the new reality was reflected by the launching on the New York Mercantile Exchange (Nymex) of the first oil future, on March 30. It was a historical turnaround for a sector that for many decades had seen the price of oil governed by more or less successful oligopolies. Though based on Western Texas Intermediate (WTI), the benchmark crude for the United States, oil futures could serve for the entire world as an objective frame of reference for any kind of oil pricing. Oil refiners, for instance, could decide to buy Arabian Light crude on the basis of its differential from WTI, or shift to more convenient oil grades. This institutional change, however, had its counterpart in the flourishing spot market, which had already become the real market for oil, one that mirrored the daily free play of demand and supply. Unchained from distorted psychology, spot transactions were now driving prices down, for sellers were keen to allocate their surplus oil and ready to offer discounts. Many OPEC countries contributed to this great sell-off, cheating on their quotas and selling their oil on the spot market. This process went on for two years, albeit with a few pauses that gave the major producers the illusion that things were about to stabilize. But they did not. By a fateful irony, it had been OPEC itself that let loose the "monster"—the market—during the 1970s, when it fell victim to its own greed shifting from long-term priced contracts to spot sales, which were then much more profitable.

Instead of facing the new reality, OPEC members blamed each other for what was happening, while the Organization as a whole blamed non-OPEC countries for exploiting the cartel's self-imposed production restraints to gain market share. The behavior of the latter became the target of an obsessive but useless campaign of recrimination by OPEC. In the words of the Organization's Secretary General Subroto, non-OPEC producers' policy was "patently unfair" and had to be stopped immediately.[9] During the OPEC conference in Geneva in December 1979, Subroto even launched an unprecedented invitation to all non-OPEC producers to cooperate with OPEC in order to stabilize the market.[10] Yet as it was OPEC itself that was first mesmerized by short-sighted policies, Subroto's appeal was like whistling in the wind.

In 1985 it became all too clear that there was no parachute, no turning back from the ruinous fall. OPEC's lack of discipline was complete. Most of its members were cheating around the quota system while the rivalry between Iran and Iraq impeded any new effective agreement on prices and quotas.[11] Only Saudi Arabia tried to salvage the situation for as long it could. As OPEC's "swing producer" it continued to slash its own production to support prices, but it was useless and painful: the kingdom saw its production plummet to an average of 2.2 mbd in May 1985 from 10.5 mbd in 1980. Tired of bearing that burden alone, the Saudis finally disassociated themselves from the Organization that they had helped create.

Almost as revenge for the failure of the other OPEC countries to heed his Cassandra-like warnings, it was Yamani who anticipated the radical turn and its consequences. Speaking at the Oxford Energy Seminar in September 1985, the Saudi oil minister said:

> Most of the OPEC member countries depend on Saudi Arabia to carry the burden and protect the price of oil. Now the situation has changed. Saudi Arabia is no longer willing or able to take that heavy burden and duty, and therefore it cannot be taken for granted. And therefore I do not think that OPEC as a whole will be able to protect the price of oil.[12]

Beneath his cryptic language, Yamani meant that Saudi Arabia was going to abandon its self-imposed production restraints in order to win back its market share, without regard for the inevitable price drop that this action would cause. Some weeks later, the kingdom began selling oil though a new price formula called "netback." Simply put, the formula fixed the price of oil at the sum of the costs of production, transport, and refining, plus a predetermined profit for buyers.* In this way, the artificial official prices were abandoned; the value of Saudi oil was now determined by market conditions, and an additional wave of oil began to sweep through the market.

In December, the whole of OPEC had to align itself with Saudi policy, in a final, futile effort to regain control of the situation. The cartel's production in 1985 was half what it had been during the golden years; its market share was little over 25 percent (despite the fact that member countries held almost 80 percent of world reserves) and its oil revenues, in nominal terms, were but a quarter of the soaring levels of 1980, when OPEC pulled in a record $275 billion.[13] This time, however, Saudi Arabia went its own way, rapidly ramping up production irrespective of OPEC rules.

At the onset of 1986, a tidal wave of oil overwhelmed the market, with consequences that no one had foreseen. Prices collapsed, dropping to below $10 per barrel or less in May 1986: at one point, Dubai's oil hit $7 per barrel.[14]

The market crash caught the world by surprise. According to all forecasts delivered until a few months earlier, oil prices were bound to defy the law of gravity because of the inexorable dwindling of global petroleum resources. The denial of such gross misinterpretations of reality, however, was so shocking as to become a major cause of concern for the United States; it sent then Vice President George H. W. Bush to Saudi Arabia to discuss the oil situation.

Reportedly, Bush's mission was to convince the Saudis to stop overproduction, as crashing prices were destroying a significant part of the U.S. oil industry. Bush himself gave credence to this interpretation, publicly remarking that he "would tell Saudi Arabia that the protection

---

* The crude oil price calculated by Saudi Aramco following the "netback" formula was the result of a subtraction: from the refined product price earned by refiners on the market, the Saudis subtracted the fixed profit they wanted to guarantee to the refiners themselves, plus the cost of refining and transport. What remained was the price at which they offered their crude to foreign companies on an FOB basis. The formula spread to other OPEC countries in 1986–1987, but later was eclipsed by the return of an "official pricing system" by OPEC.

of American security interests" required "action to stabilize the falling price of oil," and that he would persuade them to "stabilize—or even increase—the price of oil by cutting production."[15] Coming from a former oilman whose political constituency was Texas, such statements corroborated the idea that the powerful lobby of small American oil producers had once again been successful in bending the U.S. government to their will, as they had done with Eisenhower in 1959. This problem had been well known ever since.

America was full of independent, high-cost producers whose survival depended on the high price of oil. This was true, for instance, of the 450,000 "stripper wells"—wells that produce few barrels per day (as few as one or two) and are a typical feature of the country—which accounted for about 15 percent of total U.S. production in 1986, or over 1.3 million bpd. More than half of that production was shut down beginning in March 1986, as the WTI price plunged to $12 per barrel.[16] Because oil has always been prone to mythmaking, there was also a completely different version of the American diplomacy of the 1980s, according to which the Reagan administration had pushed the Saudis to induce the price collapse in order to destabilize the Soviet Union.[17]

But such conspiracy theories failed to reflect the reality of an oil system that was out of control, in which no subject, the United States included, could really influence the unpredictable course of events generated by a surfeit of competing actors. Simply put, all the main oil players could not be reconciled to a single design, as their actions were moved by different objectives, political motivations, and even reciprocal rivalries. In fact, the Bush mission seemed to achieve nothing. Saudi Arabia further increased production to 6 mbd by August 1986, and did not relent until a final agreement was reached by OPEC in December.

Probably this line of action was not even endorsed by the entire Saudi establishment, as King Fahd was an advocate of both higher prices and higher production for the kingdom. Most likely, an internal dispute over oil policy was the reason behind the sudden dismissal of the charismatic Yamani by the king himself in October 1986, after 24 years of service to the Saudi oil ministry. The truth, however, was that no one was in control of anything anymore, so that there could be no miraculous solution to quell the schizophrenic forces of a market in structural surplus. The party was over, and so was the brief golden age of OPEC.

Yet, what passed into history as the "oil countershock" was much more than that, because it shattered the rules and strategic perceptions that had governed the world oil system for the entire postwar period.

All of a sudden, the free market had displaced oligopolies, and oil seemed to have lost its status as a vital resource the tight control of which was key to global power and national security. The new credo was supported by free-market orthodoxy, by then the dominant religion of the United States and Great Britain under the leadership of Ronald Reagan and Margaret Thatcher. In compliance with it, in 1981 Reagan fully deregulated the energy sector in the United States,[18] while Thatcher initiated a vast program of privatization of public companies and services that would be imitated with varying intensity around the world. The British program did not even spare that symbol of the UK's modern quest for the survival of its once unchallenged world supremacy: BP.

For Thatcher, the company that had initiated the oil saga of the Middle East, that in 1914 Winston Churchill had brought under government control to secure British access to Persian oil, and that had eventually joined the exclusive club of the Seven Sisters, was no longer a matter of national power. Between 1979 and 1987, the British Treasury sold its controlling stake in BP on the market in successive *tranches*, without preserving any special right (the so-called golden share) to protect the company against hostile takeovers.[19] This extreme application of free-market ideology, however, was soon put in crisis.

In a sort revenge of history, after the final public offering of BP shares in October 1987—the largest the world had ever known[20]—21.6 percent of the company was progressively taken over by the Kuwait Investment Office (KIO), the investment branch of the Kuwaiti Ministry of Finance. Thus the controlling share of BP was now in the hands of the country whose oil resources and industry BP had discovered and dominated since the 1930s, together with Gulf.

Thatcher's cabinet was taken by surprise and embarrassed. It had few tools to limit the KIO's involvement in BP's management. The only possible way to do so was for the British Treasury to call for an official inquiry by the Monopolies and Mergers Commission concerning the effects of Kuwait's takeover on the public interest. In 1988, the Commission came to the conclusion that KIO's holding in BP could go against the public interest, and ordered that it be "reduced within twelve months to not more than 9.9 percent."[21] Pressed by the government, BP agreed to buy back at a premium the affected shares held by the Kuwait Investment Office, a financial blow the effects of which would last for years.

Well before the BP *affaire*, the whole oil industry paid a heavy toll for the market collapse. Like OPEC, private companies had been seduced by the castle in the air built from ever-growing prices. Small local oil companies had gotten rich quickly, joining the ranks of independent companies and bringing fame and fortune to their owners and top managers. Even the Seven Sisters had become far richer. In spite of the loss of the large oil concessions in the Middle East, their remaining and new production, along with their downstream activities, had yielded huge windfall profits that had literally buried them in cash. Their biggest mistake was to reinvest this money with bad judgment, once again demonstrating that there is nothing like a period of high prices to make the worst investment decisions ever.

In addition to vast exploration plans in costly areas and in unconventional oils, such as bituminous schist, they overspent in petrochemicals and oil refining, believing both sectors would grow at rates comparable with those of the postwar period. This was such a common impulse that in 1980, for instance, world refining capacity was close to 80 million barrels per day, while demand for oil stood at only 63 million.[22] (In 1970, oil demand and refining capacity were almost aligned.)

Yet the most devastating element of this age of irrational spending by oil companies was diversification, notably investments made in completely new sectors far from the traditional core businesses of the industry. Here the examples of foolish undertakings knew no limit. Among the most curious are Mobil's acquisition of the Montgomery Ward department store chain, or its new lines of plastic packaging and real estate; BP's aggressive entry into animal nutrition, in which it became the world leader after acquiring U.S. Purina Mills in 1986; Exxon's embrace with office automation; and Shell's development of forestry, household cleaning products, and biotechnology. This is but a glimpse of a process that blinded almost every company, reaching levels of amazing eccentricity in the case of Gulf's plunge into the entertainment sector with its acquisition of two of the most famous circuses in the world, Ringling Brothers and Barnum & Bailey.[23]

Only around 1982–1983 did this fever begin to subside. Oil companies then found themselves entangled in critical financial problems, blamed by investors for their poor investment strategies and powerless to halt the fall of their share price, sometimes to well below half or even one-third the mere value of their oil and natural gas reserves. A wave of mergers and acquisitions ensued as the weakest among them fell prey to other companies, including those unrelated to the oil business. This was the case, for example, of the acquisition of Conoco, one of the largest independent American oil companies, by chemical giant Dupont, or the takeover of Marathon Oil (another important independent player in

U.S. oil) by U.S. Steel, and the acquisition by General America of Phillips Petroleum, the company that had started the North Sea oil saga by discovering the supergiant Ekofisk field in 1969.[24]

The wave of takeovers did not spare even the Seven Sisters, striking Gulf in 1984. In 1983, before word of a possible takeover had spread, the company's market value was barely over 6 billion dollars, against a maximum valuation of its assets of 18 billion. Adding insult to injury, hostilities against Gulf were initiated not by a major corporation but by an oilman recently thrust into the limelight by his financial dynamism, T. Boone Pickens, the founder, president, and chief executive of a very small oil company called Mesa Petroleum. Through Mesa, Pickens first began buying stock in Gulf, and then formed an association of shareholders of the prey company, finally making a public offering to take it over. A legal war between Pickens and the heads of Gulf inflamed the market and triggered other takeover attempts. One was undertaken by the Gulf board itself, without success. Under siege and internally divided, the company's top management sought the aid of a "white knight"—i.e., another company willing to undertake a nonhostile acquisition but agreeable to the top management of the prey company. Help came from one of Gulf's former peers in the club of the large oil multinationals, Chevron, whose final offer of $13.2 billion beat all competition. Thus, at the end of 1984, a company that had figured prominently in the history of the modern oil industry forever disappeared from the scene.

Other companies tried winning back investors' trust by launching share buyback programs in order to sustain their market value. Even the queen of oil companies, Exxon, embraced this trend by approving the most extensive stock buyback effort yet tried, equal to $16 billion over seven years. Above all, companies had to abandon pharaonic investment projects and face painful processes of restructuring and rationalization. Everyone in the business began cutting costs, slashing human resources, focusing on "core business" oil and gas activities, reducing excess capacity in the petrochemical and refining sectors, closing marginally profitable segments, and eliminating all diversified businesses. It was only the beginning of a process that would continue for many years and would transform the oil industry's DNA into a financially driven code.

This remarkable revolution of the oil world eclipsed the catastrophic visions that preceded it. With few exceptions, most of those who had cried wolf about oil scarcity during the 1970s and early 1980s now joined the chorus proclaiming that oil had become "just another commodity," a resource that the market could easily take care of. Indeed, sometimes it is really hard to face that unsolved mystery of human nature that makes it possible for people to radically change opinion overnight, and that, eventually, allows them to happily become the staunchest apologists of the opposite view.

CHAPTER 12

A Storm in the Desert:
The First Gulf Crisis

The crisis that shattered the oil market was partially overcome in December 1986, when OPEC countries reached an understanding on certain principles. The new agreement implied neither a complete reunification nor a solid base, yet it was at least a framework for a fresh start, albeit with more modest targets. In brief, OPEC returned to an official price, this time based on the average value of a basket of seven OPEC crude oils set at $18 per barrel. All members but Iraq also accepted a new quota system, more flexible than the previous one because it provided for a revision every three months. Naturally, OPEC hoped for the return of the sun after the storm, i.e., a resurgence of prices. But it was not to be.

On average, from 1987 to 1999 oil prices fluctuated at around a modest $18 per barrel in nominal terms. The industry benchmark was now the new star of the international oil system, Brent crude, whose futures contract had been launched in 1988 on the London exchange. A Brent barrel at 18 dollars, however, meant a price of the OPEC's basket-crude of less than 17 dollars, given the better quality of Brent. Of course, over those thirteen years there were ups and downs in prices, but the overall market trend was of decline in real terms, mirroring a modest compound average growth rate of demand of 1.3 percent. In contrast, during the golden age of oil, demand increased by just under 7 percent per year. Furthermore, until 1999 oil demand was consistently outpaced by the growth of supply—over 2 percent per year.

Oil, indeed, seemed to have become "just another commodity," as the new conventional wisdom proclaimed. The new reality also withstood a dramatic test that in other periods would have derailed the system. Once again, the test occurred in the Persian Gulf. And once again, the protagonist was Iraq.

If the painful post-1986 reawakening dashed producing countries' dreams of power and wealth, it dramatically affected the imperial strategy of Iraq's president, Saddam Hussein. As soon as he had risen to dictatorial power in 1979, Saddam had embarked his country in a military buildup that was the premise of the Iraqi invasion of Iran in 1980. But the war lasted eight years with no winner, leaving Iraq in a state of financial ruin. In 1989, Iraq earned only $13 billion from oil revenues, while importing $12 billion worth of civilian goods and $5 billion of military procurements.[1] It had also run up a huge foreign debt, mainly from financing its war effort, and now owed some $50 billion to Western and Russian lenders plus a larger but undisclosed amount to Arab lenders, generally estimated at $100 billion.[2]

As early as 1990, Iraqi oil production had been temporarily rescued from damaging overexploitation with inadequate instruments, due to the Iraqi industry's lack of the necessary funds and modern tools to manage it properly. Yet prices of Iraqi crude were sliding, and by the summer hit $11 per barrel.[3] Already at odds with OPEC's inability to limit output, Saddam launched a campaign accusing Kuwait and, to a lesser extent, the Arab Emirates, of having plotted with the United States to keep oil prices low in order to condemn Iraq to starvation.[4]

Clearly there was no conspiracy against Saddam, but something in his claims was true. As for most of its history, OPEC was again having troubles with its internal discipline. Instead of respecting the cartel's ceiling set at 24 million barrels per day, its members were producing 26 mbd, with Kuwait and the Emirates considered as the main responsible (nearly 80 percent) of overproduction.[5] Adding insult to injury, Kuwait—according to Saddam—was impoverishing Iraq's oil resources by frenetically pumping crude from a supergiant field straddling the border between the countries. Called South Rumaila on the Iraqi side, it was one of the country's largest oilfields, although it had been poorly developed since its discovery in 1962. For Saddam, Kuwait's exploitation of the field was consciously intended to steal South Rumaila's oil and decrease its pressure, making it harder for Iraq to recover its underground riches.[6]

The Iraqi campaign against the small sheikhdom was also enriched with new arguments that called into question the legitimacy of Kuwait's very existence as an independent country, its sovereignty over two small islands in the Persian Gulf, and other issues.

Doubtless, Kuwait represented an appealing target for Iraq not only because of its oil but because it provided a solution to one of the historical constraints of Iraq's geography: its poor sea access. Even today, Iraq's access to the sea consists entirely of a tiny strip of 190 kilometers on the western shore of the *Shatt-el-Arab* waterway, where the Tigris and Euphrates converge to flow into the Persian Gulf; in contrast, small Kuwait has a coast of about 500 kilometers. Expanding Iraqi sea access was a major goal of Saddam's, and one that became crucial in the context of sustained development of Iraq's huge potential for oil exports. Indeed, the latter had to be transported primarily via pipeline and thus across foreign countries such as Turkey, Syria, and Saudi Arabia, making it necessary for Saddam to maintain a "good neighbor" policy with these countries and so limiting his political autonomy.

Both the Arab League and the Gulf Countries Cooperation Council tried to cope with the growing tension, but all attempts for a conciliation failed in the face of the stubborn Kuwaiti refusal to come to terms with Iraq. While meetings were unsuccessful, Iraq's propaganda grew more aggressive in a crescendo that was only a prelude to a spectacular move. At dawn on August 2, 1990, the Iraqi army invaded Kuwait with a blitzkrieg, and by the end of the day it had taken over the country. By August 28, Saddam proclaimed that his conquest was Iraq's nineteenth province.

The world looked on in dismay at the latest performance of the "New Saladin." Overnight Saddam had increased his oil holdings to more than 20 percent of the world's proven reserves. He now portrayed himself as the liberator of the Arab masses from their subjugation by both the West and the corrupt, pro-Western gulf monarchies. His emphasis on the need for a new Arab order conjured up a dreadful scenario: just past Kuwait lay Saudi Arabia and the Emirates, with the bulk of their huge oil reserves located on a narrow corridor running from Kuwait to the Arab Emirates. Within Saddam's reach was the greatest concentration of oil in the world, more than 50 percent of proven global reserves, including those of Iraq and the newly occupied lands. Even if only a dire suspicion, the idea that Saddam had not satisfied his imperial hunger by taking Kuwait was deeply troubling to many observers. In any case, it was probably the prospect of Saudi Arabia falling under Iraqi control that won the support of a U.S. Congress otherwise resistant to a major military operation against Iraq.

As Colin Powell, then chairman of the U.S. Joint Chiefs of Staff and later Secretary of State, told General Norman Schwarzkopf in the days following Iraqi invasion of Kuwait, "I don't see us going to war over Kuwait. Saudi Arabia, yes, if we had to; but not Kuwait."[7]

Another version of the same argument was spelled out by then U.S. Secretary of State James Baker in his memoirs:

> If the President had said prior to August 1990 that we were willing to go to war to protect Kuwait, many members of Congress would have been muttering impeachment. Even after Saddam had invaded Kuwait, there was little, if any, domestic support for using our military. We had to build that support painstakingly.[8]

Relatively unaffected by America's "Vietnam syndrome," however, President Bush explained to the *New York Times* the reasons why the United States could have never accepted Saddam's threat to the entire Persian Gulf, resorting to the same *fil rouge* that had linked together American foreign policy from Roosevelt's meeting with the king of Saudi Arabia in 1945 to the "Carter Doctrine" in 1980:

> Our jobs, our way of life, our own freedom and the freedom of friendly countries around the world would all suffer if control of the world's great oil reserves fell into the hands of Saddam Hussein.[9]

Yet the international reaction came slowly, while the world hoped for a peaceful settlement. On August 6, the United Nations declared in Resolution 660 a total embargo on Iraqi and Kuwaiti oil, while calling for the reestablishment of the status quo ante, namely the liberation of Kuwait. That resolution cut 4 million barrels of oil per day from international exports (the net of both countries' internal consumption), or around 7 percent of world demand, then 65 million barrels per day.

Although Saudi Arabia committed itself to offsetting the loss, and OPEC temporarily scrapped production ceilings for all its members, oil prices began rising. While in July they had fluctuated between $15 and $19 per barrel (in Brent terms), on August 22 they reached $30; then on September 24 they topped $40, after Saddam announced in a speech broadcast on Iraqi radio his determination to bomb Israel and Saudi Arabia with long-range missiles should the economic strangulation of his country not stop soon. In the following months, however, oil prices decreased, leveling off at around $30 through the first two weeks of January 1991.

The New Saladin's threats found fertile ground in Western public opinion, partly because of a flurry of shoddy analyses which, bending to the self-serving sensationalism of the media, depicted the worst scenarios possible. Iraq's military potential was magnified by reports stating that it held the fourth or fifth most powerful military apparatus in the world and possessed chemical weapons, missile warheads capable of reaching Europe, and possibly nuclear weapons—remarkable military might for a country that despite the financial backing of most Arab countries as well as Western ones had been unable to defeat the disrupted and internationally isolated Iranian army.

For his part, Saddam was a master at exploiting Western fears. Shrewd for an impenitent layman, he mingled Arab nationalism and Holy War, worldly claims of political independence and oil revenues with religious dictates against the impious unbelievers. A sort of Nasser in would-be religious disguise, with a special talent for brinksmanship, he exploited every means of modern propaganda to generate support among the Arab people and overcome the growing isolation inflicted upon him by other Arab governments, which had condemned the invasion and called for Iraq's withdrawal.

When the United Nations imposed an ultimatum on Iraq authorizing the use of force under the organization's flag should Saddam's forces not leave Kuwait before January 15, 1991 (Resolution 678, November 29, 1990), the Iraqi *rais* spared no effort to amplify his threats. Leveraging his growing support among the Arab masses, he shrugged off the ultimatum and depicted the upcoming war as the final confrontation between the West and the Arab and Islamic people, "the Mother of all Wars"—as he defined it.[10]

Henceforth, all political efforts to avoid a military confrontation failed notwithstanding the tireless shuttle diplomacy conducted by the Soviet Union to convince Saddam to step down. A huge multinational force authorized by the United Nations was formed and put in motion, in preparation for the new D-day envisaged by UN Resolution 678. It was made up of fifty-seven countries—many of them Muslim and Arab—with a total of 500,000 men, although 90 percent of them were American. America held supreme command of the whole force. And the D-day inexorably arrived.

At 2:00 o'clock on January 17, 1991 (late afternoon on January 16 for the East Coast of the United States) while night blanketed Iraq's cities and villages, the world witnessed its first television war broadcast in real time, brought by CNN to every house on the planet with its spectral green images of the massive bombings. It was the outbreak of what was dubbed "Desert Storm," an operation of U.S.-led multinational forces acting under the flag of the United Nations.

Contrary to most experts' predictions of a new war-related oil crisis, on that very day crude prices plummeted from $30 to less than $20 per barrel, where they remained for the duration of military operations despite Saddam's desperate attempts to inflame and enlarge the war by launching Scud missiles against Israel and Saudi Arabia, and the threat of the Iraqi ambassador to the United Nations on February 18 to resort to weapons of mass destruction.[11] Also Saddam's final attack of Samson syndrome, which on February 22 led him to order Kuwait's oilfields set on fire (in less than a week, Iraqi forces blew up around 800 oil wells, refineries, storage tanks, and other oil infrastructure), when he realized the war was lost, provoked no significant price rebound. How did it happen that the market behaved in this way?

Many saw a direct link between the price collapse and the U.S. government's decision to release to the market 35 million barrels of oil from its Strategic Petroleum Reserves in subsequent steps, announced the very day the war began. However, this was merely the straw that broke the camel's back. By 1990 the world was already witnessing a new oil glut, fostered by OPEC overproduction;[12] only the situation had been eclipsed by the Middle East's anxiety as it waited for the nightmare of war to materialize.

After Kuwait was invaded, many producing countries started to pump all the oil they could, with OPEC alone (excluding Kuwait and Iraq) delivering almost 5 mbd of additional oil with respect to the precrisis level.[13] Saudi Arabia's production increase was the most surprising. Before the crisis, it was generally thought that the kingdom could sustain a peak production of no more than 7.5 million barrels daily.[14] Yet by December 1990 it was producing 8.5 mbd, or 3 million more than in January 1990, and in early 1991 it even surpassed the level of 9 mbd. At the same time, while overproduction intensified, the United States, Europe, Japan, and the Soviet Union were going through an economic crisis that dampened their energy consumption.

In short, the oil crisis was, once again, a state of mind, rather than a physical one. And once expectations of a looming calamity gave way to the reality of the overwhelming U.S. attack on Iraq, promptly shown to be a paper tiger, the crisis dissipated, and the iron laws of economics dictated the course of events.

At the end of February, Iraq effectively surrendered, although the war did not formally officially end until March 2, when the country accepted all conditions imposed by the United Nations for a cease-fire. The latter included several limitations on Iraq's sovereignty, including the imposition of a no-fly zone over two-thirds of the country, UN inspections to detect and destroy all weapons of mass destruction, and, above all, economic sanctions, which included a prohibition on selling oil without specific UN authorization. Eventually, in 1991 the United Nations passed a resolution allowing Iraq to resume oil exports, but Saddam refused to accept it, protesting the unacceptable curtailment of Iraqi sovereignty entailed in the decision. A subsequent resolution (986) approved by the UN in 1995 was also dismissed on the same grounds.

Thus Iraq remained out of the international oil market through the end of 1996, its production hovering around 500,000–600,000 barrels per day in the mid-1990s (considerably less than 20 percent of the country's potential) with only a modest level of exports to neighboring countries such as Jordan and Turkey in the form of smuggling. Saddam's new brinksmanship provoked a humanitarian disaster in Iraq, starving the population but providing the dictator excellent propaganda against the punitive effects on Iraqi people of the anti-Iraq coalition's sanctions. For all the embarrassment that this policy caused Western countries, it also forced Saddam against the ropes.

After the war, he had brutally repressed any sign of budding insurgency with mass executions and systematic torture, even resorting to chemical weapons. But instability persisted within Iraq and Saddam needed to address it not only with his usual heavy stick but also with a carrot. Thus in December 1996 he accepted Resolution 986, and a moderate flow of Iraqi oil to foreign markets resumed. The new resolution allowed the country to export $2 billion of oil over a six-month period and could be renewed, while a special UN commission would manage the revenues, depositing them in an escrow account and authorizing their spending only for food and medicine. Known as the "Oil-for-Food" program, this initial step toward Iraq's reintegration into the international markets was followed in February 1998 with another resolution (1153) authorizing Iraq to export as much as $5.2 billion in oil over the usual six-month period (this too renewable). Finally in 1999, UN Resolution 1284 eliminated all ceilings on exports and expanded the range of goods Iraq was allowed to purchase, although the UN still retained control over Iraqi oil revenues.[15]

Though defeated and checked, Saddam Hussein's regime had thus survived the storm, and was even strengthened by the humanitarian disaster that hit Iraq. The U.S.-led UN military coalition chose to stop General Schwarzkopf's troops before they marched on Baghdad and administered the final blow to Saddam, fearing that that might create a political vacuum in Iraq and provoke the country's dissolution along the fault lines of its historical ethnic and religious divisions. For U.S. policymakers in particular, the risk that Iran might take advantage of Iraq's domestic chaos exceeded that of keeping in power the man who had defied the Middle East's political order, particularly now that he had been defeated and humiliated.

Indeed, the whole U.S.-Iraq relationship turned into a major case of gross political and strategic mismanagement that started during the Reagan years.[16] Its origin can be traced back to February 26, 1982, when the Reagan administration eliminated Iraq from the list of state sponsors of terrorism, a step that was legally necessary to supply Saddam with financial credits, military and civilian technologies, as well as satellite images of Iranian military sites and troop locations (Iraq was then at war with Iran). The mismanagement was compounded in October 1989, when George Bush, Sr., signed National Security Directive 26 which, in proposing "economic and political incentives for Iraq to moderate its behavior and to increase our influence with Iraq," aimed to serve U.S. "longer-term interests" and "promote stability in both the Persian Gulf and the Middle East."[17]

This awkward attempt by the United States to manipulate a potential enemy (Iraq) in order to weaken an actual enemy (Iran) was clearly short-sighted, given the personal history of Saddam Hussein and his never-disguised designs on the entire Persian Gulf area. There are few cases in history in which a policy based on the principle "the enemy of my enemy is my friend" has paid off in the long term, particularly when the chosen ally had the characteristics of the Iraqi dictator. Awkward and short-sighted in its conception and implementation, that policy quite naturally never achieved what it was intended to and later provided considerable fodder for conspiracy theorists.

Soon after the Gulf War ended, it was reported that Saddam had attacked Kuwait after meeting with the U.S. ambassador to Iraq, April Glaspie, who had given him a sort of "green light" for the military operation. In fact, Glaspie had neither been informed about the Iraqi invasion plan, nor had she supported Saddam's claims against Kuwait. All she had done after carefully listening to Saddam's claims was to inform the Iraqi *rais* of her "direct instructions from the [U.S.] president to seek better relations with Iraq."[18] No doubt, this was a little too ambiguous a diplomatic formula for a man like Saddam; however, it implied only that the United States had completely underestimated the dangers of the situation and had no clear policy concerning it. The same is true of the remarks of Margaret Tutwiler (one of the closest aides to U.S. Secretary of State James Baker), made the day before the Saddam-Glaspie meeting, that the United States had neither "any defense treaties with Kuwait" nor "special defense or security commitments to Kuwait."[19]

Only by twisting Glaspie's remarks could they be considered an indirect "go ahead" for Iraq's invasion of Kuwait. In fact, the U.S. position vis-à-vis the political equilibrium among the Persian Gulf countries had not changed since the Carter Doctrine, and while Washington had supported Iraq against Iran, it would have never have allowed Saddam to become the Napoleon of such a critical area. In this context, Ambassador Glaspie's meeting with Saddam was no more than the continuation of the blind appeasement policy followed by the United States since 1982 which, contrary to its intent, had only nurtured Saddam's imperial designs.

The ambiguous conclusion of the first Gulf Crisis and the free hand Saddam largely enjoyed in re-cementing his grip on Iraqi society could not but cast a dire shadow on any illusion of a durable pacification of the Persian Gulf.

CHAPTER 13

The Soviet Implosion and the Troubled Caspian El Dorado

At the sunset of the century, yet another epochal event seemed to confirm that there was something inexorable about the oil market's downward drive: the collapse of the Soviet Union and its oil industry.

The crisis of the USSR had already begun in the 1980s, but, paradoxically, it was the reformist attitude of the Federation's new leader, Mikhail Gorbachev, that opened the Pandora's box of its explosive underlying problems. In 1986, Gorbachev launched a plan based on two pillars, perestroika (restructuring) and glasnost (transparency), whose practical application would undermine the foundations of the Communist Empire.[1] His reforms triggered the dismantling of parts of the paralyzed economic institutions of the USSR while sparing others. However, partly because of the strong political opposition to his radical new visions, Gorbachev did not succeed in filling quickly the vacuum created by the elimination of habits and rules consolidated over many decades of total state control of every aspect of Soviet life. In a few years, thus, the USSR fell prey to economic chaos, hyperinflation, soaring government debt, and widespread corruption.

Already in bad shape, the Soviet oil industry was particularly hard hit by the crisis. The oil countershock of 1986 had drastically shrunk revenue from oil and natural gas exports, which normally contributed more than 30 percent of the USSR's budget and were essential to the survival of the industry as a whole. About 70 percent of its production was in fact sold on the domestic market, at artificially low prices imposed by the state: at the end of 1992, for instance, they were only about 5 percent of the world price in the Russian Federation.[2] Meanwhile, oil production costs soared as a result of inflation in labor and machine costs (part of the latter had to be imported, while the ruble was devalued), and the damaged state of many oilfields, which had been the victim of overexploitation and poor technical management in the past. The result was an astonishing implosion of oil production, which dropped from its record high of almost 12.6 million barrels per day in 1988 to a plateau of 7.1–7.3 mbd in 1996–1998.[3] Soviet consumption as well dived, and oil exports, which had peaked in 1987–1988 at 3.5 million barrels per day, declined by 1.5 mbd in the early 1990s.[4]

The malaise of the oil industry of the communist empire was destined to last until the end of the century, and was probably aggravated by the chaotic reorganization it went though after the dissolution of the Soviet Union in 1991 and the birth of fifteen sovereign republics within its former boundaries.

In the Russian Federation, led by Boris Yeltsin, chaos was greatest. Following the example of Poland, Russia plunged into a process of shock-therapy privatization that remade the map of the country's economic power; yet the absence of a legal framework for the process, the unpreparedness of the public for this foreign-import crash program, and the already deep-rooted corruption within the establishment turned privatization into a system for the redistribution of Russian riches into the hands of a restricted elite.[5]

This was the destiny of Russian oil industry as well. During the Soviet era, it was fragmented into about three hundred producing and refining enterprises, each governed by targets established by centrally imposed five-year plans.[6] Historically, this abstruse industrial organization had produced around 90 percent of the oil output of the USSR, or more than 11 million barrels per day at its height in the late 1980s. In the framework of the privatization process, the hundreds of enterprises that formed the Russian oil sector were reorganized into groups so as to create sufficiently large joint-stock companies, which were eventually floated on the market. From this process emerged respectively Lukoil, Yukos, Surgutneftegas, Gazprom (1993), Slavneft, Sidanko, Onako (1994), Tjumen Oil Co. (TNK) Sibneft, and Rosneft (1995). This was only the first step in the dramatic reshaping of the Russian oil industry that was to take place. Indeed, the privatization process nurtured the rise of a new aggressive breed of obscure businessmen, would-be financial investors, and speculators—profiteers of any kind that overnight took control of the commanding heights of the Russian economy, by leveraging of their political connections, financial chaos, and absence of law.

After encouraging their rise, Yeltsin called on those rampant raiders to prevent Russia's financial crash and guarantee his political survival, thereby becoming more and more an accessory and hostage to their influence. The most striking episode of the Kremlin's abdication of power to private speculators was carried out with the loans-for-shares scheme, which Yeltsin approved by decree at the end of August 1995.[7] Devised by Vladimir Potanin, one of the main exponents of the flourishing finance industry that emerged with the privatization process and shared by other members of that restricted elite, including Mikhail Khodorkovsky,[8] the plan included a Faustian pact between the Kremlin and the new masters of the economy that would have deeply changed the distinctive features of power in Russia. His philosophy was as simple as it was disconcerting.

In 1995, the Russian government was encumbered by debts and a deficit that condemned it to insolvency. The short-term situation was so dramatic that it endangered the payment of state salaries and pensions, while Yeltsin's popularity was in constant decline. To make the picture even grimmer for the Russian leader and his supporters, polls indicated that the success of the Communist Party would be a possibility in the parliamentary elections of December 1995, which could have paved the way to success for party leader Gennady Zyuganov in the presidential elections of the following year. Taking advantage of this critical situation, the proponents of the loans-for-shares scheme offered to lend the government the necessary funds to cover the state budget, and took the shares still held in forty-four partially privatized companies as collateral (the number was later reduced to sixteen).[9] Management of these companies would have been temporarily taken over by the government's creditors, which hid another objective of the understanding between the Kremlin and its financial allies: to eliminate the top management of the same companies, considered a power in and of itself, inclined to favor the opponents of the privatizations and even the Communist Party. However, the primary objective of the whole plan was much broader.

If the government was unable to repay the loans by September 1996, the shares given as collateral would be sold in public auctions in which the same creditors could participate. This clause sardonically sealed the despicable pact reached between the Kremlin and the new economic potentates, since it was rather clear that the Russian state would never recover from the debt entered into in such a short time. The only guarantee that Yeltsin demanded was to fix the due date of the loan immediately after the 1996 presidential elections. That way, he tied the beneficiaries of the loans-for-shares scheme to his political destiny, forcing them to support him in the reelections by any means: clearly enough, a communist victory would have meant losing everything. And so everything went according to plan, with some additions to the original arrangement, between November 23 and December 28, 1995.

In the oil sector, thanks to the loans paid to the state, Mikhail Khodorkovsky's Menatep financial group obtained control of 78 percent of Yukos as collateral (December 1995) for the sum of $309 million.[10] During the same period, financier Boris Berezovsky, allied with unknown young financier Roman Abramovich, obtained the controlling stake in Sibneft as collateral by lending nearly $100 million. (A few years later, Berezovsky resold the shares to Abramovich.) Instead, Vladimir Potanin obtained control of the Sidanko company, by lending $100 million. Even more astounding was the attempt to gain majority control over international colossal Norilsk Nickel (NN), which at the time of the fall of the USSR controlled 90 percent of Russian nickel production and almost 100 percent of the platinum produced in the country:[11] Potanin obtained 38 percent of the shares of NN as collateral by paying only $100,000 more than the auction starting price, which was set at only $170 million. Norilsk Nickel had recorded net profits of $1.2 billion in 1995.[12]

With respect to Potanin's original plan, not all "red managers" of the companies involved in the loans-for-shares scheme could be eliminated. The strongest and shrewdest of them, in reality, were associated with the plan, so as to take direct control of the companies they were running. So Surgutneftegaz, headed by Vladimir Bogdanov, virtually acquired control of the majority stake through his pension fund, while Vagit Alekperov's Lukoil—the largest Russian oil company at that time—allied with Imperial Bank, a Russian financial group, to take control of its capital. At that point only one obstacle loomed over the triumphant march of the rising new economic potentates, notably the 1996 presidential elections and the danger that Yeltsin's defeat would jeopardize the stake they had claimed for so little. But Yeltsin won and the pact he signed with the group of speculators could be sealed by the predictable finale. After the due date of the loans-for-shares scheme with the insolvent state, the creditors sold the share parcels received as collateral in rigged auctions, formally purchased them, and became the majority shareholders of some of the most important Russian companies producing raw materials. It was the crucial moment when the speculators came from nowhere and seized Russia's vital riches, thus becoming the effective economic "oligarchs" of the country.

The dubious sale of state assets continued after the end of the loans-for-shares period, reaching even more disturbing levels. In 1999, the government sold 49 percent of the TNK oil company to two Russian financial groups, Alfa-Group and Renova, for just $90 million, equivalent to the price of one cent per barrel of reserves.[13] In 2001 TNK purchased 84 percent of Sidanko from Potanin, which BP had already bought into in 1997 with a 15 percent share. By the beginning of the new millennium, the dividing up of the Russian oil sector had been completed. Of the large integrated companies set up between 1992 and 1993 only Rosneft remained under state control, along with Transneft—which had the monopoly of the oil pipelines in the entire country. Another exception was the gas supergiant Gazprom, which many considered a sort of state within the state. Apart from this, at this point nearly all of Russian oil was in the hands of an exclusive club of powerful new businessmen.

It is true that at the time of the privatizations, Russian oil companies were rather mysterious and in ruins and certainly could not be valued according to Western financial standards. With outdated technology and a lack of managerial culture, they were afflicted by organizational and production problems inherited from the Soviet era and plagued by an apparently uncontrollable decline in oil production. As a result of the latter, Russia had recorded a plunge in its oil production, which reached the minimum level of 6 million barrels per day in 1996, compared to the record of 11.4 million barrels per day in 1987. In addition, in most cases the low oil prices imposed on the domestic market by the state were not even enough to cover production costs, thus condemning companies to starvation. Having said that, there is no doubt that the privatization of oil companies in the 1990s had been—at best—an extraordinary deal for the buyers; at worst, which appears to be the most plausible, something very close to what was called "the sale of the century."[14] However rich with oil and natural gas reserves, a few years later each of them was worth billions of dollars: Yukos, for example, seemed headed for bankruptcy in 1995, but had a stock exchange value of $31 billion in July of 2003.

Due to the lack of a central state strong and capable enough to put the brakes on the expansion of the oligarchs, the extent of their hold on Russian society seemed to have no limits. Only at the beginning of the new millennium all this was partially destined to change, due to the rise to power of Vladimir Putin.

Things took a completely different course in another part of the former Soviet empire, a part that was to become a tangle of intense geopolitical and international oil (and gas) interests: the Caspian region.

For centuries this area had been an ethnic and religious mosaic, with dozens of nationalities living shoulder to shoulder, and the site of numerous international power plays stemming from its double nature. Indeed, it had been both a corridor for Mongolian and eventually Islamic influence on the southern flank of the Russian empire, and a natural area of expansion for Moscow, which considered it the last bastion against Mongolians and Muslims, but also the door to India and the Mediterranean. For all parties involved, moreover, the region was an essential part of a larger area, Central Asia, through which one of the most important trade routes of history—the Silk Road—runs.

Only late in the nineteenth century, oil emerged as a new ingredient of the Caspian geopolitics when Azerbaijan gave birth to the Russian oil industry, eventually assuming the unchallenged leadership of the empire's crude production for many decades. This function progressively diminished since World War II as a result of the depletion of the onshore oil deposits in Azerbaijan and Chechnya (the latter was a significant oil producer until the 1960s) and the concurrent discoveries of huge hydrocarbon deposits first in the Volga-Urals area, and later in Western Siberia. In particular, it was the emergence of the latter's hydrocarbon potential that shifted Moscow's focus to the development of this area and away from the Caspian region, which entered a prolonged limbo. As the USSR collapsed, the Caspian states were finally free to shape their own destiny, and oil played a significant role in it.

With Soviet-era geological surveys indicating vast untapped oil and natural gas potential, the whole area soon became a potential new El Dorado in the collective psychology of oilmen. This sudden euphoria was partly hyped by the governments of the newly formed Caspian republics, which sought to attract foreign companies and international interests in order to bolster their political and economic independence from Moscow. Consistent with this goal and in contrast to Russia, all the Caspian states retained total control of their oil concerns, initiating a process of concentration of the numerous enterprises then in operation into centralized state oil companies, which they allowed to establish joint ventures with international majors for exploring and developing underground resources.

Apart from the magnification accomplished by careful state propaganda, the region's oil and gas endowment was actually quite impressive, particularly in the Caspian offshore. Here there had been evidence of potentially huge oil and gas deposits since the 1950s, later corroborated by geological appraisals, but a lack of technical know-how and adequate tools had kept the Soviet engineers from bringing to the surface what they guessed was lying beneath the seabed. Thus it took a while before world's media began referring to the area as a new North Sea, at the very least, and possibly even a new Persian Gulf, as the estimates put its oil reserves alone (the area was rich with natural gas as well) at between 25 and 100 billion barrels or more. The immediate consequence of such projections was a major international competition to grab up contracts, with Kazakhstan setting the pace for foreign company penetration into the region.

In the last years of Gorbachev's rule over the Soviet Union, Kazakhstan was allowed to grant a contract to Chevron for the recovery and development (in a joint effort with one of the country's state oil companies) of the giant Tengiz oilfield, discovered in the 1970s and deemed to hold proven oil reserves in the range of 7 to 9 billion barrels. After the dissolution of the Soviet Union the deal had to be reformulated, this time through direct negotiations between Chevron and the sovereign Kazakh authorities. Ruled by the former secretary of the Kazakh Communist Party and Chairman of the Council of Ministers of the Kazakh Socialist Republic, Nursultan Nazarbaev, Kazakhstan could take advantage of its relative stability and lack of dramatic backlashes with respect to most of its neighboring countries, and thus was able to finalize the oil deal with Chevron on Tengiz by 1993. The following year, the country struck another important deal by awarding a foreign consortium of six international oil companies a contract to conduct a major seismic survey of the 100,000-square-kilometer sector of Kazakhstan's part of the Caspian Sea.

In 1994, however, it was Azerbaijan that heralded the real beginning of the new Caspian oil and gas saga.

Unlike Kazakhstan, Azerbaijan had gone through a period of dramatic instability since its independence. Several coups and violent rivalries among local elites had undermined the establishment of a strong central government; moreover, in 1993 a major crisis had shaken the country when its Nagorno-Karabakh province seceded, backed by Armenia, which led to a bloody war involving the two republics and Russia in the background, which supported Armenia. The war was the first alarm bell signaling that the fall of the Soviet empire had unleashed historical grievances and harsh ethnic and religious divides that had been previously suppressed by force and were now to bring chaos to the whole area.

Only in October 1993, with the election to the presidency of the republic of a former senior officer of the KGB and member of the Soviet Politburo, Heydar Aliyev, did Azerbaijan find its strongman and enter a phase of relative stability. Conscious that time could work against him, Aliyev soon appointed a commission to negotiate with international oil companies a deal for the development of several offshore prospects, particularly the offshore fields of Azeri, Chirag, and Guneshli—with estimated proven oil reserves of about 3.5 billion barrels. When the final contract for the three oilfields was signed in September 1994, the world press greeted it as "the contract of the century," finally carving in stone the image of the Caspian as the very last great frontier of the oil industry. More importantly, the consortium that won the deal was led by BP and involved several other Western and non-Western companies, regrouped under the flagship of the Azerbaijan International Operating Company (AIOC): thus Azerbaijan was no longer alone in confronting the difficult first phases of its troubled independence.

Indeed, as important as the Azeri-Chirag-Guneshli development was, it was nothing extraordinary by international oil industry standards: originally it involved an $8 billion investment over the life cycle of the fields to bring their production to 700,000 barrels, roughly the same amount envisaged for Tengiz.[15] However, coming after a long period in which there had been very few giant deals for the international oil industry, it was quite natural that the AIOC contract would arouse expectations of a new, epic oil venture like those that punctuated the history of oil.

In 1994, it was once again Kazakhstan that concluded a memorandum of understanding with Italy's Eni and British Gas, later joined by Texaco, for the development of the giant Karachaganak field, which promised to hold several billion barrels of oil and natural gas. On November 18, 1997, the enlarged international consortium (which included, among others, Eni, Shell, BP, Mobil, Total, and Phillips, regrouped into the Offshore Kazakhstan International Operating Co., or OKIOK), which since 1993 had undertaken a seismic survey of the Kazakh-Caspian Shelf, signed a contract in Washington with the Kazakh government for the exploration and development of an area of about six thousand square kilometers within the shelf (the North Caspian Sea project). It was in this area that, in 2000, the discovery of the largest oilfield found since the early 1970s occurred—Kashagan—holding an estimated 30 to 50 billion barrels of oil resources in place. On the same day, President Nazarbaev signed the contract for the development of Karachaganak. It is worth noting that both agreements were signed at the U.S. Department of State, in the presence of U.S. Vice President Al Gore.[16]

By that time, however, the Caspian oil rush had brought to the surface a host of problems that caused considerable disappointment about the effective development of the region's hydrocarbon potential. Most of these problems had a common root: the Caspian region was landlocked, so that getting oil and gas out of the area represented a daunting obstacle, particularly in the framework of rivalries and violent divisions straining the area itself.

In the Soviet era, Azerbaijan, Kazakhstan, and Turkmenistan (the latter richly endowed with natural gas) had all depended on the Soviet transportation system. Under Moscow's tight control, that system allowed that oil and gas from the Caspian went north and fed the Soviet energy industry, while the oil that Azerbaijan exported to the Mediterranean first entered Russia by passing through Chechnya on its way to the Russian port of Novorossiysk, on the Black Sea, from where it could be loaded onto vessels and take the Mediterranean route passing through the Turkish Bosporus and Dardanelles straits.

The first problem with that huge infrastructure was that it had been designed for central management and control, but now the center no longer existed. To make matters worse, Caspian countries had real concerns about their dependence on Russia's transportation network, because by allowing or denying access to its pipelines, as well as by setting artificially high transport fees, Moscow could hold in check the Caspian republics' future source of wealth and so continue to exercise indirect rule over their destiny. An early warning of such a risk came from the disappointing negotiations for Tengiz, whose development had been constrained by Russia's refusal to transport more than thirty to fifty thousand barrels per day via its pipeline through the northern Caucasus.[17]

While Moscow was effectively trying to reassert its influence over the Caspian, the United States also entered the picture. U.S. engagement in the region was initially promoted by the George H. W. Bush administration, which soon saw the Caspian Basin as a partial lenitive to the excessive U.S. dependence on Persian Gulf oil, particularly in the aftermath of the first Gulf War. At the same time, the White House and State Department were preoccupied by the huge arsenal of conventional weapons stockpiled in the region (Kazakhstan also had nuclear bases) and pushed for the rapid consolidation of a post-Soviet order to prevent the emergence of a dangerous political vacuum in it.

This line of action was broadened and articulated by the administration of President Bill Clinton in the second half of the 1990s, especially after a Caspian Policy Review undertaken in 1997 that entailed a new assertiveness on oil issues.[18] Beyond the stated support for a "multiple pipeline system" option, the core of the Clinton administration policy was a "dual refusal" of Russian or Iranian entry into the Caspian pipeline game,[19] which Washington considered necessary to prevent these countries from gaining a major influence over the newly independent republics. Consistent with the strategic focus he devoted to the region, Clinton even created the position of "Special Adviser to the President and Secretary of State for Caspian Energy Diplomacy," a sort of plenipotentiary whose main task was to press all Western actors involved in Caspian hydrocarbon development to support the U.S. position. Naturally, the latter process only further complicated the already thorny question of how to get oil out of the region. Setting aside its aggressive stance on the matter, Russia was the natural and cheapest corridor for Caspian hydrocarbon transport; by the same token, Iran also offered a cheap and geographically favorable route of transit.

A further complication for marketing Caspian oil was Turkey's determination to limit the bustling transit of oil tankers through the Bosporus strait, which threatened Russian oil exports to the Mediterranean. The problem was indeed serious, and grew more so. In 1994 more than 1.5 million barrels of oil daily, or 80 million tonnes a year, passed through the Bosporus bottleneck. In 2003, the figure skyrocketed to 2.8 million barrels per day, or 144 million tonnes a year. This implied a constant menace to the security of Istanbul, the heart of which directly overlooks the Bosporus and was struck several times by disastrous accidents when tankers incorrectly navigated the narrow twisting channels of the strait; consequently, the Turkish government has repeatedly announced its intention both to introduce tighter regulations and to limit the passage of oil tankers through the cursed strait.

At the same time, aiming to become a key Mediterranean hub for hydrocarbon exports from the Persian Gulf and the Caspian, Turkey had promoted the construction of land and sea pipelines reaching its coasts that would provide new routes for oil and gas coming from Azerbaijan and Turkmenistan—in both cases eluding the Russian territory (with the only exception of one gas pipeline, the so-called "Blue-Stream" pipeline).

Naturally, this strategy found an ally in the United States, and an enemy in the Russian Federation, further complicating the geopolitical game surrounding the issue of how to get oil and gas out of the Caspian.

As time passed, the latter became a maddening puzzle of several pipeline proposals, with all companies and states directly or indirectly involved showing their cards by presenting their own projects to cure the woes of that landlocked region. Pipelines stretching in every direction, some partially overlapping one another, were drawn across the map of the region creating a sort of *Pipeland*, connecting China with Turkmenistan and Kazakhstan; Kazakhstan with Iran, Russia, Turkmenistan, and (via Afghanistan) Pakistan and India; Turkmenistan with Turkey (via Azerbaijan and Georgia); and Azerbaijan with Russia, Iran and Turkey, and so on. Most of these proposals were nothing more than exercises in fiction. Actually, the major international companies stuck to the main principle informing any sound investment strategy, namely that the economic returns must justify the construction of any possible route; yet politicians of all kinds, assisted by ever-present would-be experts, spared no effort in launching bizarre projects, with local governments backed by various international powers and even signing useless memoranda of understanding concerning the realization of such projects.

Needless to say, the resulting proliferation of what were literally pipe dreams, along with political rivalries and conflicting interests excited the restless imagination of conspiracy theorists, who interpreted any dramatic event in the region as a direct consequence of a dirty quest for oil and power. The most prominent legend conjured by this kind of intellectual ferment concerned the dramatic situation of Chechnya.

After it proclaimed its independence from Moscow in 1991, the small northern Caucasian republic had to face a fierce retaliation: in 1994 Russia invaded it and began a prolonged war that intensified over time and still today is far from over. Indeed, all Caucasian republics and enclaves—from Dagestan to Ingushetia, from Nagorno-Karabakh to South Ossetia and Abkhazia—passed though dramatic phases of violence and chaos, paying a bloody toll to the reemergence of ancient ghosts. As a senior diplomat observed:

Historically, religion was the basis for the differences in the region. The Orthodox Armenians, the Georgians, and the majority of Ossets were Christian, oriented toward the nearby Christian empires of Byzantium and Russia. The peoples of the North Caucasus and what is now Azerbaijan were Muslim and received moral, economic, and military support from either the Ottoman Empire or Persia. Moreover, the relative importance of religion differed in the various cultures. The Georgians were more religiously observant than the Ossets, the Chechens and the Azeris more so than the Dagestanis or the Ingush. Some nationalities held onto their ancestral cults and pagan rituals and professed a Christianity or an Islam that were merely formal.[20]

Beyond religion, tribal identities and geographic influences played a major role in the explosion of violence, which also struck other Central Asian countries such as Tajikistan and Afghanistan. Yet the astonishing brutality and destruction of the Chechen war was unique and grabbed the world's attention. For many observers, it was clear that the major reason for the sudden outbreak of violence was Chechnya's oilfields and infrastructure. But the brutal reality that cursed the republic was written in its history, which was now reemerging in its direst form.

Indeed, the small Caucasian republic had been a fierce antagonist of the Russian central power since the time of Peter the Great in the late eighteenth century, long before oil assumed importance in human life. Albeit forcefully composed through harsh repression, the secular desire of the Chechen peoples for independence had never died, making them the soft underbelly of Russia's southern flank. During World War II, Stalin thought to solve Chechnya's issue once and for all by ordering the displacement of nearly 1 million Chechens to other parts of the USSR. Many of them died during their tremendous trip.

When the USSR fell, Chechnya's quest for independence was underestimated by Russian authorities, being only a gust of a much broader wind of freedom sweeping the entire realm of the former Soviet empire. As a consequence, Boris Yeltsin ordered the withdrawal of all Russian troops from the republic (1992) after an agreement worked out with its leadership. Later on, however, Moscow grew scared of the domino effect that the growing militancy of the Chechens could have on the whole Caucasian area, which might ultimately lead to the disintegration of Russia; moreover, Yeltsin probably saw in the war an instrument to reinforce his wavering popularity at home.[21]

As to the supposed importance of oil in provoking the Chechen wars, it is worthwhile recalling that the republic was able to produce only 30,000 barrels per day, or about one-third that of a country like Italy, which is certainly a wonderful place but an insignificant oil producer. Moreover, Chechnya's most important oil pipeline, linking Azerbaijan with the Russian port of Novorossiysk, had a reduced capacity (about 120,000 barrels per day) and was in poor shape. Its supposed importance for transporting the new Azeri oil soon proved to be exaggerated when the Russians built a line that bypassed it, rendering it useless. In sum, the modest oil resources of Chechnya, its redundant pipelines, and its outmoded refining system were only a footnote in a conflict whose roots were far broader and older. There was no oil, natural gas, or other precious resource in the former Yugoslavia, which, parallel to events in Chechnya, fell victim to one of the bloodiest spasms of ethnic carnage since World War II. Just as the awful destiny of Yugoslavia was somehow written into its DNA, so it was for Chechnya.

As for the Caspian region as a whole, additional problems cast a long shadow on an already troubled situation. In particular, no agreement was reached over the legal status of the Caspian Sea, a matter of profound divisiveness for its littoral states. According to some of them, the Caspian was (and is) an enclosed sea, whose overall jurisdiction had to be divided among littoral states following a median line of the Caspian itself. As a consequence, each state had an exclusive right to exploit natural resources under the portion of the sea assigned to it. This interpretation had solid foundations in international law, but it naturally benefited some countries more than others. In particular, it gave Azerbaijan and Kazakhstan the exclusive right to most hydrocarbon resources discovered by then (and so far) in the Caspian Sea. Strongly supported by Russia and Iran, instead, was another point of view that considered the Caspian an international frontier lake, implying that each littoral state had a limited exclusive right to the sea itself, comprising a few miles, while the rest of the sea was subject to joint disposition of all states—a sort of condominium-like approach. Thus far, the debate over these opposing views has led to no agreement.[22]

Finally, another enduring obstacle to the development of Caspian resources was the region's lack of easy access to the essential tools of the oil industry, such as marine drilling, construction fleets, and fabrication facilities—this scarcity too being a consequence of the geographic isolation and complexity of the region itself.[23]

Because of all these problems, by the end of the century only a few steps had been taken toward unlocking the Caspian's hydrocarbon potential. The only pipeline that was built was one connecting Baku with the Georgian port of Supsa, on the Black Sea, completed in 1998 with a total capacity of 150,000 bpd. A larger pipeline planned by the so-called Caspian Pipeline Consortium (CPC) for connecting the Tengiz field in Kazakhstan with the port of Novorossiysk was completed in 2002—with an initial capacity of 560,000 barrels per day, eventually to be doubled.[24] Also in 1998, Russia and Turkey—with the crucial involvement of Italy's Eni—agreed to build a natural gas pipeline, dubbed the "Blue Stream," linking both countries and passing beneath the Black Sea. In spite of strong opposition to the project from the United States, which supported an alternative pipeline from Turkmenistan to Turkey, work on the Blue Stream began in 2000. And in spite of the U.S. characterization of the project as the "Blue Dream" because of its technical and economic complexities, the pipeline was completed in 2002, setting the world record for laying undersea pipeline (at a depth of 2,150 meters). Finally, an intergovernmental agreement was signed in Istanbul in November 1999 for the construction of the long-debated and strongly U.S.-backed Baku-Tiblisi-Ceyan (BTC) oil pipeline linking Azerbaijan with the Turkish Mediterranean coast and avoiding Russian territory. It would have a planned capacity of 860,000 bpd that could be doubled. Amidst several doubts about the pipeline's commercial and economic viability, this time the United States won an initial success in its Caspian strategy. However, it took more than five years before all outstanding problems concerning the pipeline could be overcome, and only in May 2005 was its first line inaugurated.

CHAPTER 14

The Collapse of Oil Prices and Industry Megamergers

The overturning of the former Soviet oil industry and its diminished contribution to international supply left the market indifferent. Up until 1996, half of the modest growth of world oil consumption was fed by the surprising production boom of the North Sea, while OPEC countries had to play the role of "swing suppliers" of additional global needs.

The hard reality was that, once again, the world was facing excessive "oil liquidity" caused by sluggish demand, which was increasing at less than two percentage points per year, and a more than abundant supply. Oil prices reflected the actual market situation, hovering for most of the decade at around $18 per barrel or less. Given this situation, even the forced curtailment of some important oil producers' production by international sanctions—a peculiar trend of that period—did not affect the prevailing bearish perception of the market's future.

Already experiencing a dramatic plummeting of its production, Iraq became a target of the U.S. "dual containment" policy announced in May 1993 by President Clinton—a policy that also involved Iran. Since 1979, the latter had been an object of American economic sanctions, but in the early 1990s some U.S. companies had resumed contacts with Iranian authorities to explore the possibilities of oil deals. In effect, the "dual containment" strategy was aimed at impeding the resurgence of regional and military ambitions by Iraq and Iran, but they did not clearly forbid American companies from doing business with Teheran. That ambivalent position was soon to change after the U.S. oil company Conoco announced in 1995 a multi-billion-dollar agreement with Teheran to develop a giant offshore oilfield in the Persian Gulf. Pressed by Congress, Clinton then released a presidential order establishing a total embargo on dealings with Iran (April 1995). In July 1996, the U.S. Congress approved the Iran-Libya Sanctions Act (ILSA), which "threatened even non-U.S. countries making large investments in energy."[1] Though opposed by European countries, sanctions contributed to maintaining Iran in a state of great difficulty with regard to the development of its oil sector, already suffering from the long war with Iraq. As for Libya, the new sanctions only worsened the situation created by the U.S. economic sanctions of 1986, which had led U.S. companies to leave the country.

Yet despite the fact that Iraq was out of the game, the Soviet Union had collapsed, and Iran and Libya were severely constrained, the world oil supply kept on growing more than demand did, thereby putting OPEC in a paradoxical situation. In fact, OPEC's return to a policy of reined-in production and oil price support after the Iraq-Kuwait crisis made it attractive for international oil companies to invest billions of dollars in non-OPEC areas where extraction and logistic costs would otherwise have been too high. This situation nurtured three major trends in the industry: the rush into the Caspian Basin; the strategic drive toward deep and ultra-deep offshore prospects in the Gulf of Mexico, West Africa, and later in Brazil; the failed bet on the Tarim Basin in northwestern China. It is worth noting, however, that because of the bearish market situation, international oil companies put under tight control their capital expenditures and focused primarily on squeezing their existing assets base. As for the Caspian, most of their new undertakings involved long-term investment commitments but relatively modest immediate spending. Naturally, this also implied that no new significant production was coming onstream, as existing production was more than what was required.

OPEC eyed these developments with frustration and resentment, conscious that its competitors were taking advantage of its painful attempt to keep oil prices at acceptable levels. Simply put, the cartel was once again losing market share, while others joyfully filled the void created by its own policies. Almost any OPEC meeting in those years opened and closed with a ritual protest against the Organization's competitors, which met with the total indifference of international oil companies and non-OPEC producers. Actually, the oil cartel failed to reconcile itself with the survival demon that pushed oil companies to develop oilfields wherever they could and to put them onstream as rapidly as they could.

The once powerful oil cartel started creaking again, and internal in-discipline increased until it provoked an open clash between OPEC and its own "inventor," Venezuela.

After Andre Sosa Pietri was appointed chairman of PDVSA in 1990, he began questioning both the government control of the company and its association with OPEC. A former senator and member of one of the most influential and wealthy Venezuelan families, Sosa Pietri wanted a free hand in running PDVSA, which he hoped to open to various forms of interaction with foreign companies and capitals. At the same time, he did not want to be subjected to OPEC anymore, which he dismissed as "only a myth," a relic of the past that, were it to survive, would be transformed into nothing more that a "research center."[2] Taking advantage of a corrupt political system paralyzed by the impending impeachment proceedings against Venezuelan president Carlos Andres Peres, Sosa Pietri led PDVSA to become an independent source of power that could bend the will of Congress to its own aims. It was not a major effort for him, thus, to obtain the green light for making deals with foreign companies, and he drove the steady growth of Venezuelan oil production irrespective of the country's OPEC quota.

This policy was continued and pushed to extremes by Sosa Pietri's successor, Luis Giusti. Albeit initially less trenchant about the role of OPEC, Giusti was far more effective than his predecessor in taking PDVSA to the brink of an epochal break with the Organization. On January 1996, the Venezuelan company launched the largest round of international bidding on oil exploration and production rights since its nationalization in 1975, making it possible for foreign companies to return to the country and work in the oil business. Moreover, in October 1996 Giusti outlined a ten-year development plan whose target was doubling Venezuelan oil production to more than 6 million barrels per day by 2006; at the time, PDVSA was producing 3.3 million barrels of oil daily, almost 1 million over its OPEC quota.[3] Finally, entirely un-moved by OPEC's protests and warnings, Giusti grew more outspoken, openly declaring that OPEC had to "change or disappear."[4]

Only a temporary illusion prevented a final clash with the oil cartel from taking place. Between 1996 and 1997, oil prices underwent a sub-stantial increase that no analyst had anticipated. In the last two months of 1996, Brent quotations even skimmed $25 per barrel on average, spurring a flurry of mistaken analyses of the cause. Many experts argued that world oil demand was undergoing structural changes in response to the vigorous rebound of U.S. consumption, the relentless upsurge in Asia's appetite for energy, specifically the so-called Asian-tigers—China, Indonesia, Malaysia, and South Korea. What the world was still unacquainted with, however, was a persistent flaw of the oil world: poor data.

Ever since the birth of the market, the relative lack of reliable real-time figures on effective demand and supply had plagued the industry and economists, and the problem had clearly grown more acute with the globalizing of the market. Even in the 1990s, the issue was far from being resolved. Even the most qualified source of data in the Western world, the International Energy Agency, was barely able to chart the monthly or yearly movements in demand and supply, not by any fault of its own but because of the objective difficulty of obtaining statistics on developing countries' production and consumption, oil companies' inventories, and the volume of oil shipped to its final destination after being bought and sold many times by different operators. In addition, the actual levels of major producers' output capacity were more a matter of careful speculation than empirical certainty. In this situation, the risk of "missing barrels"—i.e., of underestimating or overestimating demand or supply—was always present. And in 1997–1998 that risk materialized.

In 1997, all predictions pointed toward hefty demand growth, although oil prices had started to slide. Even the effects and size of the Asian economic crisis, which began in July–August of 1997 with the crash of the Thai currency, were totally underestimated by the world's main financial and economic institutions, including the World Bank and International Monetary Fund. Asia—the refrain went—was the rising sun of the twenty-first century and would relentlessly devour increasing volumes of oil. Japan's dramatic collapse had happened just a few years earlier but had taught nothing.[5]

Moreover, both OPEC and Western experts were repeating warnings both of an impending decline of North Sea oil production—announced since the early 1980s but blatantly disavowed by the staggering output of the region—and of the consequences a tightening world oil supply would have on ever-growing U.S. imports. In sum, for the major producers there was plenty of room for moderate optimism as they looked into their future, notwithstanding their rivalry. And thus they advanced unconsciously toward their destiny.

On November 29, 1997, at a ministerial meeting in Jakarta, OPEC agreed to raise its total output ceiling by some 2.5 million barrels per day for the first six months of 1998. Through this, Saudi Arabia received a higher production quota and partial compensation for Venezuela's overproduction. But the market was saturated with oil, because OPEC was already pumping it well above its official ceiling. The market's reaction was immediate. Oil prices began to drop in what in the months to come would become a ruinous fall.

What pushed the crisis to its climax was a report by the International Energy Agency[6] stating that in the second quarter of 1998 the global oil supply appeared to exceed demand by 3.5 million barrels per day, a surprising difference that implied "a huge inventory build-up—whether real or a statistical anomaly."[7] The issue of "missing barrels" exploded: where had all that oil gone? Did there exist in the world an undetected storage capacity capable of absorbing such a vast supply, a wild divergence from the normal fluctuation band of inventories held by companies and countries? No one had a precise answer, so the simplest response was to blame IEA for its gross mistake, and even suggest it should be closed down—as some U.S. congressmen did.

But whatever the true size of oil inventories, throughout the year it had been evident that they had grown enormously, such that the Brent crude price plunged to an average level of $11 per barrel in the last three months of 1998, from $14 in the first quarter of the year,[8] and in the first weeks of 1999 it even dived below $10. Oil-producing countries and companies fell into a state of gloomy dismay, which was reinforced by all forecasts and pundits. As the *Oil&Gas Journal* pointed out at the end of 1998, the global oil industry was

> moving toward a consensus (which, in this business, is usually ample reason to be skeptical) that it is in for a sustained period of low prices...the chorus is growing louder that something of a structural upheaval—a sea-change—is under way.[9]

According to the emerging doctrine that had incubated throughout the 1990s, oil was not only "just another commodity": its consumption levels in the world were so mature that they created a permanent glut in the market and consequently a long-term decline of its price. Pessimism was so widespread that as brilliant and insightful an expert on the oil market as former Saudi oil minister Zaki Yamani observed:

> The future trends I have pointed out all suggest that oil demand growth will be weaker over the next ten years that it had been in

the last decade. OPEC's ability to produce oil, on the other hand, will continue to rise.... I'm pessimistic, as I said earlier, but this does not mean that I cannot be persuaded that there [is] some light at the end of the tunnel, and that it does not belong to an oncoming train.[10]

Many suggested that prices could even go much lower. For instance, in a lead article devoted to the future of oil, *The Economist* argued that a long-term price of $10 per barrel could prove too optimistic, and summed up its thinking by suggesting: "We may be heading for $5."[11] But nothing sounded more like a death knell for the heroic age of oil than the remarks by Crown Prince Abdullah, the de facto ruler of Saudi Arabia, at a meeting of the Gulf Cooperation Council in December:

It would be well for all of us, governments and people alike, to remind ourselves that the [oil] boom is over and will not return. We must all accustom ourselves to a new way of life that is not based on total dependence on the state. Instead, each individual, along with the state, must play a positive and active role.[12]

Before that speech, during an official visit to the United States in September 1998, the prince had already captured the world's attention by publicly inviting international oil companies to submit to Saudi Arabia development projects concerning the kingdom's oil and gas sector. At the time, most observers inquired whether that message had to be considered as the first signal of a dramatic reversal of Saudi policy, aimed at reopening the country to foreign exploitation of its oil resources.

Against all odds, however, OPEC rose from its ashes. On March 1999, the Organization reached an agreement in Vienna on cutting production, this time joined by various non-OPEC producers, such as Mexico, Norway, Oman, and Russia, which also committed to reducing their output. It took only a few months to show that this time the adherence to the cut was the highest, or almost 90 percent of the predetermined ceiling (a 100 percent adherence being virtually impossible). So oil prices quickly recovered, and by the end of 1999 the crisis had passed.

OPEC now appeared as strong as ever, capable of dealing with sudden changes in the oil price landscape. Indeed, in its March meeting the cartel had established a flexible target price of $22–28 per barrel, and decided it would intervene by increasing or cutting production should the price move outside that band for twenty consecutive days. Accordingly, in 2000 new cuts followed as prices began to slide, removing from the market more than 5 million barrels of oil per day by the first half of the new year. The architect of OPEC's resurgence was a patient, discreet, and shrewd man who had spent almost his entire life in the oil business: Ali Naimi.

Naimi started working at Saudi Aramco in 1947, when he was twelve, replacing his brother after he died. Because he was too young to work legally, he had to devise a sad story about his family's economic difficulties to convince the doctor who examined him to lie about his age. Thus he succeeded in getting hired by Aramco, starting as a coffee boy to the American staff.[13] Forty-one years later, shortly after the oil countershock, he became Saudi Aramco's Chief Executive Officer. Finally, in 1995, his superior talent was rewarded with his appointment as oil minister of Saudi Arabia. Neither a prince nor a member of the Saudi establishment, Naimi had moved up from nothing and was appreciated for his technical skills rather than his connections. Above all, he was a man of long-term vision, capable of sticking to his targets no matter what the problems and opposition might be. Over the course of 1998, there had been rumors of his looming departure from the most important oil chair in the world, due to a reported discrepancy of thinking with respect to important members of the Saudi royal family.

Clearly, Naimi believed neither that the age of oil was nearing its end nor that Saudi Arabia had to abdicate to foreign companies for work that Saudi Aramco was fully able to perform on its own. Nor did he think that OPEC was a relic of the past.

It is still a matter of speculation whether all the moves made on Naimi's advice by Saudi Arabia and Crown Prince Abdullah in 1997–1998 were part of an unannounced strategy aimed at imposing a forced solidarity on an out-of-control OPEC, as had happened in 1985–1986. Whatever the truth, Saudi Arabia's actions in those two years projected a dire scenario for all producing countries: if freed from production limits, Saudi Arabia could really cause a permanent oil glut on the market. It was probably that fear that bolstered OPEC reunification in March 1999, making it possible to overcome the oil crisis.

International oil companies, though, had not yet made it to the end of the tunnel.

Throughout the 1990s, they had fought hard to stay afloat but remained under siege. Industrial and financial restructuring initiated in the 1980s had continued, dramatically slimming their organizations and human resources. Cost-cutting and non–core asset disposal had become a permanent imperative, imposed on them by low oil prices and by the necessity of restoring their image for investors who had become highly skeptical about their capacity to create value. Worse, financial markets had generally shifted to other investment opportunities, as they too considered the oil sector a mature one with modest prospects for growth. For that reason, the oil industry had submitted to financial dictates that called for the highest possible short-term profitability. Particularly, companies allowed investors and analysts to push them into calculating value creation using a long-term oil price of about 16 dollars per Brent barrel in nominal terms. This was an unrealistic view inspired by the consolidated belief that oil had become "just another commodity" whose price was inexorably bound to decline in the long run like that of most raw materials throughout history. To make matters worse, the markets expected that in such a price scenario the companies would produce a Return on Capital Employed (ROCE) at least four to five points higher than their Weighted Average Cost of Capital (WACC), which was generally in the range of 7–8 percent. In simple terms, ROCE is an accounting indicator based on the ratio between profits and capital invested by a company and it is good for assessing past but not future performance.[14] With respect to the future, ROCE hides a demon. The less a company invests, and the more it squeezes its existing assets without replacing them, the higher its ROCE will be. Yet, in the long term, such a company is dead—or at least bound to get anorexic.

In sum, expectations of returns based on ROCE calculations shaped a short-term culture inconsistent with the nature of the long-term nature of the oil business. It is not an exaggeration to say that if ROCE had existed in the first decades of the last century, no one would have spent a dollar searching for oil in Saudi Arabia, or—during the late 1960s—in the North Sea.[15]

All this notwithstanding, companies continued to navigate in troubled waters. The more they sold off superfluous businesses and industrial sites, slashed costs, and squeezed their asset productivity, the more they realized the extent of their past mistakes and the difficulty of living with low oil prices. In addition to oil production, other traditional pillars of the industry's core business were under attack. The robust buildup of new plants in Asia and the Middle East was swelling oil refining and petrochemical capacity and blocking the desired effects of Western oil companies' massive reduction of their own capacity. As a consequence, margins in both sectors were poor, and year after year they were hit with new environmental regulations on oil products imposed by industrial countries.[16] The environment had by now become a top priority of most governments in the world, after its slow and problematic takeoff thirty years before. The symbol of the new sensitiveness was the signing of the Kyoto Protocol in December 1997 by eighty-four countries, thirty-nine of which committing themselves to reducing their own greenhouse gas emissions by 2008–2012.[17] By then, the traditional concept of an energy company had also been called into question.

A new model of doing business in the sector had enchanted the world of investment bankers, financial analysts and traders, strategic consultants, as well as the media and universities. It was the model proposed by a company that had been relatively unknown until it was catapulted overnight into the global spotlight as a superstar. Its name was Enron.

Founded in 1986 in Houston by Kenneth Lay from the merger of two companies, Enron was not properly an oil company, its activities being rooted mainly in natural gas and power transportation/transmission and trading.[18] Under Lay's leadership, it had rapidly grown by exploiting the dramatic storm that swept the U.S. natural gas and power markets in the 1980s as a consequence of Reagan's deregulation. Tackling the uncertainties created for producers and customers by the sudden revolution, Enron had been apparently successful in reshaping on its own U.S. gas market rules, thanks to a set of new instruments: highly flexible contracts, price and volume formulas capable of accommodating the needs of hundreds of different suppliers and customers, trading mechanisms based on complex mathematical equations, and a system of overall risk management that leveraged on advanced tools of derivative finance. The jewel of Enron, whose activities became the hallmark of the Texas company, was the "Gas Bank," a division established in 1989 and later to become its most important subsidiary, Enron Capital and Trade Resources (ECTR).

Enron's ascent to worldwide fame, however, began in 1990, when Lay chose as CEO of what would become the ECTR a former partner from the McKinsey consulting firm, Jeffrey Skilling. In his capacity as consultant, Skilling had worked for Enron reviewing and improving the company's business model. It was up to him to outline the idea of a "Gas Bank," which in his mind was quite simple yet revolutionary. As concisely explained by McLean and Elkind:

> Producers (acting as depositors) would contract to sell their gas to Enron. Gas customers (the borrowers) would contract to buy their gas from Enron. Enron (the bank) would capture the profits

between the price at which it had promised to sell the gas, just as a bank earns the spreads between what it pays depositors and what it charges borrowers. Everybody would be happy.[19]

Underpinning this idea was Skilling's apodictic conviction that the traditional industrial company was dead, especially in the energy sector at large. Its huge investment of capital in numerous heavy production or infrastructure projects was outmoded and had to be replaced with a completely new ability to practice the "soft skills" required by deregulated markets, taking to an extreme the use of the highly sophisticated trading tools made available by derivative finance, creative accounting, and risk management. Indeed, the development of the "Gas Bank" concept and its application to other activities by the company (power, telecommunications, etc.) was inherently connected to Enron's need to protect itself from several sources of risk it would be exposed to as a consequence of its audacious undertakings.

But for Lay, Skilling, and their financial brain, Andrew Fastow, this was probably a marginal problem. In retrospect, it is not surprising that, as a take-it-or-leave-it condition for entering Enron in 1990, Skilling asked Lay for the introduction of a new accounting system for the company called mark-to-market, which would later prove to be a major factor in Enron's ruinous fall.[20]

In itself, the mark-to-market system was not evil. Its basic principle was (and is) that of booking assets, revenues, and liabilities according to their current market value, instead of their historical value (their value at the moment in which the company had first booked them), which was the more common accounting practice. Like any other system for representing company accounts, however, mark-to-market could be used for cheating investors. For example, one could book the whole value of a ten-year contract, taking into account its future revenues as calculated according to the most optimistic projections, even if the contract had not yet translated into actual revenue. As McLean and Elkind have remarked:

Taken to its absurd extreme, this line of thinking suggests that General Motors should book all the future profits of a new model automobile at the moment the car is designed, long before a single vehicle rolls off the assembly line to be sold to customers. Over time this radical notion of value came to define the way Enron presented itself to the world, justifying the booking of profits on a business before it had generated a single penny in actual revenues. In Skilling's head, the idea, the vision, not the mundane reality, was always the critical thing.[21]

By the same token, the value of the equity participation in a subsidiary could be booked at the maximum price its shares had reached. It did not matter that mark-to-market called for a constant revision of current values: Enron specialized in hyping them through a complex and obscure system of intercompany deals, displacing onto special purpose entities (SPEs) all the risks of locking in a share price, the value of a contract, or even the debts it had accumulated. Clearly, all these SPEs were off the books, which meant that no one could trace the effective wealth of the mother company.

In any case, Enron became one of the towering protagonists of the financial bubble of the second half of the 1990s. For five years in a row, beginning in 1996, *Fortune* magazine named Enron "The Most Innovative Company in America." It was virtually impossible to find an analyst or a banker who did not praise the company, invest money in it, suggest one buy its shares, and finance its activities. The dazzling success of the Texas corporation became a case study for most university management and financial departments worldwide, and the Lay-Skilling vision became the new catechism for apprentices of the business game, as well as a reference used to blame old-style managers for their impotence in dealing with the changing paradigms of managerial capitalism and industry value creation.

Oil companies were among the preferred subjects of the new Darwinian mantra proclaiming the inevitable demise of all species incapable of adapting themselves to the dramatic evolution of the markets. Skilling himself, with a flamboyant and arrogant prose typical of one identifying himself as the messiah of a new era, did not hesitate to deride Exxon as a "dinosaur,"[22] along with all the other traditional oil companies. In contrast Enron's people were making something magical: in Skilling's words, they were changing the way the world did business, or, much more modestly, "they were doing God's work."[23]

Skilling's judgment on the obsolescence of oil companies was mirrored in the financial markets' irrational preference for the champions of the so-called new economy: high-tech, internet, financial, and telecommunication corporations. Even if the latter were only start-ups, it was enough to promise a wonderful future for them to enjoy spectacular share price increases and abnormal price/earning ratios. It did not matter that their plans projected losses for several years before breaking even: they were the future, while all the others were the past.

In most oil company gatherings in those years, analysts and consultants explained to disoriented audiences how they had to leverage Enron's experience to change their Jurassic business models and open their minds to new horizons and strategies, such as combining water, electricity, and natural gas services (the latter was already becoming a major focus of all oil companies), possibly even telecommunications and many other client-oriented services, a weird all-embracing approach that was presented as a "multi-utility business model."

Under siege on all fronts, the oil industry finally reacted with a textbook strategy: concentration.

In the summer of 1998, BP and Amoco announced their merger, actually nothing more than BP's friendly takeover of the U.S. company. December then saw a spectacular move, the merger of former Seven Sisters Exxon and Mobil. The value of the deal was 77 billion dollars, all paid in shares of Exxon—the de facto ruler of the game—to Mobil shareholders. In 1999–2001, the process continued with the mergers of Total, Petrofina, and Elf; BP-Amoco and Arco; and Chevron and Texaco. The combined value of these operations came to a stratospheric 275 billion dollars.[24] A flurry of other smaller mergers (compared with those of the oil business, which always thinks in huge numbers!) followed in the same period and even afterward.

A new class of very large oil companies emerged as a result of this sweeping concentration process, the so-called supermajors ExxonMobil, Shell, and BP. Along with them a new group of international majors appeared as well, including TotalFinaElf, ChevronTexaco, Eni, and ConocoPhillips. These seven publicly traded giants, however, were but a pale shadow of the Seven Sisters. In all, they controlled only a modest 5 percent of world oil reserves, and no more than 15–18 percent of world production.

In any case, quite ironically the twentieth century ended with a posthumous double revenge by John D. Rockefeller.

On the one hand, his view that the oil market was doomed to chaos without "cooperation"—a gentle expression that in Rockefeller's vocabulary meant competition suppression—had once again proved true. On the other, the two largest companies of his Standard Oil empire, Exxon and Mobil, were once again a single entity, after the long divorce imposed upon them and their fellow companies by the U.S. Supreme Court in 1911.

Quite ironically, the "irrational exuberance" that had marked most Western stock exchanges in the 1990s and relegated oil companies to a shadowy corner came to a shocking conclusion at the end of 2001. A sudden crash shattered global investors' expectations about the so-called new economy, triggering the ruinous fall of high-tech, Internet, and telecommunication stocks. In a revenge of history, the epitome of that huge bust was the company that more than any other had surged to be the symbol of the new energy company of the future, Enron.

Still in 2000, the Texas corporation seemed invincibly marching along its "triumph boulevard." In August 2000, its revenues topped 100 billion dollars, and its market capitalization stood at around 60 billion dollars, making it number seven in *Fortune*'s list of the 500 largest companies in the world. That same year, Fortune again proclaimed it the most innovative company in the world. But in 2001, the dark reality came to the surface.

After months of growing doubts about the real sources of the company's value generation, in October Enron had to restate its financial results and write off more than 1 billion dollars for the third quarter of the year, thus showing a net loss in that quarter of 618 million dollars. Its debts stood at slightly less than 40 billion dollars, hidden for many years in so-called "special purpose entities." The value of Enron shares plummeted from more than 90 dollars in August 2000 to less than 1 dollar at the end of November 2001. On December 2, 2001, Enron filed for Chapter 11 bankruptcy protection. The collapse of the company's paper empire vaporized around 100 billion dollars, establishing a record in the history of United States and world financial markets.

But neither Enron's nor the financial markets' crash changed investors' attitude toward oil companies or the latter's approach to investment plans. In a few years, that culture would create new problems for the oil industry and markets. But once again, oil pundits did not anticipate anything at any time before the problems actually materialized.

CHAPTER 15

The First Oil Crisis
of the Twenty-first Century

History never sleeps and never ends. Above all, it is always ready to shock everyone with its rapid shifts. So it is no surprise that all the pessimistic assumptions about the impending end of the glory days of oil-generated wealth were turned upside down in the first years of the new millennium by a new oil shock.

As in the early 1970s, there were many economic reasons that fed the crisis, whose explosion was mainly the result of many years of inadequate investments in new oil production capacity and infrastructure. However, as in the 1970s, a new political drama anticipated the crisis and shaped its psychological background. Once again, that drama erupted from the unsolved problems of the most oil-rich region of the world, the Middle East; this time, though, it took an unprecedented and dreadful form, through the terrorist attacks to the soil of the United States that, on September 11, 2001, destroyed New York's World Trade Center, a wing of the Pentagon in Washington, and the lives of nearly 3,000 people.

All of a sudden, the threat to global security seemed to have no border, no headquarters, no traditional army, no formal rules of engagement. Its elusive nature was held together by a messianic call for "Holy War" against all the alleged oppressors of Islamic values and culture, a war to be conducted by any group of "good Muslims" wherever in the world they felt oppression took place. Based on sometimes loosely connected or completely unconnected military cells devoted to martyrdom, this borderless guerrilla-like organization had its apparent source of inspiration in the alarming figure of Osama bin Laden, the founder of the al-Qa'ida terrorist network, who claimed responsibility for plotting the

September 11 attacks. As spectacular and sudden as it appeared on that occasion, the threat brought forward by bin Laden had a complex background, which cast a grim shadow over the possibility of a peaceful living together of different civilizations, as well as on the security and free availability of the entire Middle East's oil resources.

The evisceration of Islamic terrorism paralleled the rise and fall of hope for a peaceful Israeli-Palestinian settlement in the mid-1990s, and was partly fed by it.

Initially fostered by the Oslo and Washington peace agreements of 1993 between Israeli premier Yitzhak Rabin and PLO leader Yasser Arafat, those hopes were soon frustrated by Arafat's strategy of trying to squeeze as much as possible from the new situation, and the growth in Israel of a new anti-Arab radicalism supported by ultraorthodox groups. The latter rejected any agreement with the Palestinians based on land transfers, particularly of the settlements Israel had established in the Palestinian territories. Rabin himself fell victim to Jewish radicalism, murdered by an ultraorthodox young man in November 1996. Quite rapidly, the situation degenerated.

For most analysts, the turning point that set off another season of violence was the provocative stroll by the future Israeli premier Ariel Sharon (then the aspiring conservative candidate for the premiership) to the al-Aqsa Mosque esplanade in Jerusalem, a holy place for all Muslims, on September 28, 2000. With that gesture Sharon wanted to reaffirm, as a Middle East's expert pointed out, "the Jews' ancient claim to Jerusalem and took the first step in a deliberate strategy to undermine the logic of peace that had been built into the 1993 Oslo Peace Accord."[1]

The Israeli-Palestinian conflict, however, was just one of the many justifications invoked by Osama bin Laden for al-Qa'ida's terror strategy. Another key factor prompting bin Laden to wage a global "Holy War" was the U.S. military presence in the holy land of Saudi Arabia during and after the First Gulf War, an act of impiety as well as a symbol of Muslim subjugation to American and Western interests. That subjugation, according to bin Laden, would not be possible without the alliance between the United States and what he called—resorting to Nasserist prose—the "corrupt" and "unholy" monarchies of the Gulf, first among them that of Saudi Arabia. For the al-Qa'ida founder, the unacceptable U.S.-Arabian marriage was based on trading cheap oil to the West in return for American protection of the Saudi and other Gulf monarchies. In an open letter to the American people, Osama was particularly outspoken as to the significance of that marriage:

You [the Americans] steal our wealth and oil at paltry prices because of your international influence and military threats. This theft is indeed the biggest theft ever witnessed by mankind in the history of the world.

In December 2004, a man identified as the founder of al-Qa'ida, speaking on an audiotape posted on an Islamic Web site, repeated this charge, and urged Muslims to stop it by sabotaging oil infrastructure and fields in Saudi Arabia.[2] By then, the terror war launched by Osama's network had succeeded in changing the mood of the world by spreading fear and insecurity.

In January 2002, President George W. Bush had proclaimed the new doctrine of "pre-emptive war" against "rogue states" deemed to constitute a potential threat to the security of the United States. Even before its formal unveiling, that doctrine had already been tested in the massive 2001 attack by U.S.-led forces against Afghanistan, the long-time headquarters of Osama bin Laden. Eventually it was applied to Iraq (March 2003), in what became the Second Gulf War.

Events are still too fresh to offer a sober judgment on the new military operation against the Saddam regime. What we know for sure is that, in the post-9/11 climate, the Bush administration quickly put Iraq in the dock, charging it with supporting al-Qa'ida and possessing weapons of mass destruction that would enable Saddam to launch a large scale offensive upon the entire Middle East. Later revelations demonstrated that neither charge had been proven, and that both had been based on inaccurate interpretations of intelligence, but also to some degree on manipulation of incorrect information concerning Iraq's acquisition from Niger of enriched uranium. This charge was used by President Bush in his State of the Union address in January 2003 to demonstrate that Saddam Hussein could possess weapons of mass destruction, but it later proved untrue.

In any case, the post-9/11 chain reaction, labeled as a "War on Terror" by the Bush administration, led to the fall of Saddam's regime in May 2003 and the occupation of Iraq by U.S. military forces, along with multinational troops. Saddam himself was captured and imprisoned in December 2003. But instead of heralding the advent of a new era of peace and stability, the conclusion of the war was the beginning of a new nightmare for the Arab country and its oil sector.

Terrorist attacks throughout the Iraqi territory became a common daily occurrence, many of them directed against the oil infrastructure. Already curtailed by years of war and mismanagement by Saddam's regime, Iraqi oil production capacity was thus further reduced and put at daily risk, depriving the world of an important cushion of crude supply.

As many have predicted, if there was one option that truly ran against the self-interest of the large international oil companies it was the war itself. Saddam's Iraq had cultivated good relations with virtually all of the world's oil companies, opening its immense deposits to joint development by local and foreign companies. The regime had signed many contracts and memorandums with foreign companies, to take effect after UN sanctions were lifted. The war, and the removal of Saddam, turned them all into worthless scrap paper. In other words, for the more cynical oil companies Saddam and his regime were an economic Godsend; by converse, as oil expert J. Robinson West explained in the summer of 2003, "even with Saddam toppled, people such as ExxonMobil's CEO Lee Raymond warned that it would take some years to develop confidence in the stability of the political and commercial regime in Iraq before any large investments are made."[3]

Albeit on a smaller scale, attacks to oil infrastructure have also hit Saudi Arabia since 2001, showing that the interruption of the oil flow from the Persian Gulf was a key objective of radical groups. Bin Laden's propaganda against the Saudi monarchy, moreover, was matched by a growing attack by many American opinion and decision makers against the kingdom's supposed support of radical fundamentalism, facing the striking evidence that fifteen out of the nineteen terrorists who had carried out the 9/11 massacre were Saudis, that bin Laden himself belonged to a prominent Saudi family (albeit he had been disclaimed by his own family and deprived of Saudi citizenship), and that the kingdom had continued to lavish money to extremist Islamic organizations. Whatever the perspective, thus, Saudi Arabia appeared as a weak and unreliable link of the world's oil chain, whose political and economic system was blamed both by terrorists and important Western circles.

With most of the Persian Gulf region a prey to apparent chaos, pro–Western Arab oil monarchies in check, and no easy prospect for a way out, the issue of oil security again gained momentum, right in a period when the first oil crisis of the twenty-first century was taking shape.

Actually, just before the specter of Islamic terror fell on the world, prices of oil and energy in general had began climbing well above the modest levels registered since 1986. After having plunged to $14.4 on an average annual basis in 1998, the barrel of Western Texas Intermediate

(WTI) rose to $19.3 in 1999, $30.3 in 2000, and then slightly decreased to around $26 in 2001 and 2002. Only after the war against Iraq, oil prices started an unstoppable march that led U.S. benchmark WTI to average $31.1 per barrel in 2003, $41.4 per barrel in 2004, and a stellar 56.5 in 2005. The most important international benchmark, Brent crude, followed the same trend, with a discount to WTI oscillating between 1 and 2 dollars per barrel until 2002, which eventually increased to more than 3 dollars in 2004 and 2005.[4] On August 30, 2005, oil reached its all-time record in nominal terms at $71 dollars for a barrel of WTI.

These values were still far from the temporary peak of $42 per barrel in 1980 for a barrel of Arabian Light, comparable in the United States to something between 80–100 dollars per barrel in 2005 dollars— depending on the inflator used. Moreover, the impact of oil on the world's economies had drastically fallen since the 1970s. For example, while the production of 1,000 dollars of GDP in the United States in 1980 required 1.8 barrels of oil, by 2004 it required only 0.6 barrels.[5] In other words, starting with the most oil-addicted country in the world, the oil intensity (the amount of oil needed to produce a unit of GDP) of the global economy was drastically diminished.

However, these arguments could not placate the psychological demon of the crisis. Not only was the world's oil supply by then insecure under the menace of terror and war, but there were many concrete elements that seemed to justify the direst visions: global spare oil production capacity was narrowing to its lowest levels in many decades, hovering around 2 mbd according to most estimates, or less than 3 percent of world's consumption. Production in traditional Western areas outside the influence of OPEC (like the United States and the North Sea) was dwindling, while the trend of declining new discoveries begun in the early 1960s continued, with yearly findings of completely new oil representing only around a quarter of the petroleum consumed every year— or about 7 billion barrels in 2000 against consumption of 27 billion barrels (even if after 2001, new discoveries jumped again).* Discoveries of giant new oil fields were a rarity, lending further support to the doomsayers' contention that only small and marginal deposits remained to be located. In sum, everything supported the idea that the world was running out of oil.

---

* See Chapter 18.

As happened so many times before, prophets of doom got the upper hand in what became a heated debate. The man who distinguished himself as the leading advocate of an imminent supply crunch was a geologist who had spent part of his life working for oil companies, Colin Campbell.

Campbell had not discovered a magic device to know the unknowable, namely the real amount of oil hidden in the earth's subsurface. His crystal ball was made up of statistical and econometric models based on the intuitions of another geologist, Marion King Hubbert, who in the late 1950s had correctly predicted the peak of oil production in the United States. But the empirical evidence on which Hubbert had based his original forecast was not exact science, and Hubbert himself failed to replicate his only success when he tried to apply his method to other countries and the world as a whole. Campbell's upgraded model offered no absolute truth either. His changing predictions for the peak of global oil production and its subsequent depletion turned out inexorably wrong, yet he insisted on updating his predictions with a lack of embarrassment over his poor past performances, which had gone generally unnoticed or ignored. His stubbornness was finally rewarded. In 1998, he succeeded in finding a wide audience thanks to the publication of one of his articles (written with Jean Laherrére) in the prestigious magazine *Scientific American*. Entitled "The End of Cheap Oil,"[6] the article summed up and updated Campbell's views and suggested that world oil production would reach its peak in the first decade of the new millennium.

Thus, when oil prices began climbing in 2000, and even overcame what was then considered the unsustainable barrier of $35 per barrel, Campbell's storm clouds turned into an impressive downpour: for many of the world's opinion makers, Campbell became a long unheeded Cassandra whose tragic premonitions had finally came true. Books and cover stories in prestigious journals echoed and elaborated the geologist's message with an intensity that could not but increase in 2004–2005, when oil prices skyrocketed and repeatedly surpassed sixty dollars per barrel.[7]

In this climate, other worrisome analyses found a vast audience. In 2004, for instance, a Texas-based investment banker with a reputation as an oil expert, Matthew Simmons, captured world attention by saying that Saudi Arabia's oil reserves had been overestimated, and faced faster-than-expected depletion.[8] Simmons's warning echoed one of the favorite theses of oil depletionists, namely that all OPEC countries had artificially inflated the extent of their oil reserves in the 1980s to obtain higher production shares within the cartel.* Despite a massive Saudi Arabian campaign to debunk Simmons's analysis and the flaws the latter bore, the notion that the richest oil producer in the world was itself facing peak-production penetrated the already confused thinking of pundits.

In fact, there were plenty of practical reasons that could account for what was happening.

More than anything else, high prices, reduced spare capacity, and limited discoveries of new oil were the result of years and years of low investments by major producing countries. Having absorbed the lessons of history, the latter had sought to avoid generating excess production capacity, the cause of recurring oil market crashes since the time of Rockefeller. Following this logic, in the previous twenty years they had aimed at mere reserve replacement, limiting exploration and the development of new oilfields. Moreover, U.S. economic sanctions against oil countries such as Iran, Iraq, and Libya during the 1980s and 1990s further frustrated their ability to adequately replace their oil production capacity, or to increase it.

Oil production capacity in the world was also restrained by political events that occurred in two major oil countries, Venezuela and Russia.

In 1996, Venezuela had shocked OPEC with its plan to double its production capacity by 2006, which was then considered a mortal blow to an already too liquid oil market. The plan, however, was soon cancelled after the election of Hugo Chávez to the presidency of Venezuela in December 1998—a major shift in the politics of the South American country that also heralded a dramatic change in its oil policy.

A radical nationalist and socialist who had grown up in the military, Chávez had unsuccessfully tried to overthrow the government six years earlier through a military coup. Arrested and released from prison after two years, he had then decided to use political and constitutional means to come to power, setting a program that promised Venezuelans to free the country from poverty and corruption, and to break the traditional concentration of wealth in the hands of a small elite. During his first three years in power, Chávez took bold steps to overhaul Venezuela's institutions and attempted to redress the country's social ills, justifying the hopes of many Venezuelans who had supported him. But his policies also provoked economic problems, and a backlash from those who were

---

* See Chapter 18.

losing their power. Reelected in 2000 with wider popular support after having changed the constitution and increased presidential powers, Chávez thus faced growing and militant opposition, which brought the entire country to a state of chaos.

On April 11, 2002, he was ousted by hostile military factions during a general strike against his supposed authoritarian politics, but he was freed two days later by military units that had remained loyal to him and returned to power. Then, in December of the same year, another major strike began, in response to Chávez's decision to change the management of the state oil company PDVSA. As a response, 18,000 PDVSA workers who had joined the protest were fired, and Venezuela's oil production plummeted from around 3 million barrels per day to less than 1 mbd in the first months of 2003. The Venezuelan leader's opponents had underestimated Chávez's willingness to sacrifice oil to politics. The United States in particular had rushed to bless the new military regime that had temporarily removed Chávez from power in April, only to discover soon that the Venezuelan president enjoyed far stronger support than expected. Indeed, despite a crippled economy, Venezuela and Chávez resisted all the hits. Yet an already suffering world oil market was further damaged by the radical leader's eventual strategy of privileging a political use both of oil and of PDVSA in order to cement his domestic power and international alliances, while making it harder for foreign investments and technologies to flow into the country and help restore its wounded petroleum sector.

The other major blow to the world's oil supply was the dramatic day of reckoning that took place in Russia against the "oil barons," just when oil prices were relentlessly soaring.

With Vladimir Putin's election to the Russian presidency in 2000 it was clear right from the preliminary actions of the new leader of the Kremlin that the golden age of the "oligarchs" was destined to come to an end. Shortly after the elections, in June of 2000, telecommunications magnate Vladimir Gusinky was arrested. Released from prison, he sold all of his publishing business to the natural gas monopoly Gazprom and chose to go into exile. In November there was the fall of Boris Berezovsky, another great oligarch of the Yeltsin era.[9] In March of 2001, the government changed Gazprom's top management, which many considered the most influential center of power of the whole of Russia.[10]

Yet the most sensational act occurred in October 2003, when Russian authorities arrested the Mikhail Khodorkovsky—master of the then top Russian oil company, Yukos—as he was getting out of his private airplane in Siberia. The charges against him and his company were fraud and tax evasion, and soon ended up including the whole of the activity carried out by Yukos and its control group (Menatep) between 2000 and 2004. Through subsequent audits the Russian authorities ordered Yukos to pay the total sum of $28 billion, and in the meantime froze the company's accounts and payments to third parties. It only took a few months for Yukos to find itself on the brink of failure, while its hydrocarbon production plummeted.

Putin then stepped up the realization of an oil and power strategy by proceeding in the summer of 2004 to confiscate the crown jewel of Yukos, Yuganskneftegaz, as partial compensation for nonpayment of overdue taxes by the parent company. It was not long before the government announced its plans to sell Yuganskneftegaz in an international auction, a threat that materialized in December.[11] Meanwhile in August 2004, the government increased the crude oil export tax with the aim of taking away from oil and gas companies a part of the extra profits earned when crude oil prices rose to over $30 a barrel. In September 2005, Gazprom announced the acquisition of the controlling stake (75 percent) of Sibneft, the fifth largest Russian oil company. In the following months, the company started what appeared to be an imperial policy of expansion outside Russia, particularly toward Western European natural gas markets.

All this seemed to prove that Putin was carrying out one of the main economic and political objectives he had outlined before becoming Russia's master, in a dissertation prepared to earn a diploma in Economic Sciences from the State Mining Institute of St. Petersburg.[12] In that document, he had clearly explained that in strategic sectors of the Russian economy—and particularly in the energy sector—the government had to take back control of the commanding heights, aiming at establishing major industrial groups capable of competing with Western multinationals without losing their national identity. As a part of this design, at the end of 2004 the Russian government presented a new law at the parliament (Duma) that limited participation to national companies—with 51 percent of the capital controlled by shareholders and companies registered in Russia—in auctions for the awarding of strategic hydrocarbon reserves.[13] In 2005, then, it began questioning and revoking some exploration licenses and development of hydrocarbon previously assigned to international and Russian companies.

The new resource nationalism of the Putin government preoccupied international investors, but above all stopped the pace of the Russian oil production growth, which after the 1990s crisis had registered an un-expected boom, soaring from 6.1 mbd in 1999 to more than 9.2 mbd in 2004, a dramatic jump of more than 3 mbd that was mainly due to higher efficiency and more advanced techniques for extracting oil that the "oligarchs" had imposed upon their newly controlled companies.[14]

If limited spending and political problems of the most oil-rich coun-tries had constrained the world's petroleum supply, international oil companies' past strategies had also in some way contributed to that out-come. Desperately searching for value creation and short-term financial performance, long had they ignored many investment opportunities whose returns were below the levels required by investors, thereby re-fraining from undertaking expensive and uncertain exploration cam-paigns. Apparently, this strategy paid out. In the first years of the new millennium, oil companies were awashed with cash and stellar profits. ExxonMobil emerged as the number one industrial company in the world in terms of 2004 and 2005 net profits— reaching in 2005 a stag-gering $36.1 billion figure—the world's all-time record of profits for an industrial company; in the first months of 2005, it outpaced General Electric as the world's largest industrial company in terms of market capitalization, at more than $380 billion. As a whole, between 2000 and 2004, the six largest international oil companies* registered cumulative operating cash flows of around $500 billion, net profits of more than $300 billion, and returned their shareholders about $200 billion in the form of dividends and share buybacks. In the same period, they em-barked on capital expenditures of $400 billion—a sign that they were recovering from the crisis of confidence of the previous decade.

Their past prudence on investements, however, had partially eroded their capacity to replace reserves—i.e., to find new oil to replace their daily production. Shell was the first victim of this unsustainable policy of short-term financial expectations.

In January 2004, the Anglo-Dutch Group (which was then the second largest oil company in the world) shocked experts by announcing a down-ward revision of its "proven" reserves by more than 4 billion barrels of oil and gas, equivalent to about 20 percent of the company's total. In the following months, Shell made three additional cuts, albeit much smaller ones, and in February 2005 it announced another 10 percent reduction in its proven reserves.

---

* ExxonMobil, BP, Shell, TotalfinaElf, ChevronTexaco, and Eni.

Whatever the complexities and the purely economic approach behind the issue of reserve booking in the oil industry,* the amazing revelation by Shell and the growing problems of other oil companies in replacing their hydrocarbon reserves made that issue one of the hot-button problems of the oil industry, bringing grist to the oil doomsayers' mill. In fact, oil industry problems had nothing to do with a physical scarcity of oil, but rather with their financial prudence as well as the impossibility of accessing the largest and cheapest reserves in world. Few recalled that global oil reserves virtually accessible to international companies were less than 25 percent; in most countries where they could operate, moreover, they had the local government as their major partner, and the latter took most of those very reserves. The plain truth was that from the 1980s on, the world's private oil companies controlled no more than 8 percent of the world's oil reserves. At the same time, any new frontier of petroleum exploration and development was technically and environmentally challenging, and above all costly.

Looking at the other side of the oil supply-demand equation, the picture appeared even more worrisome, because demand for oil showed no sign of easing.

Petroleum consumption had grown steadily throughout the twentieth century, increasing from 500,000 barrels per day in 1900 to more than 10 million barrels in 1950, 75 million in 2000, and more than 83 mbd in 2005—even though the rate of growth had slowed significantly after the oil shocks of the 1970s. Yet, a large part of mankind still consumed less than one-tenth of the per capita energy used by an American or a Western European, and was expected to increase its future need for oil. This was apparently confirmed after 2003 by a sudden jump of global energy consumption. While in the first years of the new century, the annual increase in world oil consumption had been around 600,000–700,000 barrels per day (even less than the modest average of little more than 1 mbd during the sluggish 1990s), in 2003–2004, global consumption growth reached respectively 1.8 million barrels per day and slightly less than 3 mbd. The possibility that such a leap forward reflected a cyclical rebound in demand—a reasonable explanation—was soon dismissed because of a long-term, worrisome shadow looming large on that growth: China.

The country's economic growth and oil consumption gave fresh meaning to Napoleon's comment "Let China sleep, for when she awakes

---

* See Chapter 18.

she will shake the world." After having registered an annual average growth of more than 9 percent since 1978, China's GDP increased by 15 percent in 2004, spurred by an investment boom that was transforming the physical landscape of a significant part of the country. By then, China was already consuming 40 percent of the world's cement, 30 percent both of its copper and its steel, and was the major consumer of many other resources. This implied a massive demand for every kind of energy, including oil. In 2004, a leap of around 17 percent in demand (or more than 900,000 barrels per day) raised consumption to more than 6 million barrels of oil per day, compared with 4.7 mbd in 2000 and 2.3 mbd in 1990.[15] China had now become the second largest oil consumer in the world after the United States, and its future appetite for oil seemed bottomless, considering that each Chinese still consumed less than 2 barrels of oil per year, compared with just under 13 for each European and 25.5 for every American.[16]

It was thus inevitable that the combined effects of Islamic terrorism, tight supply, heavy consumption, and the assumed threat of an approaching collapse of the world's oil resources would spur global anxiety about the future of "black gold," and affect the oil market psychologically. Since the Second Gulf Crisis in 2003, "black gold" prices incorporated a fear factor difficult to quantify. Saudi oil minister Ali Naimi estimated it at 10–15 dollars per barrel in November 2004, when oil prices were around 45 dollars per barrel.[17] Others correctly responded that there was no reason for prices to exceed 30–32 dollars per barrel. They pointed out that overall finding, development, and production costs of the most expensive oil in the world—or the last barrel needed to supply effective demand—did not surpass that level. But fearing the next disruption, physical and financial operators were ready to pay high premiums to hedge against an uncertain future, or simply to speculate on a worse one.

During 2005, several banks and financial institutions began offering hedging contracts to oil companies that locked in the price of a part of their future production to the five-year future contracts. Never used before as a credible reference for long-term prices, by then the latter indicated an oil price higher than 55 dollars per barrel from 2006 to 2010. Financial analysts as well—the ones who during the 1990s calculated oil companies' profitability against a long-term scenario of 16–18 dollars per barrel—were now bullish, envisaging a permanent upward shift of oil prices. A report by investment bank Goldman Sachs on March 30, 2005, suggested that the world could be facing a "super spike" scenario that could see oil prices skyrocketing as high as 105 dollars per barrel.[18]

Adding insult to injury, Murphy's Law "if anything can go wrong, it will," was relentlessly at work. Whatever the effective consistence of global oil reserves and production, the world was increasingly incapable of processing the qualities of crude oil made available by producing countries. New specifications on fuel quality, introduced in all the industrial countries and around the world, made it hard to refine heavy and high sulfur oils, due to the inadequate state of global refining. In this case, there was a worldly explanation. Years and years of excess capacity, low margins, environmental restrictions, and hostility by local communities have prevented companies to make significant investments in oil refining, finally putting it under stress. The United States was by now a net importer of gasoline and other oil products, its refining capacity being lower than demand for fuels of its domestic market. Europe was long on gasoline production, whose quality was not adequate to immediately enter the U.S. market, but was short on diesel production, due to the structure of its refining system: and diesel was by then the most popular fuel in Europe for transportation. In Asia the situation was worrisome as well, the refineries in the area being generally outmoded and thus inadequate to deal with medium-heavy, high sulfur crudes available in the market.

All this pushed up demand for high quality crudes, driving up their price relative to heavier, high sulfur oils. As a result, while "good" oils such as Brent and WTI broke the sixty-dollar barrier, around 50–60 percent of worldwide oil supplies lagged far behind, because of poorer qualities. The price difference between WTI and Maya crude (the main Mexican variety), for instance, reached 18 dollars at its peak in 2005, providing a windfall for refiners able to transform Maya into the quality oil products required by the market. In the 1990s, the WTI-Maya differential had been around 5.5 dollars. By the same token, the differential between Brent and Russian Ural oil had remained under one dollar, but in the first months of 2005 it reached 10 dollars. The paradox in this situation, thus, was that while the world appeared to be adequately supplied with oil, it was not of the quality demanded by refiners. Lower-than-necessary investments also explained other infrastructure shortages such as the inadequacy of oil rigs, transport vessels, and so on—all elements that put pressure on oil prices.

So many negative elements could not but stretch the oil market to its limits, both physically and psychologically, turning any minor disruption in supply into a fatal blow. August 2005 was a terrible milestone in the unfolding of the crisis.

On August 1, King Fahd of Saudi Arabia died, ten years after having been substantially incapacitated by a stroke. His half-brother, Crown Prince Abdullah,[19] was immediately pronounced king. The quick and smooth succession kept an already nervous oil market from going crazy, even though prices soared during the month to $67 per barrel.

Actually a new, age-related source of anxiety spread among world's policymakers, even though Abdullah's succession was largely expected and his political and economic ideas were well known and tested. The new king had been effectively running the country for a decade due to the poor health of Fahd. And despite being critical of U.S. policy in the Persian Gulf after September 11, and particularly of the 2003 war against Iraq (he had denied the Americans the use of Saudi bases for the war), he was a staunch defender of the strong alliance between the Saudi monarchy and the White House that has shaped relations between the two countries since World War II. Indeed, Saudi officials moved quickly to reassure the markets that there would be no change in their country's oil policy.[20]

Nonetheless, the fact that King Abdullah was over eighty, and the quickly named crown prince, Sultan (another son of ibn Saud), was in his late seventies, lent some credence to the possibility "of a crippling tribal feud within the House of Saud, or a nationwide movement of discontent among average Saudis,"[21] at the moment of Abdullah's death. In addition to this medium- to long-term risk, several analysts also saw the new king as "one of the architects of the higher prices policy" within the Saudi establishment,[22] in spite of all reassurances to the contrary.

Then on August 29, Hurricane Katrina hit the Gulf of Mexico, provoking a devastation never recorded before in the history of American natural calamities. Together with the human tragedy it brought, the hurricane upset the critical energy production and infrastructure system of the Gulf, leading to the temporary curtailment of about 1.5 mbd of oil production in the area, along with the shut-down of 16 percent of U.S. natural gas production, and a 10 percent cut-off of the country's oil refining capacity.[23] Oil prices moved accordingly, and by August 30 the price of WTI surpassed the $70 per barrel level.

The "perfect storm" that had upset the world oil market since 2003 showed neither sign of relief nor turning point. Although preliminary data on 2005 pointed both to a significant slowing down in the rate of oil consumption growth to 1.2 mbd—a drastic fall with respect to the increase of 3 mbd registered in 2004,[24] and to an unanticipated jump in world's refining capacity of about 2.7 mbd,[25] for most pundits they did not seem to demonstrate that the market was somehow reacting to stellar oil prices. Short-term factors that could aggravate the situation, leading to further disruptions of oil supply, overwhelmed by far all elements that could ease the market in the longer term. As a consequence, the direst predictions on the future of oil seemed to be confirmed. Our planet has indeed entered the age of a Copernican Oil Revolution, and nothing could stop the latter from shaping the energy future of mankind.

Apparently, the most important recognition that the situation was indeed serious came in the form of a surprising swing of political consciousness by President Bush. In his January 31, 2006, State of the Union address, Bush declared, "America is addicted to oil, which is often imported from unstable parts of the world." To curb that addiction, Bush dramatically set the goal of replacing 75 percent of U.S. oil imports from the Middle East by 2025.

Coming from a man known for his pro-oil sympathies, the message was received by the world as a sort of smoking-gun—the final proof that a Copernican Revolution was really occurring in the realm of energy. Few recalled that more than thirty-two years before, Nixon had launched a similar message to the world's public opinion in a desperate public relations exercise aimed at calming it down, while his own staff knew very well that the message was empty and its goal—energy independence—impossible. Moreover, few pointed out that, if George W. Bush were to be really serious, the very day of his speech he would have increased the federal taxes on gasoline and other oil products, in order to start mitigating the "American addiction" with oil. To the contrary, as *The Economist* pointed out, his controversial 2005 Energy Act "handed out tens of billions of dollars in subsidies to every imaginable source of energy. But it did nothing to promote carbon trading; it did not mention carbon taxes; it had no tightening of vehicle fuel-economy rules."[26]

This is the unpleasant ending of our oil history, so far. But because history never ends, before accepting as unquestionable truths what our recent past and current experience project as future trends, a careful analysis of some mystery and misperception of "black gold" is required.

PART TWO

Misperceptions and Problems Ahead

CHAPTER 16

Are We Running Out of Oil?

As we have seen during our brief journey through the history of the oil industry, crying wolf about the availability of oil has been a recurring theme in times of crisis. In the twentieth century, there have been at least three major cycles of oil catastrophism: the first one started during World War I and ended with the tremendous oil glut of 1930; the second one erupted in the United States during World War II, and was few years later denied by the growing petroleum overproduction marking the world up to the end of the 1960s; the final cycle came with the beginning of the 1970s, culminated in the two oil shocks, and was dramatically reverted by the oil countershock of 1986.

However, past experience has failed to warn against such fears and past oil floods after phases of apparent scarcity have not instilled prudence in analysts' judgment, while price collapses following skyrocketing oil values have failed to teach the basic rule governing this peculiar market: its unreliability.

It should thus come as no surprise that at the dawn of the new millennium, a new wave of oil doomsayers predicting an imminent era of petroleum scarcity has gained momentum, with an increasing number of books and newspaper articles reflecting their dire vision.* Like a rising tide, the media's amplification of any voice predicting the earth's inevitable oil insolvency has swept away any reasoned opposition to that unproven notion, making the public debate about oil seem at ease only at the extremes, dominated by a prose rich in superlatives, phantoms, and conspiracy theories.

---

* See Chapter 15.

Yet it would be incorrect to trivialize the fears of a world short of crude, given the intensity they have touched in the 2000s and the serious concerns they have disseminated in the world. What is more, the effective endowment of our planet's oil resources is such a central issue in any analysis concerning the future of energy and its political and economic implications that coping with it is a necessity. In trying to dispel the major oil dilemmas looming large on the next years, then, we will start with that topic.

For all those who are not acquainted with oil matters, an initial warning is necessary. Today's petroleum doomsday visions have been made much more esoteric and convincing for the casual observer through the intensive use of formal statistical and probability models that seem to penetrate the unresolved mysteries of our subsoil. But in fact they do not. Even in our day no one knows how many treasures the earth's subsoil holds in its depths, and no acceptable method has been devised either to assess them or to calculate the extent of future oil recoverability from the already-known reservoirs. In simple terms, searching for the ultimate figure about the earth's oil endowment is like searching for the Holy Grail—a never-ending rush with several people claiming to have discovered what in effect remains a mystery.

After this clarification, we can begin our walk through the current oil dilemmas by examining the arguments of oil doomsayers. Whatever their predictions, all of them have a common denominator or, better, a mantra: the so-called Hubbert model.

As we have seen, Marion King Hubbert first made public his intuitive model in 1956; from a conceptual point of view, it was relatively simple.[1] Complying with an empirical rule of thumb followed by the first petroleum engineers, Hubbert observed the production curve over time in a known oil province resulting from the successive start-up of the fields discovered in the area: starting from zero, production grows over time until it peaks when half of existing recoverable resources have been extracted—the so-called mid-point depletion. At this stage—according to Hubbert—production tends to decline at the same rate at which it grew. In other words, the symmetrical rise and fall behavior of oil production may be represented by a bell curve: the area below the curve shows the cumulative production of an oil province, or the "ultimate recoverable resources" (URR) it holds.

The shape of the Hubbert curve results from the sum of oil production from individual fields brought onstream successively, following the discoveries from the main fields to their satellites. As a consequence, Hubbert maintains, if an oil basin has been sufficiently explored, it is possible to reasonably forecast when it will achieve peak production and when it will run out of oil.

In order to predict the ultimate oil resources of a country, or the entire world, the Hubbertians have refined the original intuitions of their school's founder and introduced some additional considerations. The first is that the geological structure of our planet is already well known and thoroughly explored, so it is highly improbable that there are any completely unknown oil deposits left.[2] However, forecasting the future oil stocks of a country, or the world, is complicated by the erratic distribution of thousands of oilfields, by more recent discoveries, and by possible future new findings, each one having different characteristics.

The Hubbert followers seem unconcerned by these problems because they resort to mathematical applications whose fundamentals are similar to the one expressed in the "Central Limit Theorem," which states that the sum of a large number of erratic variables tends to follow a normal distribution, assuming a bell-curve shape.[3] For those unfamiliar with statistics or probability theory, this notion may be easily understood by making reference to the "normal" behavior of those leaving for weekend holidays. Also if it is impossible to know for certain the single decisions of all movers, empirical observation and statistics say that most of them will leave during the so-called rush-hours (generally on the afternoon of Friday, from 4:30 to 6:30), while only a wiser minority will avoid that hellish concentration. If one represents with a graphic that situation, it will be a bell-shaped curve.

Given these features of Hubbertian models, one needs to accurately assess past production and discovery trends, as well as geological data pointing to the potential for new discoveries and their order of magnitude. Taken together, these factors will permit the models to forecast the future behavior of production in a specific country or the world at large. Think, for example, of an intensively explored country. Its production keeps growing, but the rate of new discoveries is declining despite intense drilling activity, and geological surveys show that all the great oil basins have been tapped. What may remain is limited and scattered. The logical consequence to be drawn is that the country is moving toward its midpoint depletion (or has already passed it) and it is probably squeezing its existing reserves without either replacing them or having the possibility to do so in the future. The Hubbertian model estimates when peak production will be achieved, and then the game is relatively simple: the exhaustion of existing reserves will follow the declining side of the bell curve, which is the mirror image of the rising side.

Thanks to his original model, in 1956 Hubbert succeeded in precisely predicting the peak oil production point of the forty-eight contiguous U.S. states (which do not include Alaska and Hawaii), indicating that the critical year would be 1965 or 1972—depending on a best- or worst-case scenario—plus or minus one year. In fact, it turned out to be 1970.

Such a success is relatively easy to explain, even without the support of a formal and complex model. Simply put, the United States was (and is) by far the oldest and most intensively known, explored, and aggressively exploited area in the world. The knowledge of its subsurface outpaces that of any other region of the world except western Europe by a factor of 100. Consider, for example, that in Texas alone nearly 1 million wells have been drilled, against 2,300 in all of Iraq, and that today there are more than 560,000 producing wells in the United States as against slightly more than 1,500 in Saudi Arabia.[4] Nonetheless, without all restrictions imposed by the Texas Railroad Commission, "the U.S. production would have effectively reached a maximum not in 1970 but in 1957."[5] In fact, one of the major problems of the Hubbert model and of the entire art of forecasting is their inherent incapability of predicting political decisions affecting production, change of habits affecting consumption, price trends and technological evolutions affecting both production and consumption, and so on. Furthermore, more than anything else resource estimates are totally uncertain, because there is no method of ascertaining even their order of magnitude.

In sum, Hubbert's success with the United States was partly a piece of luck, but an isolated one. As one of the Hubbert disciples has rightly underscored, in choosing his symmetrical curve to predict the future production of the United States, Hubbert followed one of the oldest traditions in science, the one based on the so-called Occam's razor: "try the simplest explanation first."[6] The method worked in his case, but only because of the maturity of the U.S. oil industry, the extensive knowledge of its subsoil, and the unexpected help that U.S. oil policies gave Hubbert himself. Indeed, the American geologist initially did not represent his original model as anything other that what it was, a clever empirical intuition. As he pointed out:

In my figure of 1956, showing two complete cycles for U.S. crude-oil production, these curves were not derived from any mathematical equation. They were simply tailored by hand subject to the constraints of a negative-exponential decline and a subtended area defined by the prior estimates for the ultimate production.

Only later, after his forecast for the United States proved correct and he became a kind of folk hero, his self-complacency led him to the conviction that he had devised a method applicable to the entire world. All this is not intended to diminish the importance of Hubbert's intuitions. But even Sir Isaac Newton made mistakes, and Hubbert was not a latter-day Newton.

In fact, the Hubbert curve has been partially validated only for mature and intensively developed areas where knowledge of the subsoil is the highest and available technologies have been fully exploited. But the increase in subsoil knowledge, the spread of technological progress, and the advancement of drilling—along with political decisions and oil price changes—have shown time and again that peak production can be increased and delayed, so that the decline phase of the bell curve can be shifted to the right, limiting the applicability of Hubbert's theory. Even the highly mature United States still bears within its depths huge volumes of oil that are simply not recoverable today for economic or technical reasons.

Further, hydrocarbon exploration in the world is still far from being complete. Oil and natural gas may be found in sedimentary basins. Of the already known basins of this kind existing on our planet, only about 30 percent today produce oil or natural gas, and a part of them still needs appropriate exploration with advanced tools (Iraq is an example). Of the remnants, 39 percent have been tested with only moderate success: nevertheless, in many cases the results obtained in those basins cannot be considered the last word on their potential, due to the modest spending and poor technology applied to their scouting. Finally, more than 30 percent of global sedimentary basins is still unexplored.[7]

Hubbert underestimated the difficulty in setting up a model able to gauge the size of the world's ultimate recoverable petroleum resources because the world is not the United States. Its subsurface knowledge is scattered, and in many cases very limited. Overall, he underestimated the dynamic nature of many variables affecting the evolution of oil resources. Thus, when Hubbert tried to broaden the scope of his research by estimating the world's total oil resources, he grossly failed. In the early 1970s, for instance, he projected that the world would reach peak oil production in the mid-1980s at the latest, followed by a sudden decline to only 34 million barrels per day in 2000. In fact, the actual figure for 2000 was around 75 million.[8]

The major problem is not Hubbert himself, but the majority of his followers, who claim to have succeeded where he failed. Consequently, Hubbert's basic mistakes have not been corrected, and his disciples have

continuously pointed to an imminent oil crisis that has never materialized.

As we have seen, the most famous among these disciples is geologist
Colin Campbell, whose gloomy views about oil resources have been
disseminated through the press and have influenced many opinion makers.[9] He is also the founder and chairman of the Association for Study
on Peak Oil (ASPO)—that has acted as a resonance box for many oil
depletionists. According to Campbell and others, the world will achieve
its peak production point some time during this decade, and will then
face a rapid depletion of its oil reserves, causing prices to skyrocket and
triggering an urgent need to develop alternative sources of energy.

However, Campbell made subsequent revisions of his own estimates
of ultimate recoverable petroleum resources—respectively in 1989, 1990,
1995, 1996, and 2002—each time increasing it; once his predictions
proved wrong, he simply moved forward his doomsday projection of
peak oil production. Currently, he estimates URR at around 2 trillion
barrels. In 1989, his number was 1.57 trillion.[10]

Now, each of us may have his own ideas about any subject, including
oil. But none of us must mislead the public by claiming to hold an
objective truth in our hands. One may fear the exhaustion of oil, but
cannot claim to possess a scientific instrument to predict it, as the advocates of oil exhaustion do.

All of this still does not respond to our basic question *"are we running
out of oil?"* but serves to neat the camp from the idea that a doomed
scientific answer already exists. An appropriate answer to that question
requires a brief journey through the fundamentals of oil resources,
quality, production, transformation, and consumption. For the casual
reader, it will be an eventful journey, because there are few economic
issues that are plagued by uninformed knowledge like that of oil.

CHAPTER 17

The Inner Secrets of Oil

Oil resources are finite; this is irrefutable. But it is equally true that no one knows just how finite they are. And trying to assess their order of magnitude is a very complicated puzzle.

Contrary to most people's understanding, oil is not held in great underground lakes or caves, a situation that would make evaluations far simpler. Unfortunately, it is trapped in porous subsurface rocks (the reservoir), sometimes in cells so minuscule that they cannot be seen by the human. In these deposits, oil is commonly associated with natural gas and water, the three elements arranging themselves according to the geological characteristics of the reservoir: gas is at the top, oil in the middle—with natural gas sometimes melted in it—and water at the bottom.

Nature worked for millennia to create these formations. Originally, they were huge accumulations of living organisms, decomposed and covered by successive strata of rocks that pushed them deeper and deeper into the earth, until they reached a level where an impermeable rock stratum stopped them from sliding even lower. Upward pressure and high temperatures exposed those organic sediments to a chemical reaction, which over time transformed them into today's oil and gas. Like a cork, another impermeable rock stratum sealed up the reservoir and entrapped oil and gas in their millenary underground prison.

It should be noted that the exclusively organic origin of oil is challenged by some sources, and even today it is not universally accepted by the scientific community as an absolute truth. Since the 1950s, the Soviet school of geology supported the theory of oil's inorganic roots or, in other words, that oil could be made without the help of living organisms deep in the crust of our planet. Called *Abyssal Abiogenic (or Abiotic)*

Theory, this fascinating and unorthodox view led to the discoveries of oil in non-sedimentary, crystalline basement rock scores of hydrocarbon.[1] Orthodox geologists have generally rejected with scorn that theory, explaining the presence of oil in inorganic basins as a result of a migration from other rock strata. Yet an additional hint that may support this theory is that methane (the most important component of natural gas) can be found in many parts of the universe, including planets of our solar system where there is no trace of past or current life. Methane, like oil, is a hydrocarbon, and in most cases is trapped with oil in the same reservoir, so it seems reasonable to assume that oil may have similar nonorganic origins. Once again, orthodox scientists contend this view on the ground that the isotopic composition of our planet's hydrocarbon is consistent with that of organic elements, and much lower than the one present in inorganically generated methane.

In 2004, however, a team of prominent American scientists (among them the Nobel laureate Dudley Herschbach of Harvard University) produced new evidence that the Soviet school theories had some foundation.[2] In the words of Russell Hemley, a coleader of the team, the experiments he and his colleagues carried out pointed "to the possibility of an inorganic source of hydrocarbons at great depth in the Earth—that is, hydrocarbons that come from simple reactions between water and rock and not just from the decomposition of living organisms."[3]

Whatever the future of this upsetting research, until now geology, geophysics, and exploration have indicated that there is an "oil-window" beneath the surface, situated in a range between approximately 500 and 7,500 meters in the subsoil, or 1,500 and 22,000 feet. These depths correspond to subsequent geological eras, starting from the "youngest," Pleistocene, following with Pliocene, Miocene, Oligocene, Cretaceous, and finishing with the "oldest" (in terms of oil formation), Jurassic. Average production depths generally deepened with increased age, although the bulk of oil reservoirs—so far—has been found in the median strata of these geological formations. In any case, only in that "window" pressures and temperatures are consistent with the presence of "organic" oil. At depths below that, temperatures may exceed 200°C, or 400°F, and generally tend to rise by about 30°C for each additional kilometer, creating an environment that resembles the hallway to hell (also are reported many cases of nonlinear progress of temperatures as depth increases, i.e., of colder temperatures in deeper strata).[4] That is why no one has ever taken into account what the inner depths of our planet could contain.

Once a reservoir is drilled into, something resembling the uncorking of a champagne bottle takes place. The reservoir's internal pressure, fed by natural gas and water, is freed from its upper rocky cap, so that its content of hydrocarbons is allowed to vigorously rise to the earth's surface. Of course, this simple representation hides many snares. Wrong or inaccurate drilling may cause severe damage to oilfields, and even ruin an area of the reservoir by consuming its internal pressure too rapidly, or causing major accidents such as explosions and implosions. All of these setbacks may permanently reduce the amount of recoverable oil from a field.

For many decades since the birth of the oil industry, wildcatters literally wasted many fields because they did not know anything about the physical laws that affected their work. The first pioneers of drilling simply made as many holes as possible, sometimes so close to those of their competitors that a man could hardly walk through the dozens of derricks dotting the first oil fields. Photographs of the first oil regions in western Pennsylvania or Azerbaijan have given us images of hellish overcrowding, which doomed the future of the fields as intensive and foolish drilling rapidly exhausted underground pressure and destroyed the delicate structure of the reservoir, thus making it impossible to recover additional volumes of hydrocarbons.

The notion of *recoverability* is crucial to the oil industry. Given its complex nature, the reservoir will always entrap a part of the hydrocarbons it retains, even after very long and intensive drilling. This means that fields that no longer produce oil and are considered exhausted still contain more or less ample volumes of hydrocarbons that simply cannot be recovered with existing technologies.

Beyond internal pressures and technology, other physical factors contribute to make oil recovery simpler or more difficult, among them the porosity of the reservoir rocks, their pay thickness, and water saturation within each stratum. Today, the world's average recovery rate for oil is about 35 percent of the estimated "oil in place," which means that only 35 barrels out of 100 may be brought to the surface. As always occurs with statistics, these figures hide huge disparities. In many Persian Gulf countries and in the Russian Federation, for example, the recovery rate is less than 20 percent; by converse, in the United States and in the North Sea—where advanced technologies are widely used by private companies—the indicator may exceed 50 percent.

Before drilling, it is necessary to assess the probability of finding oil and gas in a certain area, and this is not a simple exercise. In the early days of the industry, improvised wildcatters were guided by an almost magical approach, relying on divination, dowsers, spirit mediums, smellers, and would-be inventors of improbable machines for locating oil.[5] Others concentrated their activities next to cemeteries, simply because many cemeteries stood on hilltops, which had turned out to be promising places to find oil. Subsurface analysis based on geophysics began to be widely applied only by the 1920s, along with many other instruments that revolutionized the search for petroleum.

Today, the most advanced tool for assessing the probability of finding hydrocarbon reservoirs is three- and four-dimensional seismic (3- and 4-D seismic) prospecting, a method that was first commercially applied in the late 1970s.* Thanks to its steady evolution, this technology today allows the gathering of plentiful data about the subsurface. Processing it through sophisticated computer-modeling software develops three-dimensional representations of a reservoir, and—thanks to 4-D seismic—also to assess its dynamic qualities and behavior under production. Yet 3-D seismic can provide only a reasonable hint, not the ultimate truth about the existence and recoverability of oil. Its most important contribution to oil exploration is that it guides drilling activity as precisely as possible, but only exploratory and appraisal wells will confirm whether oil is or is not present, along with giving experts well logs, core samples, and other elements whose careful study helps understand the inner nature of underground rocks and their hydrocarbon deposits.

The exact boundaries of an oilfield with recoverable oil and gas, however, are never known with complete confidence, until years or decades of successive geophysical analysis and adequate drilling have gone by. A reservoir may extend through tens or even hundreds of square miles and, at the same time, may have a vertical depth that is initially unknown. Consequently, during the first years of exploration and production, estimates of hydrocarbon resources contained in an oilfield tend to be incomplete, and thus conservative.

All of these elements point to a fundamental concept: knowledge of already discovered oil resources is not static, but increases over time through the expansion of scientific understanding of the fields. This explains why resources may increase over time—in tandem with increased knowledge—though a dynamic, ongoing process. In other words, they

---

* For the basic principles governing seismic prospecting see Chapter 4.

are not a once-and-for-all fixed truth. A brief example from the prominent oil economist Morris Adelman is illuminating:

> In California the Kern River field was discovered in 1899. In 1942, after forty-three years of depletion, "remaining" reserves were 54 million barrels. But in the next forty-four years, it produced not 54 but 736 million barrels, and it had another 970 million barrels "remaining" in 1986. The field had not changed, but knowledge had.[6]

Another good example is offered by the Troll field in Norway, as reported by the International Energy Agency:

> Troll, originally a gas field, contains oil in thin layers, making it hard to extract. At one time, it was not thought that any oil could be recovered profitably. But the incremental deployment of various techniques increased the field's oil reserves fivefold between 1990 and 2002. The recovery rate increased to 70% over this period.[7]

These are but two of hundreds of cases reported in oil-related literature that underscore the inherently dynamic nature of oil reserves. Think, for example, of the most recently discovered great oil frontier in the world, Kazakhstan, and its major element, the gigantic Kashagan field.

Geological estimates about the general area where Kashagan is located—the Kazakh North Caspian Sea Shelf—have existed since the Soviet era, but they only indicated the possibility of vast hydrocarbon deposits. In 2002, after the completion of just two exploration and two appraisal wells in the Kashagan field, official estimates were raised to 7–9 billion barrels of producible reserves. In February 2004, after four more exploration wells in the area, estimates were increased again, to 13 billion barrels.[8] And this is still only the beginning, because the area in question spans over 5,500 sq km, an area as large as the state of Delaware, and six exploration wells are a very modest indicator of its future potential.[9]

At this stage of our examination of the mysteries of oil, another complication enters the picture: the difference among "resources," "recoverable resources," and "reserves." Even within the industry there is still considerable disagreement over the meaning of the last two categories.

In an attempt to simplify the general terms of the issue beyond the strict boundaries of the oil sector, it is worth pointing out that the term "resources" should describe the overall stock of a certain mineral in plain physical terms, without any associated economic value and/or estimate of the likelihood of its being extracted. In other words, there may be large quantities of a given resource that cannot be technically recovered or may be too expensive to be extracted, and that nonetheless exist: this is, for instance, the case of the gold dispersed in the oceans. Only a part of existing mineral stock is both technically and economically exploitable, and this should be considered as "recoverable resource." Finally, only a part of the latter may be produced and marketed right now, and this is regarded as "reserve."

In the oil world, the prevailing classification system for assessing reserves has been devised by the Society of Petroleum Engineers (SPE)[10] and the World Petroleum Congress (WPC),[11] who outline three categories:

- *Proven Reserves*—defined as the amount of oil and gas in place in known reservoirs that can be estimated with "reasonable certainty" to be commercially recoverable under current economic conditions, operating methods, and government regulations. The concept of "reasonable certainty" is associated with a probability of profitable recovery of at least 90 percent.
- *Probable Reserves*—the probability of profitable recovery falls to 50 percent.
- *Possible Reserves*—profitable probability of recovery no less than 10 percent.

Almost all countries in the world have adopted the SPE/WPC system to assess the level of their oil and gas reserves, with the Russian Federation and Norway being the most notable exceptions. Statistics on reserves that are commonly published by newspapers and journals reflect this methodology.

Today, all major sources estimate that the world's proven oil reserves waver between 1.1 and 1.2 trillion barrels.[12] From a geographical point of view, they are highly concentrated. Nearly 65 percent are found in five countries in the Persian Gulf area: Saudi Arabia, Iraq, Kuwait, United Arab Emirates, and Iran. Outside the Gulf region, only two countries have large proven reserves, Venezuela and the Russian Federation (see Appendix 1). Oil reserves in most OPEC countries are directly controlled by state oil companies, but also other great producers such as Mexico and Norway run their hydrocarbon endowment through national oil companies. As a result, the percentage of proven oil reserves held by private companies, generally referred to as international oil companies, is strikingly modest, amounting to less than 8 percent of global reserves. In other words, the good old times of the Seven Sisters are gone forever (see Appendixes 4 and 5). Moreover, with few exceptions (one is the United States) the property of oil reserves is always held by the host country—also when private companies operating there book a part of its reserves as their own. Thus, private companies directly own only a very thin portion of world's reserves: what they book as such mainly derives from the calculation as present volumes of future productions that companies are entitled to receive for twenty or thirty years in return for their exploration and development contracts.

Since yearly oil consumption overcame 30 billion barrels yearly in 2005, or more than 83 million barrels per day, the ratio between current reserves and consumption suggests that the life span of proven oil reserves is about thirty-eight years.

Pessimists argue that this projection is misleading because future demand will be higher than today's, thus shortening the effective longevity of today's reserves. This argument is itself flawed since it assumes that only consumption grows, while reserves and resources are forever fixed. As we will soon see, this is untrue, since both terms of the equation will change over the long term. In fact, the life-index projection for proven reserves has increased in the last decade despite growth in production (as discussed in detail later on).

As we have already observed, proven reserves are only a small part of a much greater stock of global oil resources. According to the U.S. Geological Survey, in 1996 the world held more than 2.3 trillion barrels of remaining recoverable oil resources.[13] This figure includes estimates of 891 billion barrels of existing reserves ("reserves remaining," a definition resembling that of "proven reserves"), and probabilistic assumptions as to future additions. The latter are divided into two categories: 732 billion barrels of "undiscovered conventional oil" (oil still to be discovered), and "reserve growth" of 688 billion barrels, which reflects the assumed increase of reserves due to improved knowledge of existing ones. Overall, the survey fixed the world's original oil in place at around 7 trillion barrels, from which one must deduct the 710 billion barrels that our planet has already consumed ("cumulative production").

Again, it must be stressed that all of these figures are only estimates, which will evolve over time due to multiple variables.

To start with, all of the above-mentioned figures concerning remaining recoverable resources have been calculated assuming a long-term oil price scenario of 18 dollars per barrel, which induces an under-estimation of their effective size because it excludes all more expensive oil resources. Secondly, all current evaluations of world resources and reserves do not take into account the so-called unconventional oils, such as bitumen-like ultra-heavy oils, shale oils, or tar sands. The world is rich in these neglected hydrocarbon deposits, with more than 4 trillion barrels of resources in place.[14] Conservative estimates put recoverable resources of unconventional oil in Canada and Venezuela alone at 600 billion barrels. Other large unconventional resource holders, such as the Russian Federation, are not included because of the absence of reliable data.

So far, these unconventional oils have not been factored in to official oil statistics because of their high production costs relative to conventional oils, but also for some specific features that make them differ from liquid hydrocarbons. For example, Canadian tar sands may be mined like coal, while most ultra-heavy oils are quite solid, and so on. But things change over time. Venezuela and Canada are now producing higher volumes of synthetic oils derived by thermal and chemical treatment of their nonconventional resources. In 2001, their output was 270,000 and 654,000 barrels per day respectively.[15] By 2010, they will be able to deliver more than 2 million barrels daily. In the near future, thus, the importance of these kinds of hydrocarbons will grow as technology reduces their production costs. A significant share of these oils is already profitable with oil prices at 16–18 dollars per barrel, and a much greater part is profitable with oil prices in the range of 25–30 dollars per barrel. For this reason, one the most important observers of the oil industry, the *Oil&Gas Journal*, has recently started to include a tiny fraction of nonconventional oil resources as part of its own estimate of proven oil reserves—lifting the overall figure to much more than 1.2 trillion barrels.[16]

Peak oil advocates make it a point to dismiss the potential contribution of unconventional oils to global supply, basically on the ground that they are too costly and too messy to produce. But this argument openly contradicts their own world view. Should conventional oil become scarce and its price dramatically and permanently increase, then it follows that unconventional oils would become economical to extract and market. The executive editor of the *Oil&Gas Journal*, Bob Williams, has correctly remarked that such a scenario "would argue for a concerted effort by government and industry to further commercialize and expand supply from these resources beginning now."[17] As a matter of fact, in the last few years Canadian tar sands have been a target of an investment boom as a consequence of soaring oil prices.

Another factor that dramatically affects the size of existing reserves is their growth over time, even without new discoveries. As we have seen, the U.S. Geological Survey devotes a specific chapter to the "reserve growth" phenomenon in its classification of recoverable resources, a phenomenon we have already touched on in the case of the Kern River oilfield in California and in that of the Troll field in the North Sea. Its extrapolation as a peculiar category stems from the observation that, historically, "additions to proven recoverable volumes" of hydrocarbon have been "usually greater than subtractions," also without any new oil discovery.[18] One of the latest demonstrations of this phenomenon on the world scale has been recently carried out by two prominent geologists from the U.S. Geological Survey. According to their extensive analysis, the estimated proven volume of oil in 186 well-known giant fields in the world (holding reserves higher than 0.5 billion barrels of oil, discovered prior to 1981) increased from 617 billion barrels to 777 billion barrels between 1981 and 1996.[19] This reflects neither magic, nor anything resembling the mythical cornucopia, but more trivial factors.

As we have already pointed out, the first explanation is the steady expansion of knowledge about existing reservoirs. But there are other elements that play a crucial role in expanding the availability of oil and gas even in the absence of new discoveries: technology, price, and political decisions.

Technological advances can dramatically increase the recoverability of oil from its reservoirs, as well as the knowledge of inground resources, and thus reduce the cost of exploration, development, and production. On average, only about 15 percent of oil in a given reservoir can be recovered from its rocky prison by relying on natural pressure alone—in a process known as "primary recovery." Over the decades, many new technologies have evolved for bringing as much additional oil as possible to the surface, and others are under development. Since the 1920s, it was understood that by injecting natural gas and water (water-flooding) down into a reservoir, its pressure could revive and push oil up. These techniques are generally referred to as "secondary recovery," even if today they are applied from the early stages of development of an oilfield. From the late 1970s on, a technological revolution swept the oil industry, with the advent of 3-D seismic prospecting and horizontal drilling, which permits tapping of oil deposits through their lateral walls, instead of only through vertical holes, as had been done since the birth of the industry.* This made possible a dramatic increase in the amount of oil that could be recovered. Other equally important tools have been developed to manage reservoirs more effectively and thus maximize oil recovery—leading to the stage of "tertiary" or "enhanced" oil recovery. The set of techniques that have made it possible range from pumping steam or chemical fluids into the reservoir to increase the viscosity of natural gas and water (thus reinforcing their power to push oil up) to injecting or producing heat in the oilfield—thus reducing the viscosity of oil or helping it overcome the low porosity of the rocks where it is trapped. Thanks to new technologies, the recovery rate from world oil-fields has increased from about 22 percent in 1980 to 35 percent today. Given current production levels, an increase of a single percentage point in the recoverability factor can result in additional reserves of between 35 and 55 billion barrels, equivalent to one or two years of global consumption.

This partly explains why the life-index of world reserves has constantly improved over the years. In 1948, the ratio between proven oil reserves and current production (R/P ratio) indicated a remaining life of 20.5 years for existing reserves. By 1973, the life-index had risen to 32.7 years, and 32 years later—in 2005—the same index pointed to reserves lasting around 38 years.

New technology also helps slash production costs and makes it possible to develop fields once deemed too expensive to exploit. For example, new technology introduced over the last two decades has enabled exchange-listed American oil companies to reduce their finding and development cost per barrel of oil equivalent (oil plus natural gas) from an average of about 21 dollars in 1979–1981 to under 6 dollars in 1997–1999 (in 2001 U.S. dollars).[20]

Naturally, technology does not follow a linear pattern. And because of its unpredictable nature, no one can be sure of what its next leap forward will be. But sooner or later it is bound to come, when external conditions make it worthwhile to invest more money in research. Plenty of new techniques are under test today that may bring about fundamental

---

* Actually, the first experiments in horizontal drilling were made in the 1930s, but it took many decades before such techniques could be improved and generally adopted.

innovations tomorrow. Some of them may appear a little exotic, like the use of microbes that feed on hydrocarbon to increase oil recoverability (in fact, some microbes are already used to strip oil from polluted areas),[21] or the artificial ignition of a sort of blaze within the reservoir, and others.

Oil doomsayers downgrade the importance of technology by arguing that the hydrocarbon sector has already reached its technological plateau, so that we cannot expect any further help from that corner. This argument echoes the positions of obscurantist ideologues who deny the future possibilities of science, and is perfectly consistent with the philosophical attitude of those claiming that "the end is near," whatever the object of their impending doomsday may be: history, humanity, relations among civilizations, economics, and so on—oil included.

Price expectations are another fundamental pillar as far as resources discovery and development are concerned. Generally, high price expectations stimulate spending on new technologies and exploration campaigns, while low price expectations have the opposite effect, making companies and producing countries risk-adverse. The high prices of the 1970s, for example, made possible what was probably one the greatest capital investment efforts in the oil history, the development of the North Sea, which was then considered a frontier area in terms of technological and environmental challenges. This implied that its development costs were so high as to make a great part of it uneconomic under the oil price conditions existing before the first oil shock. The same pattern in now repeating in Canada, where low prices and inadequate technologies have long delayed the development of the country's huge unconventional oil resources.

The political behavior of producing countries is also crucial as a driver of investment. Sometimes, price and political expectations are closely connected. For example, the nationalization of oil reserves by many OPEC countries in the 1970s not only encouraged Western oil companies to undertake the North Sea venture but also pushed them to increase their efforts to develop cost-effective technologies to exploit resources in new and much more difficult areas of the world. More frequently, politics plays a crucial role as a setter of the regulatory and fiscal climate within which oil companies must operate, since their investment decisions depend heavily on the contractual and fiscal frameworks set by the producing countries. Usually, a more favorable attitude toward foreign investment triggers an oil rush, while rules that threaten profitability and protection of capital keep investors away and delay the development of a country's resources. For more than thirty years, private oil companies have seen a deterioration in these key conditions—prices, politics, access to exploration, and contract terms—in most of the world, partly justifying their prudent attitude toward embarking on vast exploration campaigns.

The only modest conclusion we may derive from this picture is that the dynamic nature of hydrocarbon resources makes forward-moving targets of the size of ultimate recoverable resources. In this context, even the assumption of a probabilistic figure for "reserve growth" does not mean that once growth actually occurs, its previous estimate decreases because a part of it has now become a proven reserve—thus leaving unchanged the overall figure for ultimate recoverable resources. If knowledge and technology also increase, as they relentlessly do, then all numbers will move upward. The same is true for the category of "undiscovered resources": once they are discovered, there is no deterministic process that simply leads to subtracting them from their probabilistic limbo. That limbo grows over time too, for the same reasons we have just explained.

Indeed, thus far all estimates of ultimate recoverable resources have always increased—including those calculated by the oil doomsayers. Yet the latter are especially bothered by those who insist that no resource estimate should be considered as a firm figure cast in stone, even if their dogmatic attitude to the evaluation of ultimate resources has always produced embarrassing failures. Thus it is hard to understand why any "latest estimate" they deliver should be considered as the final truth: what about all the previous ones? The simple fact is that there is neither absolute truth in their methodology nor exact science.

Serious oil experts such as Professor Adelman even consider "irrelevant" any conjecture about the total stock of resources. In his seminal works *The World Petroleum Market* and *The Economics of Petroleum Supply*,[22] the most important oil economist of our age suggested that one take into account only the flows of additions to proven reserves as an acceptable indicator of the future supply potential. Probably, this is the most correct approach. But once again, it is an approach that involves many traps, because oil pessimists have made the issue of proven reserves another battlefield for their dire ideologies.

CHAPTER 18

The Puzzle of Oil Reserves and Production, and the Quest for Their Control

The fact that today the life-index of proven oil reserves is higher than it was 30 or 60 years ago sounds like heresy to the ears of pessimists. For them, all current estimates of reserves are distorted by omissions and accounting tricks that inflate the real figures and hide the truth about the world's oil supplies.

These accusations are generally supported by citing four major issues: the global decline in new oil discoveries since the 1960s, the sudden upward revision of proven reserves by several OPEC countries in the 1980s, the "reserve replacement" problems encountered by many international oil companies in the first years of the new millennium, and, finally, the dramatic drop in the world's spare oil capacity.

All these issues are serious, and deserve careful analysis. In order to guide the casual reader, however, it may be useful to anticipate that each of them can be explained through established facts, rather than by the myth of an impending exhaustion of the world's oil supplies. Let us see why this is so.

There is no question that the peak of new oil reserves additions from the discovery of new fields was registered in the mid-1960s, and that major discoveries of oil have dwindled since then, albeit following an irregular pattern. As a result, new oil discoveries in 1998 replaced only one-fourth of annual global consumption, or 7 billion barrels out of 27 billion. Today, yearly increases in proven reserves depend largely on upward revision of existing stocks, deriving from increased knowledge of their size or, more simply, thanks to the availability of advanced technologies that allow for their commercial exploitation. Nonetheless, this situation stems from historical, political, and price factors.

As we have seen, after the 1986 oil price collapse, an atmosphere of risk aversion swept the industry and restricted oil exploration worldwide. Most OPEC countries became obsessed by the fear of creating new excess capacity that would further depress prices; this attitude was reinforced by their desire to preserve oil resources for future generations, a concept that gained currency in the 1970s, to which they still subscribe, and which in many cases has been enshrined in their legislation. Thus, producing countries preferred to focus on developing their existing fields rather than searching for new ones. At the same time, the international oil companies' investment plans were constrained both by their financial prudence and by their lack of access to the world's largest oil deposits.

All this partly explains the low level of oil discoveries registered until 1998, and also the slowdown in research and development of new technologies over the same period.

Nonetheless, since 1999 the dire assumption of a constant declining trend in new oil discoveries has been turned upside down. In 1999 and 2001, major new findings of oil exceeded 18 billion barrels each year, while in 2000 and 2003 they were higher than 10 and 14 billion respectively.[1] This significant reversal of the historical trend proves once again that there is no ultimate truth in the world of oil, and one may assume that new discoveries will occur in the future, as many national and international companies increase their exploration budgets in the face of higher oil prices.

All of this said, however, it is probably true that our planet does not hide many more gigantic, yet-to-be-discovered deposits of conventional oil. But the discovery trends we are witnessing today can make this irrelevant. Indeed, a process of "deconventionalization" of reserves is taking place that will probably make the future supply of oil the result of a mosaic of many increments, many of them relatively small, coming from both new and traditional producing countries, and from unconventional sources such as gas liquids, ultra-deep offshore deposits, ultra-heavy oils, shale oils, and tar sands. For example, many African countries such as Chad, Sudan, Mauritania, Ivory Coast, Senegal, etc.—once excluded from the global oil map—are now emerging as significant producers of oil, thanks to the relentless effort of oil companies to find new hot spots for replacing their reserves. All this without taking into account the huge potential for new reserves additions in the Persian Gulf and the Russian Federation, which remain hostage to political decisions and—in many cases—outmoded technologies and poor management.

In any case, and in spite of all political restrictions, so far the flow of additions to proven reserves—be it from revisions of already discovered fields or from completely new ones—has always compensated for the oil produced and consumed every year.

Let us now come to the charge that some OPEC members made unjustifiable upward revisions to their proven oil reserve figures during the 1980s. Indeed, between 1984 and 1988, oil reserves in five Persian Gulf countries increased quite suddenly by 237 billion barrels—a move that was generally seen as the result of a fierce struggle among the major producing countries of OPEC to obtain higher production quotas within the cartel.[2] Actually, the explanation is far more complex.

Those revisions occurred after many OPEC countries had nationalized the oil concessions on their territories, which had long been the exclusive domain of the Seven Sisters. As we have already observed, the Sisters' self-interest from the 1950s onward was to cap oil output to avoid overproduction in an already flooded market, a policy that also led them to underestimate official reserves in order to resist the producing countries' strong pressures to increase output so they could pocket more revenues.[3] Once the OPEC countries had stripped the largest Western companies of their oil concessions, that artificial measuring system collapsed and a more consistent and realistic assessment of reserves became possible. At the same time, in the early years of their liberation from foreign companies, several great producers undertook exploration campaigns, which yielded new discoveries at the end of the 1970s or during the 1980s. Coupled with much higher oil prices—that allowed for promoting in the category of proven reserves what were previously possible reserves—all these factors led to and justified the upward revisions of the Persian Gulf countries' reserves.

Since then, they have substantially restrained their oil investments, limiting themselves to maintaining a steady level of production capacity through reserve replacement. Looking beyond these self-imposed restraints, the potential for growth in many OPEC countries remains great.

Despite its long history as an oil producing region, the Persian Gulf is still relatively virgin in terms of exploration. Only around 2,000 new field wildcats (wells made for exploring the presence of hydrocarbons in the subsoil) have been drilled in the entire Persian Gulf region since the inception of its oil activity, as against more than 1 million in the United States.[4] Even today, more than 70 percent of exploration activity is concentrated in North America (United States and Canada), which holds less than 3 percent of world's oil reserves. By converse, only 3 percent of wildcat wells drilled between 1992 and 2002 were in the Middle East, which holds more than 70 percent of world's oil.[5] Drilling activity as a whole (which includes new field wildcats, appraisal, and development wells) today, follows the same trend. In the first half of 2005, there were more than 1,300 active drilling rigs in the United States, as against around 230 in the whole Middle East, with respect to a global figure of 2,435.[6] By the same token, from 1995 to 2004, less than 100 new field wildcats were drilled in the countries of the region, against more than 15,700 in the United States.[7] Finally, the expulsion of Western companies from exploration and production activities in the region has led to rapid obsolescence of techniques and technologies available there, further frustrating the development of new resources, and posing a real, major challenge to the development of the huge productive potential of the whole area.

Let us consider Iraq, for example. Despite its long history as a producer, the country is largely unexploited as far as oil development is concerned. Since production began at the dawn of the twentieth century, only 2,300 wells (for both exploration and production) have been drilled there, compared with about one million in Texas.[8] A large part of the country—the western desert area—is still mainly unexplored. Iraq has never implemented advanced technologies, like 3-D seismic exploration techniques or deep and horizontal drilling, to find or tap new wells. Of more than eighty oil fields discovered in the country, only about twenty-one have been at least partially developed.[9] Furthermore, 70 percent of current Iraqi production capacity derives from just three old fields: Kirkuk, discovered in 1927, and North and South Rumaila, discovered in 1951 and 1962 respectively. Yet, even at this early stage, Iraq's current proven oil reserves exceed 110 billion barrels (or more than 10 percent of the world's total), third only to Saudi Arabia's and Iran's. Given this state of underdevelopment, it is realistic to assume that Iraq has far larger oil reserves than documented so far, probably about 200 billion barrels more. These numbers make Iraq—together with a few others—the fulcrum of any future equilibrium in the global oil market.[10]

Even the most oil-rich country in the world, Saudi Arabia, still has a large potential to express. Despite a flurry of doubts raised about the effective size of its reserves in the last few years—in a renewed attempt to discredit the country's role as the world's Central Bank for oil—the kingdom will probably continue to defy skeptics for decades to come. Currently, its 260 billion barrels of proven reserves, more than a quarter of the world's total, represent nearly one-third of the original oil in place estimated by Saudi Aramco,[11] which says its measurement does not take into account potential future advantages brought about by enhanced recovery techniques. According to the Saudis, the kingdom had more than 700 billion barrels of original oil in place (OOP), an all-encompassing definition used by geologists to cover already produced oil, proven reserves, probable reserves, possible reserves, and yet-to-be-found reserves (the latter being a probabilistic assumption).

Saudi oil minister Al-Naimi estimated in 2005 that over the next two decades the evaluation of original oil in place could reach 900 billion barrels, an increase of 200 billion barrels. According to the *Oil&Gas Journal*, this increase is "even less than the estimate of undiscovered oil resources that the U.S. Geological Survey assigned to Saudi Arabia in 2000."[12]

More than half of Saudi oil production, or about 5.5 million barrels per day in 2005, comes from a single field, al-Ghawar. It was discovered in 1948, came onstream in 1951, and remains by far the world's largest. Yet, according to Aramco, only half of the field has been developed so far. Altogether there are about twenty fields already developed in Saudi Arabia, while more than fifty await development. Less than 300 new wildcats were drilled in Saudi Arabia between 1936 and 2004.[13] From 1995 to 2004, the kingdom carried out less than 30 new field wildcats, less than 20 appraisal wells (made for testing the production capacity and features of an oilfield), and less than 1,500 developments wells (drilled to bring an oilfield to the production stage)—insignificant numbers compared with the frantic activity in mature areas such as the United States and the North Sea. Indeed, over the same period in the United States more than 15,700 new field wildcats were drilled, more than 12,300 appraisal wells, and more than 250,700 development wells![14] The unexpressed potential of Saudi Arabia makes credible Aramco's long-term goal to raise its production capacity from the current 10.5 mbd to 12.5 mbd in 2009, and, eventually, to 15 mbd, and to maintain that rate for 50 years.[15]

Most suspicions arisen about Saudi potential in the last few years concern the relatively high *water cut* in Ghawar and other oilfields—i.e., the share of water produced with oil. The *water cut* typically increases in all oilfields as they get older, after several years or decades of production and particularly after water injection techniques have been applied to sustain their internal pressure (and thus their production capacity). In the United States, plenty of oilfields have a water cut of 90 percent, while the average world's water cut associated with oil production is around 25 percent. In 2000, Ghawar's water cut reached 37 percent, meaning that for every 63 barrels of oil produced, 37 barrels of water were also produced.[16] Since then, however, Aramco has succeeded in curbing it to 33 percent using more sophisticated technologies (which until a few years ago the Saudis did not use), and the company projects reducing it even more to around 27–28 percent in the next several years. Indeed, a water cut higher than 30 or 35 percent is not per se an indicator of peak production; once again, numbers must be interpreted considering how the reservoir has been exploited, what kind of techniques and technologies have been applied during its productive life, and so on.

The situation is even worse at Burgan, the largest oilfield of Kuwait and the second largest in the world. Discovered in 1938 and put on stream in the early 1940s, Burgan has produced so far more than 28 billion barrels of oil, and still delivers 1.7 mbd today. Yet all these results have been achieved by maintaining "the equipment that was installed by the Anglo-Iranian Oil Company (now BP) and Gulf Oil (now Chevron) in the 1940s and the 1950s..."[17]

Outside OPEC, there are countries whose potential is also largely undervalued. The most important among them is the Russian Federation, whose effective potential is estimated by most experts in Western oil companies at several times the official figure of more than 50 billion barrels of proven reserves. The world's leading reserve auditor, De-Goyler and MacNaughton, estimates Russia's proven, probable, and possible (3P) reserves at 150 billion barrels.[18] But there is another striking element that must be taken into account: the poor technological state of the Russian oil sector. Today, the rate of oil recovery in the country is a very modest 16 percent, as against more than 50 percent in most parts of the North Sea where Western technology has been deployed in all its strength. This poor performance base partially explains the success of the Russian company Yukos under the leadership of Mikhail Khodorkovsky. After hiring the U.S. engineering company Schlumberger, Khodorkovsky succeeded in raising the recovery rate of Yukos's oilfields from 9 to 26 percent, dramatically boosting his company's production with no help from new discoveries.

As to the charges that oil companies have inflated their proven oil reserves, some technical background is needed to give the reader a comprehensive picture.

Strict rules imposed by the U.S. Securities and Exchange Commission (SEC) on oil companies listed on Wall Street add a layer of complexity to the assessment of those companies' "proven reserves," which are the only category of reserves accepted by the SEC.[19] In an ill-advised effort to turn estimates into quantifiable certainties, the Commission requires that reserves gauged on the basis of the SPE/WPC system be booked in the "proven" category only if—in the arcane language of regulators—they "produce a positive net present value when evaluated by discounted cash flow using year-end prices," by applying a 10 percent discount rate. It also requires that a company make a formal commitment to "spending the funds required to recover the volumes and has necessary government approvals."[20]

These rigid rules have no connection with the physical reality of reserves. What is more, they are largely outmoded. When they were first defined in 1982, many technologies and techniques now in use did not exist. Moreover, the adoption of the 10 percent discount rate in the same year, on the recommendation of the Financial Accounting Standard Board (FASB), was consistent with the high interest rate (around 13 percent) then offered by U.S. Treasury Bonds in a period of high inflation. Those conditions no longer apply and U.S. Treasury Bonds yield around 5 percent. Yet, for oil companies, the 10 percent discount rate fixed in 1982 has never changed. Finally, the use of the year-end price as a reference for calculating future cash flows—for a period of fifteen or twenty years—has no significance, and may induce huge distortions.

Putting it simply, discounting—a technique popular with investors—establishes the current valuation of an asset by factoring in the future stream of cash flows it will produce over the expected life of the investment. Each cash flow is discounted by a factor, increasing exponentially with time to take into account that a dollar tomorrow is worth less than a dollar today. The discount factor depends on the cost of money (inflation, investors' expectations, and so on). One of the most brilliant and witty simplifications of this methodology was offered by Burton G. Malkiel, a renowned stock trader and later a professor of economics at Princeton University. Branding it "a fiendishly clever attempt to keep things from being simple," Malkiel explained:

> Discounting basically involves looking at income backwards. Rather than seeing how much money you will have next year (say $1.05 if you put $1 in a savings bank at 5 percent interest), you look at money expected in the future and see how much less it is currently worth (thus, next year's $1 is worth today only about 95 cents,

which could be invested at 5 percent to produce approximately \$1 at that time).[21]

Thus, the higher the discount rate, the lower the current value of the asset. In the case of oil companies, a high discount rate implies an underestimation of hydrocarbon reserves.

The main goal of the SEC is to provide investors in publicly listed companies as precise as possible a short- to medium-term valuation of the assets a company claims to hold, in an attempt to represent them as an inventory of finished goods of an ordinary firm. This goal, though, is quite difficult to achieve, given that oil reserves are far from being like an inventory of wine bottles or computer software.

As CERA has described it, estimating oil reserves is "an integrated activity cutting across several technical and commercial disciplines, including geology, geophysics, petrophysics, advanced mathematics and statistics, engineering, and economics. It is a process of continual learning, dialogue, sharing, and consultation."[22] Given this complexity, it is common for the judgment of one geologist or petroleum engineer to differ from that of others as to the potential of a given reservoir. Even companies engaged in jointly exploring or developing the same oilfield often differ strongly in measuring and booking its reserves.

Whatever the effort at achieving some reasonable certainty, there will always be a fundamental paradox in adopting the term "proven" to describe an estimate. The only acceptable method to establish a reasonable degree of control over the evolution of private companies' proven reserves is to invite them to periodically disclose reserves field by field—as is already the rule in the mining sector—and not to present them as an aggregate, making possible a continuous scrutiny of the basis for each evaluation. Otherwise, the SEC's stringent and absurd rules only create an artificial and parallel reality that fails to provide investors and experts with a fair and well-founded indication of a company's hydrocarbon endowment.

These and other subjective factors generate a grey area that allows companies to be conservative as well as overenthusiastic in stating the amount of their proven reserves. As we have seen, for instance, the Shell Group upset the oil community in 2004 by announcing a drastic downward revision of its proven reserves. Probably, the company's previous aggressive reserve booking was driven by the desire of its top management to cover unsuccessful exploration and production campaigns, and, above all, to spur short-term returns and profitability by squeezing the company's existing assets while under-spending on new projects. Yet those reserves still exist, and sometime in the future they will reappear as "proven," whereas today they only belong to the limbo of "probable" or "possible" reserves.

In fact, constant revisions are the norm among oil companies. For example, if there is a change in the year-end price of oil required by the SEC as a reference for calculating proven reserves, the size of the latter will shift accordingly, because of the inherent variations in future cash flows, as well as the effects of contractual agreements between companies and producing countries. All contracts, indeed, are subject to different fiscal regimes that are heavily influenced by the price of oil. For instance, in Production Sharing Agreements (PSA), one of the most common contractual formulas in use today, companies are rewarded by host countries for their successful exploration and production investments with oil (and gas) volumes for the entire duration of the contract—which is on average—twenty to twenty-five years. The company accounts for the total of future oil volumes it assumes to receive as proven reserves. Yet those volumes grow if the oil price decreases—and vice versa. As a consequence, every year a company must recalculate the future volumes it is entitled to receive from its PSA according to current prices, and this can lead to significant changes in its proven reserves.[23]

Nonetheless, although some reserves may formally disappear from year to year under the hatchet of the SEC's rules, they remain physically in their reservoir, and companies will continue to develop them if price and production costs justify it.

All the elements I have tried to describe so far have conspired to limit the growth of the world's proven reserves of oil, as well as its production capacity.

In 2005, the latter was estimated at less than 86 million barrels daily, in the face of average daily consumption of more than 83 million barrels. This left a "spare capacity" of around 2 mbd, the lowest level since 1973; at its peak in 1985, the world's spare capacity had been more than 12 million barrels. As noted at the start of this chapter, oil pessimists see the drop in spare capacity as further evidence to support their theory about the impending decline of oil reserves. But once again, that is not the case.

The decision to minimize excess production capacity represents optimal economic behavior for any producer of any good. As Western economic textbooks teach, it is simply absurd to spend money to create something that will not be sold, and will probably induce a general fall of the price of that very product. In the oil world, this attitude has become the norm for the OPEC countries, particularly after the two overproduction crises and price collapses of 1986 and 1998. Desperately trying to avoid recreating an oil glut, OPEC has pursued an approach similar to that devised by Japanese carmakers in the 1980s: just-in-time inventories. The problem with this approach in the oil sector is that it takes a long time to put onstream new production when it becomes necessary, so that inevitably a razor-thin spare capacity generally turns into higher prices, and make any sudden supply disruption or consumption peak a lethal blow.

Only the skyrocketing levels of oil demand and prices of 2004–2005 have finally convinced those countries to loosen the purse strings, and to develop new production capacity that will require some time to come onstream. International oil companies have also become more confident. Although they do not believe in the long-term sustainability of oil prices higher than thirty dollars per barrel, spurred by their need to find new reserves they have significantly increased their exploration and development budget over the last several years. As mentioned earlier, robust development of new and traditional fields worldwide points to a mosaic of many different future sources of oil that will keep the global supply growing.

One of the most detailed, field-by-field projections of 2010 production capacity—carried out by CERA—has put it at more than 101 million barrels per day, more than 15 million barrels higher than in 2005.[24]

Several non-OPEC producers are bringing new production onstream: Angola, Azerbaijan, Brazil, Canada, and Russia top the list, followed by Ecuador, Kazakhstan, Sudan, and many other small producers. The whole Caspian basin, moreover, is still in its infancy in terms of new production. The huge investment wave started in the 1990s is continuing, but its effects will not be felt until 2010–2015, when the combined production of the area could exceed 7 million barrels daily, more than three times its current level. Thanks to the contribution of those areas, non-OPEC output will continue to dominate future supplies as it does today, at least until 2015.

Indeed, while most people think that OPEC has a stranglehold on oil production and prices, its eleven members* today supply less than 40 percent—or 31 million barrels daily—of world production. At the apex

---

* OPEC members today are (in alphabetical order): Algeria, Indonesia, Iran, Iraq, Kuwait, Libya, Nigeria, Qatar, Saudi Arabia, United Arab Emirates, and Venezuela.

of their power in the 1970s, on the other hand, they controlled roughly 60 percent of global production. Saudi Arabia, for example, controls only 13 percent of global oil output, although it remains the largest oil producer in the world, with a production capacity of around 10.5 mbd. Iran, which is the second largest producer in OPEC, currently produces around 4 mbd and is far outpaced by the Russian Federation (9.3 mbd), and even the mature United States (7.5 mbd). Other producers who outstrip most OPEC countries are Mexico (3.8 mbd), China (3.6 mbd), Norway (3.2 mbd), and Canada (3.1 mbd) (see Appendix 1). The real strength of OPEC rests on its ability to export most of its production, while most of the major non-OPEC producers consume what they produce, and must import oil to supplement their domestic output. Only in a distant future will OPEC's unexpressed potential endow it with a critical role in fulfilling mankind's growing demand for oil.

There is no single country or group of countries, thus, that is in command of global supply. By the same token, no single company or group of companies is capable of performing such a role. The only successful example of oil oligopoly in our times, that of the Seven Sisters, is only a faraway memory, while OPEC's difficulties in running properly its internal discipline and to gain a larger share of the market is the additional evidence of how complicated the oil world is. Actually, today the quest for the control of reserves and production is as competitive as ever, and sees the participation of very different actors moved by different strategies and targets.

Let us start from one specific category—that of the international oil companies (IOCs). This broad category is generally referred to as including all publicly listed companies, whose majority stake is owned by private shareholders. Here we find the heirs of the "Seven Sisters," like ExxonMobil, BP, Shell, and ChevronTexaco, as well as once state controlled (or quasi-controlled) giant companies, such as TotalfinaElf, Eni, international independents such as Conoco-Phillips, Repsol, Occidental, Amerada Hess, Anadarko, and many others. Finally, to this heterogeneous group belongs also a myriad of small, independent companies that generally play a niche role in specific markets. As we have seen, the whole of these companies controls today around 8 percent of global oil reserves, and around 30 percent of global oil production. Reserves replacement is thus the core problem of this wide and apparently powerful category, which is cash rich but opportunity poor. And this may resemble a nightmare, if one thinks that—on average, each single *SuperMajor* (ExxonMobil, Shell, and BP) has to find around 1.5

billion barrels of oil and natural gas every year in order to maintain a steady production and reserve base.

Indeed, IOCs may have been wrong in the past to abdicate to the severe diktats of financial markets, to privilege short-terms results to long-term accomplishments, and to renounce good opportunities fearing the charge of value destruction: yet the real Sword of Damocles pending on them was and remains their lack of access to the largest and cheapest reserves in the world, those in the Persian Gulf and, partly, in the Russian Federation. Paradoxically, it is the very attitude of most OPEC countries to limit their production increases for the sake of maintaining high prices, to unlock new opportunities for the IOCs—a phenomenon that is already taking place in the Caspian Sea region, sub-equatorial Africa, Asia, Canada, and in the ultra-deep offshore. As long as prices remain high, thus, most IOCs will find a strong incentive to invest in highly difficult oil regions or themes—such as unconventional oils; by the same token, super-independents or small-scale companies, which may encounter several hindrances in going global, may play an important niche role in developing specific areas and marginal assets that do not interest larger operators.

Unfortunately for them, IOCs are apparently challenged in their survival strategy by a new breed of dynamic national oil companies of fast-growing consuming countries (CNOCs) such as China and India that in the last few years have begun competing with traditional majors for available reserves by leveraging two mighty advantages. First and foremost, their expansion policies are supported and underwritten by their governments, which seek new sources of supply to fuel their economies' growing needs. Secondly, those companies do not have to please shareholders with ever-improving quarterly financial results and higher dividends. Sometimes, this has allowed them to outbid their international rivals by offering contractual terms that would not pass the rigorous scrutiny of private investors. Moreover, they can gain leverage from bilateral agreements between their own governments and those of producing countries, including giants such as Saudi Arabia and Iran and smaller players like Sudan and Myanmar. Usually, those bilateral agreements are portrayed as a "win-win" situation: the consumer—let us say China—secures a stable source of supply by signing a long-term treaty with—let us say—Venezuela; it also gets red carpet treatment for its own national oil company in terms of access to new projects in the producing country. Conversely, Venezuela gets a good source of investments at low cost and a guaranteed, long-term market for its oil.

Finally, unlike the increasingly constrained international companies, the CNOCs are free of government and investor pressures to stay out of blacklisted countries that do not comply with Western human rights and labor standards. China's CNPC (China National Petroleum Corp.) and its publicly listed subsidiary Petrochina, CNOOC (China National Offshore Oil Corp.), and Sinopec (China Petrochemical Corp.), India's ONGC (Oil and Natural Gas Corp.), and Malaysia's Petronas are among the leading practitioners of this new strategic pursuit of far-flung reserves. In recent years, their efforts have caught the attention of the experts, and brought many a sleepless night to the leaders of the major internationals.

Flush with cash and supported by the political clout of their government, the Chinese have been especially aggressive. They have seized opportunities to invest in reserves wherever they could be found, in Kazakhstan, in Libya, in Algeria, in Saudi Arabia, in Iran, in Venezuela, and even in Chad. Often they have outbid their major Western rivals with irresistible offers in the race both for reserves or exploration rights, and for the construction of oil infrastructure. In 2005, CNOOC even tried to takeover Unocal—one of the largest U.S. oil independents—by outbidding a previous offer made by Chevron. The latter finally won the game, but it had to make a new, pricey offer.

Naturally, the third broad category of oil companies struggling for reserves and production is that formed by the national oil companies of great producing countries (PNOCs), whose strategies and policies coincide with those of their own governments. In this category, it is correct to include also most Russian companies that, albeit formally publicly listed, do not share with IOCs either the later degree of market transparency, or their real autonomy from governmental decisions.

There are at least two new facts concerning this group of companies that require a specific focus. First, the oil crisis has revived and given fresh vigor to their governments' *resource nationalism*, making PNOCs much more self-confident and assertive in devising their strategies of growth also outside of their national boundaries. Secondly, some of them are effectively able to do pretty much what the internationals do, and thus have increasingly less need for their cooperation. The state companies that reached this level are still relatively few, and include Saudi Arabia's Aramco, which boasted an extraordinary level of technical and management proficiency, Brazil's Petrobras, the world leader in deep-offshore production, as well as Malaysia's Petronas and Norway's Statoil. These companies have the financial strength and technical expertise to handle exploration and development, and are also capable of handling the technical and financial risks of major infrastructure projects. In any case, they can contract outside experts when they lack specific resources. The IOCs also do this, as outsourcing has led them to shed a significant part of their core expertise in exploration, drilling, reservoir management, as well as project and risk management, for the sometimes irresponsible sake of cutting costs.

Thus a usually ignored category of oil companies—the one of the oil service companies—has assumed a prominent role in the last decade as real master of the development of new oil prospects worldwide: "contractors" such as Halliburton, Schlumberger, Saipem, Snamprogetti, and many others, today do the job for large producers, as well as for IOCs and CNOCs. With the support of oil contractors' technical expertise, it is probable that PNOCs will continue to go their way in developing their own resources, and maybe in seizing some opportunity abroad.

This quite Darwinian picture of hypercompetition between international and national companies may be worrisome for many of its actors, but it is good news for the world's need of more oil. Indeed, as long as oil prices remain sufficiently high—even at much lower levels than today's—all of those actors will have a strong incentive to invest in new oil projects, some of them to survive, some of them to feed their countries' oil thirst, some of them to avoid losing market share and potential future revenues.

Adequate production capacity alone, however, cannot reassure the world, and our times offer tangible evidence of it. Actually, another apparent extravagance of the oil market is that it may simply refuse certain qualities of petroleum, thus restricting the "usable" supply. This is an additional complication in the already intricate world of "black gold," that requires us to prolong our journey into its secrets.

CHAPTER 19

The Problems with Oil Quality, Price, and Consumption Trends

Not all petroleum is created equal. Like wine, it comes in a staggeringly wide range of types and qualities. Some high quality barrels command high prices, while others barely find a market. Even the most common image of petroleum, which portrays it as black gold, is off the mark. In fact, as I write this, I look at a pair of miniature glass barrels sitting on my desk, one filled with bright yellow Libyan crude, the other with dark brown Nigerian oil. Some varieties look almost green. (It goes without saying that also the reference to the "barrel" is more symbolic than real, because oil actually never goes into a barrel, like it did in the early days of the industry. Today, the barrel is only an unit of measurement.)

Every kind of crude embodies different physical and chemical characteristics, including different quantities of various dissolved metals that establish its market value. The two most important variables are density and sulfur content. (See Appendix 3 for a list of the main qualities of oil.) Density reflects the relative thickness or lightness of the oil, and is measured on a scale established by the American Petroleum Institute (API): the higher the API number, the lower the density of the oil. High API numbers represent the lightest oils. The API scale for crude ranges from $10°$ (that also equals the density of water) for the heaviest conventional oils to more than $45°$ for the lightest. Within these extremes, there are specific categories indicating clusters of crudes, but unfortunately the definition of each category varies from source to source. In order to simplify the issue for the reader, here I will use a basic and broad generalization, dividing the main kinds of oils in three categories: "heavy," that covers crude up to 25 on the scale, "medium" crudes range from 26 to 34, and "light" crudes are those rated above 34. Brent crude, for example, is a light crude rated at 38.3 API, as is West Texas Intermediate (WTI), at 39.6°.

Sulfur content has its own classification scale. So-called sweet oils contain less than 0.5 percent sulfur by weight, while oils with between 0.5 and 1.5 percent sulfur are called medium sour and those with more than 1.5 percent are defined as sour. The terminology goes back to the early days of the oil industry, when wildcatters literally tasted their crude to judge its sulfur content and found that the more sulfur the nastier the taste. The lighter oils tend to have the lowest sulfur content as well. Brent and WTI, for instance, combine lightness with low sulfur content, but geography alone does not determine either lightness or sulfur content; some countries produce a combination of light, medium, and heavy oils, with varying permutations of sulfur content.

There are other characteristics that distinguish different kinds of crude and help determine its ultimate use, the most important being it pour point temperature and viscosity. Both affect the ease—or difficulty—with which it can be transported by ship or pipeline. Crude oil with a high pour point is harder to transport because it tends to solidify at ambient temperature, especially in the case of crudes with high wax content. Viscosity measures the oil's flow resistance: the higher the viscosity, the less easily it flows.

Today, nearly 60 percent of the world's crude consists of medium density petroleum with medium to high sulfur content. Light crude accounts for about 20 percent of global supplies, and half of that also has significant sulfur content. The bulk of Persian Gulf production consists of medium crudes often containing high levels of sulfur. The most common Saudi oil, Arabian Light, contains average levels of sulfur, but struggles to keep its place in the "light" column, hovering at the 34 level on the API scale.

The available qualities of crude do not follow a linear pattern, depending on different oilfields coming onstream in different periods of time. For some decades now, experts have pointed to an inexorable process of heaviness of crude (i.e. of progressive dwindling of the production of light oils), but actually such a process has not taken place. A detailed field-by-field study of production capacity to 2010 carried out by CERA, for example, shows a significant increase in the availability of light crudes, which will represent by then around 26 percent of a much higher production capacity of more than 100 million barrels per day.[1]

The vast mix of different crude oils challenges the downstream side of the industry, since each type of oil delivers different volumes and qualities of final products, in keeping with a basic underlying principle: the better the quality of oil, the more the quantity of value added oil products (like gasoline and diesel fuel) that a refiner can obtain.

Crude oil can be refined into more than a hundred different end products generally grouped into four major categories: light products (like gasoline and virgin naphtha); middle distillates (mainly kerosene and diesel), residual fuels (like fuel oil), and specialties (like bitumen, lubricants, coke, etc.). Gasoline is used primarily as an automotive fuel, while virgin naphtha is used as feedstock for petrochemical products. Kerosene is mainly used for jet aircraft fuel, and diesel for transportation, space heating, and farm and industrial equipment. Residual fuels are used for power generation, but also for the production of asphalt (bitumen), and lubricating oils for reducing friction in the operation of machinery.

In order to understand the different yields of different kinds of crude, let us look at what can be obtained from similar quantities of Western Texas Intermediate (39° API, sulfur 0.2 percent) and the Russian benchmark Urals (32° API, sulfur 1.3 percent). Given the relative quality differentials of such crudes, a simple refinery can produce around 16 percent of gasoline from a barrel of WTI, but only 12 percent from a barrel of Urals. At the opposite end of the value chain, residual oil from WTI and Urals represents respectively 35 percent and 45 percent of the output. Residual oil is the lowest value-added product derived from primary distillation of crude oil, and a growing problem for all refiners, because its market is narrowing worldwide.

A simple refinery usually consists of distillation (topping), desulfurization, and reforming units, and is referred to as a "hydroskimming plant." While topping simply separates the components of crude by heating it at different temperatures, desulfurization purifies oil products of sulfur, and reforming increases the octane number of gasoline, improving its antiknocking qualities. A simple refinery like this, however, cannot change the standard yield of a given kind of oil so that from a barrel of Urals it will always be left with nearly 50 percent of residual oil.

The addition of more sophisticated processing units to a refinery can improve its capacity to "convert" the same quantity of crude into more gasoline and diesel and other high value-added products, while drastically reducing the amount of residual oil. In other words, an advanced refinery can manipulate the raw material to squeeze more high-quality value-added products from it. That is why the structure of the world's refining system is crucial for getting the most out of the available types of oil.

The basic process to carry out "barrel manipulation" is "cracking," whose first version was patented in 1913 by William Burton. This consists of breaking the larger hydrocarbon molecules of residual oil through high pressures and temperatures to obtain smaller molecules that can be reprocessed to obtain more gasoline, diesel, and other middle distillates. Today there are several other versions of cracking and deep cracking, mostly based on catalysts (hydrocracking, visbreaking, coking, etc.). Refineries with cracking units are referred to as "high-conversion" ones.

The higher the conversion capacity of the refinery, the greater its flexibility in using a wide range of crude oils to deliver the products needed in its market. Coming back to the previous example, a high-conversion refinery can get around 25 percent of gasoline, 50 percent of diesel, and only 16 percent of residual oil from a barrel of cheaper Urals—i.e., many more value-added products than a simple refinery can get from a barrel of more expensive WTI.[2] This allows a sophisticated plant to exploit price differentials between light-low-sulfur and heavy-high-sulfur crudes. And because our world is demanding more and more light products with low sulfur content, this is a fundamental factor of competitiveness and survival.

Just as there is no single variety of crude, there is no universal price level for either crude or refined products. Just as there is no single country or group of countries that is really in command of supply, so it follows that no one can really control the price of oil. In spite of the uninformed view expressed by the media, OPEC can only indirectly influence the price of oil by fixing its own production quotas, but even this power is limited by its members' lack of discipline in respecting the ceilings—a behavior that has historically hampered the cartel's effectiveness. In recent history, a major factor behind oil overproduction and subsequent price collapses has been the OPEC producers' attempt to cheat the organization's rules by selling more oil "under the table" with respect to official quotas.

Basically, prices vary depending on quality, on destination of exports from a given country, on market conditions, and on refining structures. But this is only one side of the moon. As for any other commodity, oil prices are influenced by expectations about future supply/demand that affect both physical and financial transactions of petroleum.[3]

Most oil sales/purchases are made up of medium- to long-term contracts between producers and oil companies, the latter relying primarily on this type of agreement to cover the bulk of their needs. However, the effective price at which these transactions are concluded depends on two different references, spot markets and future contracts, and their interaction. On the spot market, commercial buyers and traders may acquire a single cargo, the contents of a tanker, as it leaves a loading platform, and immediately resell it to another buyer who may in turn sell it again to yet another buyer, creating a long chain of buying and selling for a single lot of oil. The spot market chain, however long, covers actual quantities of oil, and makes up around 30 percent of all oil traded every day.[4] Traditionally, spot markets are based on benchmarks—such as WTI, Brent, Dubai, etc.—that set the relative values of all other crudes, depending on their different qualities and points of origin with respect to the benchmarks themselves. Due to the small volumes of current benchmarks' production (see Appendix 3), though, their spot markets have become very thin, and even illiquid. As a consequence, few and small spot transactions may have an immense influence on the whole oil market, and even distort it dramatically if—for example—a single cargo is bought at an abnormal price.

Thus, over time, oil future contracts have become much more significant in setting oil prices. These contracts do not involve a physical delivery of actual crude, but financial transactions (traded in lots of 1,000 barrels) that take place mainly among commercial operators— sellers and buyers—as well as purely financial operators, who try to speculate on the price movements they expect. For these reasons, the oil barrel involved in such transactions is called a *paper barrel*. Expiration dates for future contracts start from one month, and run to three years in London and as long as five years in New York, but the vast majority run for one, two, or three months.

The prices of oil on the futures market include a discount or a premium over the spot market, reflecting the buyers' and the sellers' expectations about which way the market may be heading. A seller may agree to a discount on the spot price to protect himself from the danger of a more significant price drop; similarly a buyer may agree to a premium over the spot price to cover himself against a larger price increase. This kind of hedging plays a major role in the working of the futures market, whose main raison d'être is offering commercial operators instruments for managing risks. Each contract is often traded hundreds of times in a single day, much like a corporate stock, and is ultimately settled through the payment of price premiums. As for stocks, however, wrong indications from the physical spot market, bearish and bullish sentiments about future trends, as well as purely speculative operations, may distort price reality. The fear of oil disruptions or the perception of prolonged imbalances between supply and demand—for instance—may amplify upward and downward price movements. Actually, in the last few years financial speculation on oil futures has been held partially responsible for skyrocketing oil prices—a charge made by many observers, and among them OPEC. More than speculation per se, however, it is market psychology that tends to inflate or deflate prices. Financial operators in the oil market represent only a small quota of all traders, the latter being essentially commercial ones. When the shares of financial versus commercial operators grows—which occurs only rarely—there might be a hint that a speculative phenomenon is under way. Contrary to the general perception, the positions of noncommercial operators in the oil future market decreased during 2004, just as oil prices began soaring.

When the newspapers report that prices have gone above $60, or dropped to less than $20, they refer to the future contract on specific type of crude, usually WTI or Brent (also, in the case of Brent, they generally refer to the *Dated Brent*, set by the spot market); other, lower-quality crudes may be selling at $10 or $20 less at the same time.

As we saw in chapter 15, for example, in April 2005 the price differential between a barrel of WTI and one of Mexican Maya (22° API, sulfur 3.3 percent) was slightly less than $20, after hovering around $5–6 throughout the 1990s. In such a situation, a refiner able to squeeze more valuable products from Maya thanks to the conversion capacity of its refinery can make a lot of money. On the other hand, a simple refinery that can obtain only 10 percent of gasoline from Maya, and cannot cope with its high sulfur content, will be forced to use the much more expensive WTI or similar crudes. OPEC's crude reference basket, which is formed by eleven different qualities of petroleum,* trades at a discount to WTI. When the latter oscillated around $20 per barrel, such

---

* Crude oils forming OPEC's reference basket are (as of August 2005): Saharan Blend (Algeria), Minas (Indonesia), Iran Heavy (Islamic Republic of Iran), Basra Light (Iraq), Kuwait Export (Kuwait), Es Sider (Libya), Bonny Light (Nigeria), Qatar Marine (Qatar), Arabian Light (Saudi Arabia), Murban (UAE), and BCF 17 (Venezuela).

discount is about two dollars, but the price differential usually increases as far as prices go up. Thus in 2003 and 2004, the differential averaged five dollars per barrel; and when on August 29, 2005, WTI overcame the $70 per barrel barrier, the OPEC basket stood at slightly more than $60 per barrel.

The complexity and interrelationships of crude qualities, their derivatives, and their relative prices give rise to a fundamental and somewhat confusing aspect of the oil market: even crude supply levels higher than demand may not necessarily be adequate to meet that demand. This can happen if world refining capacity becomes insufficient, or if it is unable to handle certain kinds of crude.

This phenomenon developed slowly in the 1990s and has grown since 2000, largely as a result of the ever more stringent quality specifications imposed on gasoline, diesel, and other oil products worldwide. The global pressure to reduce lead, sulfur, and other polluting elements in fuels for environmental reasons has led to increased demand—and higher prices—for high quality crudes that yield higher volumes of high quality end products. At the same time, this has put pressure on many simple refineries whose owners failed to upgrade because of a deep-rooted fear of never recovering their investment.

In fact, for more than two decades the world's refining system has been the weak link in the oil industry, strained by overcapacity created in the early 1980s, when refining capacity reached more than 80 mbd, against consumption of less than 64 mbd. Hit by very low or negative margins, and restrained by growing environmental regulations that heavily increased costs, most refiners concentrated on cutting capacity rather than increasing it. Moreover, the so-called "Nimby" (Not in my backyard) or "Banana" (Build absolutely nothing anywhere near anything) syndromes—i.e., the negative attitude of local communities against the buildup of intrusive and environmentally "suspected" industrial plants—made it impossible even to think of creating new capacity in industrial countries.

All over the world, refining problems have assumed different forms, producing a patchy picture. While global nominal refining capacity is apparently in line with demand, the available conversion capacity lags far behind the actual needs of each regional market. And oddly enough, the weakest link in this articulated system is the United States.

Even though the country boasts of the most advanced high-conversion refining system in the world, it has nonetheless a refining capacity of only 17 mbd, with respect to a consumption of about 21 mbd. The negative consequences of this gap reverberate over the global market, given the U.S. position as the world's number one consumer of petroleum products, accounting for around 25 percent of total consumption.

This absurd situation stems not only from low investment in new refining capacity, but also from a federal system that delegates regulatory autonomy to the individual states. Because each state has the right to set its own product specifications, an archipelago of independent gasoline submarkets has developed, each alien to the other. In 2004, there were eighteen gasoline blends in the United States, with differing summer and winter specifications. Each of these blends is approved for sale only in a specific region, state, or even a metropolitan area within a state, but not in the rest of the nation.[5] As a result, American companies are compelled to import additional quantities of gasoline (more than 1 mbd in 2005), or to concentrate their demand for oil on specific crude grades suitable for their refining systems, making its price soar on the international markets, irrespective of the effective overall supply of crude.

These problems have even pushed President Bush to propose the building of new refineries in areas under federal control, previously used by the military: indeed, a largely theoretical solution, given the isolation and long distance of such areas from consuming markets, which would make highly expensive the building and transporting of oil products.

In Europe, demand for transportation fuels continues to shift away from gasoline to diesel, whose 2005 sales outpaced gasoline for the first time. Yet European refining capacity lacks the ability to deliver all of the low-sulfur diesel required, while still producing too much gasoline and fuel oil. Asia has enough nominal refining capacity, but it cannot handle heavier and high-sulfur crudes, while all the countries in the region are adopting new fuel standards following the Western pattern.

In sum, the solution to refining problems depends entirely on two things: investment and the regulatory system of each country. Resulting shortages of refining capacity can last for several years, testing the international markets for both crude and refined products, and driving up prices for both. But sooner or later the markets react to take advantage of the situation, and producing countries move to develop their own refining capacity to make use of crude oils that they could not otherwise sell.

Quite unexpectedly, in 2005 worldwide refining capacity increased by 2.7 mbd, the largest such increase since the early 1990s, despite a decrease in the total number of operating refineries. What's more, large producers and developing countries have now a strong incentive to build up their refining capacity to meet their domestic needs, and to export value-added refined products. Thus, while Saudi Arabia, Iran, and other major producers are building or modernizing large refineries, a flurry of new or enlarged refineries is under construction all over Asia—where India has planned to build the world's largest refinery by enlarging an existing one, and bringing its total capacity to 1.2 mbd. At the same time, countries that are neither major producers nor major consumers, but are well placed geographically to export refined products, are building up their refining capacity—especially in North Africa. Moreover, refinery creep alone (the additional production deriving from marginal investments in existing refineries) is estimated to add around 4 mbd by 2010.

Finally, new technologies to squeeze additional high quality product from lower quality crude are also being introduced, including those that will handle "ultra-heavy" oils and tar sands.

This is a very important evolution for making much more oil products available from existing or future supply. In fact, the refining process itself leads to relatively large "losses" through the production of "poor" derivatives: currently the difference between the amount of crude entering the system and the amount of valuable refined output represent about 10 percent of the 85 million barrel per day global production of petroleum, nearly as much as all of Saudi Arabia's average output.[6]

Thus, by 2010 the world's refining system could find again a sound equilibrium. But of course, oil supply and refining capacity alone mean nothing if they are not related to expected levels of consumption.

Before examining this aspect, it is worth pointing out that since the beginning of the oil age analysts have tended to underestimate supply and overestimate demand. The major flaw in such an approach has been to consider that while oil resources are finite—so that their production is constrained as time goes by, and their price inevitably soars— their consumption is essentially independent of price considerations, or inelastic, because people will not give up their cars, stop heating and lighting their homes, or refrain from traveling. As a consequence, it has always been easy to assume an ever-growing demand in the face of a limited expansion of supply, resulting in an unavoidably gloomy view of the energy future of mankind. This general attitude becomes all too evident when an extraordinary jump (or fall) in consumption takes place: in this case, forecasts of future consumption tend to be "straight-line," extrapolating future behavior from the current situation. That is why in periods like ours, it is hard to find anyone who believes that demand may also fall significantly or remain flat. Yet the world is incoherent, and cannot be defined by deterministic curves.

As we have seen, the history of oil is characterized by peaks and valleys, including dramatic collapses in demand due to severe economic recessions or unsustainable price levels. So a word of caution is necessary before speaking of any revolution in oil consumption trends. After all, it was only few years ago—in 1998–1999—when oil prices hovering at around $10 per barrel, and an apparent shift away from oil, moved most analysts to predict a permanent oil glut—and even the end of the age of oil. Has anything dramatically changed in such a short time? Let us start with some basic figures.

In the last twenty-five years, oil demand has registered a modest annual compound growth rate of 1.6 percent, whereas for all of the twentieth century up to 1973, demand doubled every ten years, a progression implying annual growth of about 6.5 percent. Also, oil's share in the world's energy consumption has decreased. By 2004, it represented only 34 percent of the total, after having peaked at 45 percent in the 1970s. Finally, energy intensity—the quantity of oil needed for each dollar of GDP—has been cut by more than half in all industrial countries. In 2005, the world consumed around 83 million barrels of oil per day (see Appendix 2). Fifty percent of this is absorbed by the transportation sector alone, a percentage that dramatically increases in the industrial countries. In the United States, for example, transportation absorbs more than 70 percent of oil consumption, compared to 38 percent and 32 percent respectively in China and India. According to the International Energy Agency, two-thirds of the incremental demand for oil up to 2030 will come from the transportation sector alone,[7] but it is likely that transportation's share of oil consumption will rise much more than current models indicate. Why? Because a process of oil substitution has already began in many sectors of our life, eroding its supremacy. And the more advanced the economic structure of a country, the lower the share of oil used in sectors other than transportation.

Even if oil remains the main energy source of mankind, there are other sources that can effectively and economically replace it in many end-uses, from power generation to heating, and even in petrochemical production—where natural gas can displace it as the main source of feedstocks. Only in the transportation sector will oil remain substantially irreplaceable for some decades to come. This general trend will ease the pressure on oil in the future, much more than is predictable today through the most sophisticated of econometric models.

Beyond the "improper" use of petroleum (i.e., the use of petroleum in activities where it can be replaced by other, economically competitive sources of energy), consumption growth of the last few years is based on a series of factors that cast a shadow over its sustainability, especially in Asia, whose economic boom has powered this growth.

Between 2003 and 2005 several structural problems have played a role in driving oil consumption to new records in many Asian countries. In Japan, for instance, the temporary shutdown of sixteen nuclear plants for maintenance has required more oil than normal. In China, increased demand over the last two years has largely represented an adjustment from the stagnation of previous years, and has been driven partly by inventory buildup, and by one-off needs—mainly due to the delays in the construction of coal-fired and water-powered electric generating plants.[8] Indeed, even after the recent buying rush, China still accounts for only slightly less than 8 percent of global demand, and even sustained consumption growth would have only marginal effects on an otherwise normal petroleum market.

At the same time Asia's thirst for oil has been heavily subsidized by local governments or encouraged by price regulations, making oil products much cheaper than on the international market. This is the case of China, for example, where the price of gasoline and diesel in April 2005 was 44 percent lower than on the open Asian market.[9] Worse, artificially low prices have encouraged waste and inefficiency throughout the region, making it much more oil intensive than the West.

Asians, as *The Economist* has pointed out, "consume more oil per unit of output than Europeans or even gas-guzzling Americans. Thailand and China, for example, use more than twice the rich country average, while India burns through almost three times as much."[10] Direct and indirect subsidies have made oil consumption throughout the region largely unresponsive to oil price spikes, straining the budgets of several countries facing multibillion-dollar energy bills.[11] Starting in mid-2005, however, most Asian governments announced plans for the gradual elimination or reduction of state subsidies.

But this will not get at the root of the biggest source of concern about future oil demand in Asia: the very low per capita consumption in the region, especially in China.

As we have seen, a Chinese consumes only 1.55 barrels of oil, as against 25.5 burnt by an American and 12.7 by a European,[12] and the rapid growth of automobile sales points to increased consumption down the road. More than 2 million new cars were sold in China in 2003, as against 200,000 in 1999, and there are still only ten cars for every thousand inhabitants, compared with 770 in North America and 500 in Europe,[13] a situation that is only slightly better in other Asian countries. Thus there is a vast market for automobiles, even though mass motorization, entailing a ratio of at least two hundred cars per thousand inhabitants of a country, is still many decades away.

Over the next decade, the gradual expansion of automobile ownership in China, India, and other developing countries is not likely to push the growth of petroleum consumption beyond the rates we have experienced in the last fifteen years. Indeed, increased consumption in the transportation sector will probably be offset by reductions in other areas. Both China and India have, or soon will have, access to petroleum substitutes. Both have coal and nuclear power generating plants, and China enjoys a huge hydroelectric potential. Natural gas, only marginally developed today, will also play an increased role in the future of both countries.

The inevitable increase in oil consumption in the developing countries will be moderated by this process of substitution in the non-transportation sector, as well as the stagnation and decline of demand in the more developed industrialized countries. Most developed countries have already experienced their phases of oil maturity. In Europe, Japan, and Australia, oil demand is poised to decline in the medium-to-long term, despite occasional upward spikes due to temporary phenomena. Several factors underlie this trend: the aging of populations, which makes distances covered and traveling time by car shorter; a sweeping shift from oil to natural gas in power production; increasingly stringent environmental regulations; heavy taxation of oil products; and, finally, continuing improvements in car efficiency.

The United States is the only exception to the consumption decline trend in the industrial countries. Here, a still robust demographic growth, coupled with a lack of sensitivity to car efficiency and very low taxes on petroleum products, have caused demand for oil to soar through the 1990s, reaching nearly 21 million barrels per day in 2005: an astonishing figure, equivalent to slightly less than the total consumption of Asia.

Despite all these considerations, it would be wrong to assume that there is no hope for change even in the United States. The main culprit of the relentless upsurge in the country's oil demand from the 1990s has been the American consumer's love affair with the highly popular Sport-Utility Vehicles (SUVs) and light trucks. From 2000 to 2004, these gas-guzzling vehicles have represented about half of all vehicles sold in the United States, reaching a peak of slightly less than 60 percent in 2004. In a country where gasoline alone accounts for about half of all the oil consumed, it is easy to understand that the American problem with oil is largely one of foolish consumption habits.

Yet by the end of this decade, the absurd legal loopholes that have allowed SUVs and light trucks to proliferate will expire;[14] moreover, high oil prices have already struck a first blow to the sale of SUVs and light trucks in 2005, while the U.S. consumer has begun to display interest in hybrid cars, buses, and other more energy efficient and less polluting vehicles.

These are hopeful signs. In 1978 the United States consumed about 18.5 million barrels per day of oil, or 32 barrels a year per person, the highest level of per capita consumption ever reached. Today, per capita consumption has dropped to less than 26 barrels.[15] In the early 1980s, the United States succeeded in curbing its oil addiction by more than 3 million barrels per day.

In sum, we must look at future global demand for oil as a dynamic equation in which growing demand by developing countries in Asia and other emerging regions, such as Latin America, will be partly offset by reduced consumption in industrial nations. This is a long-term trend that doesn't take into account the possibility of a sudden backlash on the demand side. This possibility can never be excluded, because the effects of high oil prices penetrate slowly but deeply into the world economy. Actually, early data for 2005 reveal that the sky-high prices reached during the year have already substantially cooled demand for petroleum products, and according to IEA, growth in demand dropped from 3 mbd in 2004 to about 1.2 in 2005.[16]

Also if it is too early to derive a comprehensive assessment from these data, one should never forget that, day by day, high oil prices alter economic fundamentals such as inflation, the production cost of most goods, family incomes, and many others. They may have a much more modest impact on the world's GDP than they had thirty years ago, but the more they rise, or remain high, the more they become like eutrophic algae that kill all vital organisms around them by slowly consuming all the oxygen in the waters where they take root.

CHAPTER 20

Flawed Forecasts, Foggy Alternatives to Oil

All of the elements I have tried to outline so far show that I do not believe there is a Copernican Revolution sweeping the world oil market, and that the stage we are passing through is a perfect storm that will not endanger the capability of our planet to supply the oil the world requires. But no one can predict how long this situation will last, or if and when oil prices will substantially decline.

Indeed, the oil market is shaped by so many complex and sometimes contradictory forces that it is impossible to produce reliable forecasts. Making matters worse, the work of forecasters is made harder by several flaws and holes in their set of information. Some of these are particularly tricky.

The first hole is the problem of poor data, an issue that has plagued the oil world since its inception and is far from resolved. Consider, for example, that China and many other developing countries have never adopted a comprehensive statistical collection system for data about their oil and energy consumption, inventories, and so on. By the same token, most major producers consider their current and future production capacity figures a matter of national security. As a result, even the International Energy Agency—the most important data source on global supply, demand, and inventories—is oftentimes bound to fail in its attempt to release reliable sets of data that the Agency delivers on a monthly basis through its extensively used *Oil Market Report*. Month after month, the Agency is obliged to revise its previous forecasts, and sometimes to change them significantly. And yet few sources are better than the Agency itself, because the vast majority of them use the data supplied by IEA with small modifications. Oddly enough, OPEC itself relies on secondary sources, even as far as its own production is concerned.[1]

The second hole is the intrinsically dynamic and usually unpredictable nature of oil spare capacity. Shrinking spare capacity props up prices, but at the same time it stimulates investments in new production capacity. Because of the time lag in developing new oilfields, the bulk of new production may well reach the market when demand is already shrinking—due to high prices. This happened in the early 1980s with the oil coming from the North Sea, Alaska, and Mexico. Spare capacity may also silently grow due to a phenomenon of "production creeping," a higher than expected production from existing oilfields, that may come as a result of improved management, adoption of more sophisticated and costly technologies, or simply because of previous miscalculations.

The major dilemma for all forecasters, however, is the timing of market forces—and particularly the lag before demand for oil responds to a significant change in price. A result of consumer inertia, this allows high prices to survive much longer than generally thought possible, and for unpredictable periods. Empirical observation has shown that sometimes it can take several years before high oil prices lead to a reduction in consumption. That is why I commented earlier that oil prices run slowly but penetrate deeply in the circulatory system of the global economy, and sometimes act as eutrophic algae, sapping its oxygen. The real problem with the issue of demand reaction to oil prices is that the notion of *expensive oil* is highly elusive, and thus difficult to define.

Here a warning is necessary. The price of oil products paid by the final consumer in most parts of the world—the one that directly influences demand for oil—is only slightly connected with the price of crude petroleum, as well as with its transportation and refining costs.

Taxes of various kinds add substantially to the price of the finished product, and often greatly exceed the cost of the original petroleum, giving rise to serious disputes between OPEC and the majority of consuming countries.

For many years, OPEC has accused European countries in particular of being the main beneficiaries of the oil wealth, because all over western Europe gasoline taxes are so high as to make up around 70 percent of the final price paid by the consumer, depending on the circumstances. Indeed, because the main fiscal component of the gasoline price is made up of excises—fixed taxes on each liter produced—when the price of crude oil grows, its share of the final price of gasoline (and other oil products) slightly grows as well, while the fiscal drag lowers. Considering that part of the final price covers refining and retail margins, what consumers pay for the raw material is only a fraction of the price at the pump, ranging from 10 to 25 percent. In other words, the price paid by the European consumer for a barrel of oil originally sold at $60 is something between $200–$250—taking into account that the same barrel also yields other products subject to different tax rates.

In the United States, on the other hand, taxes are generally low, averaging 20–22 percent, and when American consumers complain about skyrocketing prices of gasoline, they do not realize how lucky they are compared to their counterparts in Europe and the rest of the world. (In Japan taxes on oil products represent nearly 50 percent of the price at the pump.) In May 2005, for instance, the retail price of premium gasoline in the United States averaged $2.39 per gallon, while in the Netherlands it sold for $6.36 per gallon, in the United Kingdom at $6.09, and in Italy at $5.91.[2]

Finally, in many developing countries oil product prices are distorted both by high public subsidies and state-administered prices, so that consumers are insulated from the reality of the marketplace.

Only in the case of the United States, thus, the price of crude oil has a direct impact on the price at the pump of gasoline, diesel, etc. Yet oil consultant John S. Herold demonstrated in June 2005 that even at a price of $50 for a barrel of WTI, the price of gasoline was still a great bargain in the United States compared to most consumer items, when relative price increases since 1982–1984 were taken into account. Over that period gasoline (weighted at an average price of $2.24 per gallon on April 29, 2005) was 67 percent more expensive in nominal terms than in 1982–1984. Yet many other items and services have increased much more: food by 89 percent, housing by 93 percent, personal care services by 103 percent, fruits and vegetables by 134 percent, medical care by 219 percent, tuition and school fees by 330 percent, tobacco and cigars by 396 percent.[3]

Another curious analysis made by the Herold consulting firm found that if other commonly used products were also priced by the barrel, they would outstrip—sometimes dramatically—the price of crude oil. A barrel of Coca-Cola, for instance, would cost $119, a barrel of Perrier water $426, a barrel of Budweiser beer $410, a barrel of Bertolli olive oil $1,165, a barrel of Stop & Shop Dandruff Shampoo $1,469, a barrel of Jack Daniel's Black Label Whiskey $4,460, a barrel of McIlhenny's Tabasco sauce $4,542.[4] These relative values call into question several philosophical issues, rather than economic ones. Why do we Westerners believe it is our divine right to have oil—along with water and other precious gifts of Nature—for nothing, or almost nothing, while we are ready to borrow money to pay incredible prices for less important things? Why are we ready to cry and call for political intervention if the price of gasoline soars, while the prices of the rest of everything around us skyrocket? Why are we so eager to buy the newest and most expensive SUV, and then protest if the fuel to make it go costs a few dollars more? The simple answer is that we are so accustomed to take for granted the "low" price of certain items that we have lost any perception of their intrinsic value.

Economists have argued for two centuries about the "intrinsic value" of a good, without coming up with anything more plausible than what the Romans had already discovered twenty centuries ago: *res tantum valet quantum vendi potest*—a good is worth as much as it can be sold for. Thus if people are willing to pay the highest prices for gasoline in two generations, and if demand for oil keeps increasing, this simply means they value oil as much as they pay for it. And, at the very least, they should not complain about it. But how long can this last? Are we still witnessing consumers' spending inertia that will sooner or later disappear, or are higher prices now affordable because of both the lower impact of oil on the broader economy, and the parallel growth of the world's wealth?

Unfortunately, even such traditional indicators as the ratio between economic growth and oil demand—or oil-to-GDP ratio—have been totally upset in recent years, so that they cannot tell us if and for how long higher oil prices will be sustainable. As Yamani's Center for Global Energy Studies has noted:

> During the 1990s, the world economy grew at an average rate of 3.5% per annum while global oil demand grew at 1.2% a year—implying an oil-to-GDP ratio of just over a third. . . . Although world economic growth has risen to an average level of 4.2% in 2003 and 2004, global oil demand growth is now expected to average 2.8%—an oil-to-GDP ratio of nearly two-thirds, which is about twice the average level of the 1990s.[5]

The best studies devoted to the oil-to-GDP relation have failed to reach any final conclusion beyond confirming that prices still matter in determining long-term oil consumption trends.[6] The general slowdown of oil consumption growth in 2005—relative to 2004 and 2003—suggests that the long wave of market reaction to abnormal prices could be already moving on. If that is the case, then one might think that if oil prices continue to soar because of politically driven production restraints—particularly by OPEC countries—the way should open for the replacement of oil with other, cheaper sources of energy.

This is another way to judge whether the price of a good is expensive or not, at least in relative terms. As noted in the previous chapter, a process of oil substitution has been under way for many years in most sectors of economic life, except transportation. Even at $50, and probably $60 and $70 per barrel, no alternative source of energy can challenge oil's role in moving people and things. Despite a flurry of articles and books, and aggressive pronouncements by politicians and futurologists touting hydrogen fuel cell vehicles as the new frontier for mass transportation, that future is not around the corner.

Let us start with fuel cells. Their basic technology is quite old. It was first used at the end of the 1950s in specialty vehicles like tractors, and developed further in the 1960s for use in the aerospace industry. At the cells' core is a device called a proton-exchange membrane stack that converts a mixture of hydrogen and air into water and electricity. The electricity in turn powers the car's electric engine. Theoretically, such a process offers the promise of a totally clean, low-noise, easy-to-maintain vehicle. But several major stumbling blocks make it prohibitively expensive to achieve such results.

Although hydrogen is the most abundant element in the universe, it does not exist in nature in a readily usable form and must be extracted from other sources. Ninety-six percent of the hydrogen used in the world today comes from fossil fuels such as coal, oil, and natural gas, while only 4 percent derives from water.[7] The reason for the overwhelming role of fossil fuels in producing hydrogen is simple: they are by far the cheapest source for it. Yet they are not cheap enough.

When the price of oil is about thirty dollars per barrel, the cost of producing and delivering one kilogram of liquid hydrogen, which has roughly the same energy content of a gallon of gasoline, is about 4–5 dollars, if derived from natural gas.* At the same price of oil, though, the cost of producing and delivering one gallon of gasoline is only about

---

* Still today, the price of natural gas is directly linked to the price of oil in most markets of the world.

1.2–1.3 dollars.[8] Even worse, the process for obtaining hydrogen does nothing to improve air quality, since it produces the same emissions as those resulting from the direct combustion of natural gas.

Water is the only source that could fulfill the promise of a completely clean, zero-emission fuel, albeit at a very high cost. But this, too, is somewhat of a mirage because deriving hydrogen from water involves a process called electrolysis, which requires vast amounts of electricity. This, in turn, is the product of burning coal, oil, or natural gas, whose unfortunate and unavoidable by-products are air pollution and green-house emissions. The carbon dioxide emissions generated by the full-cycle hydrogen process, from its production by electrolysis to its final use in a fuel cell vehicle, equal those produced by a modern gasoline vehicle, but are 20 percent higher than those from an advanced diesel engine, and 100 percent higher than those produced by a hybrid vehicle using natural gas.

Theoretically, one could obtain emissions-free electricity for electrolysis by using renewable sources of energy such as wind or solar power, or even nuclear power. But this is only wishful thinking in the foreseeable future.

Renewable sources are extremely costly, and would make the final cost of hydrogen twenty or thirty times higher than the equivalent quantity of energy delivered by burning gasoline. As for nuclear energy, its development is politically difficult because of fierce opposition by local communities to having nuclear plants in their neighborhoods. Also, the claimed economic competitiveness of nuclear energy is questionable. Historically, many producers have failed to factor into their overall costs future spending for nuclear waste management and the shutting down of plants once their life cycle has expired. The exit costs of nuclear plants have generally been underestimated because no one envisaged the enormous environmental and safety problems involved.

The problems of hydrogen production, though, are only a part of a multifaceted and messy puzzle. A still insuperable hurdle is how to store, transport, and deliver hydrogen, given its high volatility, low energy density, and safety risks. And no one has yet come up with an effective solution to the problem of storing sufficient onboard quantities of hydrogen in a vehicle. While plenty of solutions have been suggested, all have foundered on the shoals of cost. For example, the building of a hydrogen distribution network in the United States would cost about 100 billion dollars.

In the final analysis, to produce hydrogen and make it available for cars at the service station pump may cost from four to thirty times the cost of producing oil and delivering gasoline, with the cheaper options being as polluting as directly burning natural gas. And the list of complications is not complete yet. Current commercial fuel cells still have problems of size (too large for cars), durability (unacceptably limited mileage), and cost: prototypes may cost from 500 to 2,500 dollars per kilowatt produced, while an internal combustion engine costs 30 dollars per kilowatt.[9] Considering all these unresolved issues, it is unrealistic to assume that fuel cell vehicles will have a significant impact on the market in the next thirty years.

A much brighter future is at hand for hybrid vehicles. The mechanical and engine configurations of available hybrids vary significantly, but their common feature is that they use both a traditional internal combustion engine fueled by gasoline or diesel, and an electric one powered by a battery that is charged by the conventional engine. This combination reduces fuel consumption and emissions, with the greatest reductions coming from those hybrids able to operate in an electric-only mode. They can get 45–60 miles per gallon of gasoline (mpg)—or 4 liters per 100 kilometers, more than doubling the mileage of traditional cars of the same size.[10]

The hybrids' success in slashing emissions is also admirable. Most of them meet California's rigid "Super Ultra Low Emission Vehicle Standards" and are the cleanest category of cars available today and probably in the near future. According to a major 2003 study by the Massachusetts Institute of Technology, "the projected gasoline hybrid for 2020 has roughly the same life-cycle greenhouse gas emissions as the hydrogen fuel cell vehicle."[11]

Initially, hybrids were very expensive relative to traditional vehicles, but their cost-effectiveness has improved as they have gained market success. This has enabled hybrid-vehicle producers—initially only Toyota and Honda—to realize economies of scale that reduce their price. In July 2004, Honda estimated the price differential between a hybrid-vehicle and a gasoline or diesel car of the same size at about 3,000 dollars.[12] In 2003, with the price of WTI averaging $31 per barrel, the very popular Toyota *Prius* hybrid sold for 3,500 dollars more than an equivalent conventional gasoline car. These price differentials may be partially offset by savings in fuel costs over the life of these highly efficient vehicles. These savings will increase if gasoline prices go up, making the cars that much more attractive from an economic standpoint. That is why the high oil prices of 2004 and 2005 have prompted a hybrid mini-boom in the United States and Japan, giving rise to waiting lists as buyers outnumber production.

But hybrids still represent only a tiny share of the cars sold in the world: in 2004, about 130,000 were sold worldwide out of a total of 55 million vehicles sold. About half of the hybrids were sold in the United States. However, the market is expanding and new hybrid models will be introduced by Ford, Lexus, Chrysler, General Motors, and Mercedes-Benz during the 2005–2007 model years. In September 2004, Ford became the first automaker to launch a hybrid Sport-Utility Vehicle (SUV) on the American market.[13] The increase in the existing hybrids' market penetration, along with the emergence of new competitive models by other automakers, will tend to reduce further the price gap relative to conventional cars.

In any event, hybrid cars will also have to find their way by competing against the constant evolution of internal combustion engines—which are by no means in their sunset years.

All the talk about new technologies challenging traditional ones ignores the fact that no technology competes only against itself, and combustion engines are rapidly evolving, too.[14] Over the last several years, they have continued to improve their efficiency in terms of fuel consumption and reducing greenhouse and other air polluting emissions.[15]

Diesel engines lead the innovation race among conventional vehicles, delivering fuel efficiency similar to that of the hybrids, but at lower prices for consumers, offering 30 to 40 percent better performance than gasoline engines.[16] At the same time, the process for producing and delivering diesel fuel releases 30 percent fewer greenhouse emissions than that of gasoline, although diesel combustion generates more particulates (soot) and oxides of nitrogen. Thanks to these beneficial features and to tax incentives, diesel has become the fuel of choice in Europe during the last few years. As *Newsweek* has pointed out, "largely as a result of embracing diesel, Europe's cars get on average 50 percent better mileage than U.S. autos."[17]

In the meantime, research continues to make both gasoline engines and diesel engines cleaner and more efficient. The improvements they will achieve in the near future will raise the competitive bar for the new nontraditional engines.

Given all these variables, one thing is clear: public policy, not consumer choice, will determine the future of mass transportation. Unfortunately, cars are a status symbol throughout the world, a tool for showing off one's social and economic progress. It is still rare for people buying cars to think in terms of social responsibility, opting for cleaner vehicles instead of looks, options, or power. If politicians were serious about environmental concerns and reducing dependence on oil, the solution would be at hand: a mix of tax increases on oil products, more rigid mileage and tighter emission standards for automakers, incentives for buyers of "cleaner" cars, and a ban on older and more polluting models.

With proper preparation of the cultural and economic ground, as well as advance notice of target dates, all these measures could be put in place relatively fast—let us say four years from now. Their imminent introduction would stimulate automakers to step up research, development, and production of new and more efficient cars, while ensuring a manageable transition toward a less intensively carbon-driven society.

In this desirable scenario, oil will remain the king of transportation, even at prices that today we consider very expensive. However, this combination of public policies and greater efficiency would lead to some unwelcome side effects on oil demand, and thus on oil prices, that most observers usually underestimate.

Because of the so-called paradox of efficiency, the more a good is produced or used more efficiently, the more its unit costs decrease; in other words, a consumer pays less to have the same quantity of that good than he enjoyed before. The consequence of this phenomenon is that the overall demand for that product will tend to increase, once again causing its price to rise.

Because of the extreme difficulty in assessing the sustainable price of oil, serious economists and investors usually take into account another approach to the whole issue, one based on marginal production costs.

According to the theory, in a free and open market the price of a product like oil should equal the overall cost (including the remuneration of capital employed, i.e., a margin for the producer) of the last unit of that product that is needed to satisfy demand. Hence the adjective "marginal." In the case of oil, the marginal cost should grow as consumption increases, because—the theory says—the last barrel has to come from the less efficient and more expensive oilfields, after all cheaper reserves have been so exploited that they cannot supply any additional barrels. As a consequence, the price of oil would rise, ensuring the producer of the "last barrel" a thin margin, and providing great margins for the most efficient producer. In the longer term, however, technology, new production methods, and other factors should push down the costs of even marginal production, thus dragging down that reward differential. Yet for most of the oil history, things have not precisely worked this way.

First the Seven Sisters, and later OPEC, limited their ample, low-cost production in order to sustain the price of oil. This policy permitted less efficient producers to fill the production gap by developing otherwise uneconomic oilfields. This situation persists even today, as the cautious attitude toward new exploration and development campaigns by the richest oil countries in the world allows international oil companies to undertake risky and costly projects. Figures speak for themselves.

To find, develop, and produce a barrel of oil in Saudi Arabia costs approximately $2, while in most Persian Gulf countries it costs between $2 and $4. Also the cost of oil transportation from those countries to a final market is quite modest, oscillating between one and two dollars on average. As a consequence, if those countries had developed their full potential in the past, their low-cost production would have left no room for many other producers in the world. In the 1990s, for example, the total costs for finding, developing, producing, and marketing a barrel of new Kazakh oil were estimated at between $14 and $18, threatening the viability of all projects there when world oil prices hovered at those levels. And most Canadian tar sands—among the most expensive sources of oil production to date—require long-term prices of around $28–32 per barrel over at least fifteen years to ensure decent double-digit profitability. That is why high prices of oil have created a momentum for a rush into unconventional oils since the beginning of the new century.

Following the theory of marginal costs, and assuming the cost of development and production of unconventional oil as the highest marginal level, today's and tomorrow's prices should not exceed $30–32 per barrel. (Naturally, this hypothesis assumes that OPEC continues to ration its production potential in the future; otherwise marginal costs would be much lower.) It is true that prices may be propped up by unexpected demand, by the inability of the refining system to cope with available oils, or by disruptions due to political and natural events. But because these factors are temporary in nature, in a long-term perspective prices that far exceed marginal costs of available sources of production are by definition unsustainable. Or, putting it another way, too expensive.

Those prices should allow plenty of otherwise uneconomic production to come onstream in the medium-to-long term, filling the demand gap and provoking a new competitive scenario in which traditional, low-cost producers see their market share shrink in favor of new producers. As time goes by, the latter become more efficient and thus less costly, putting further pressure on the incumbents. At the same time, when demand is fully satisfied, prices begin to decrease, tending to match marginal costs, and even to fall below them in case of overproduction.*

Over the short run, the concept of marginal cost may appear to be simply a theoretical abstraction in the face of a market moved by such factors as political tensions and expectations, psychological issues, and flawed analyses. Nevertheless, over the medium term, a radical departure from marginal cost calculations can yield dramatic results. In particular, it is always wise to bear in mind a very simple rule of thumb, which has proven its validity throughout the history of the industry: demand always responds to extreme and prolonged variations of oil prices with respect to their marginal costs. It may take a long and tortuous path before doing so, but it does. Unfortunately, when that reaction occurs, prices usually do not decline, but simply collapse. By the same token, their sudden crash even to below marginal costs will eventually pave the ground for new price spikes in the future, as investments drop just when low prices allow consumption to recover—as happened in 1998–2000.

Without monopolistic control such as that created by John D. Rockefeller, the oil market is bound to remain prey to volatility, maintaining its characteristic cycles of booms and busts, of expensive and cheap oil.

Rockefeller feared booms no less than busts, because he realized clearly that the booms were only the prologue for the busts, and vice versa. Few things appear to have changed since then.

---

* The use of marginal costs as a long-term reference for forecasting is always a wise choice. Yet it deeply changes the very nature of forecasting itself, leading to the setting of "normative scenarios" (as opposed to forecasts) that describe reality as it should be, instead of trying to figure out how it will be. On the basis of a normative scenario, each forecaster may then simulate the impact on his budget/plans of significant variations of prices.

CHAPTER 21

Arabs, Islam, and the Myth of Oil Security

Despite the vast abundance of oil, the fear of a politically driven curb on supplies casts a long shadow over the oil market. This fear has a long history.

Referring to British behavior during the 1956 Suez crisis, historian Hugh Thomas observed:

> Ever since Churchill converted the Navy to the use of oil in 1911, British politicians have seemed indeed to have had a phobia about oil supplies being cut-off, comparable to the fear of castration.[1]

This "castration syndrome" has been a common preoccupation of all advanced or rising economies facing the emotionally driven prospect of losing their oil supplies. Sometimes, the mirror image of consuming countries' hysterical oil insecurity has been an increasing "resource nationalism" on the part of oil producers.

Historically, these two camps have sometimes come close to clashing, with the former struggling to retain access to a key resource for their economic survival, and the latter trying to leverage oil to increase their own economic and political status. But even as the collective imagination has always seen disaster looming behind these conflicting pressures, the laws of the marketplace—however imperfect—have always prevailed, keeping disaster at bay.

As we have seen in our voyage through the history of black gold, a shortage of crude cannot last forever; the flexibility and broad articulation of the market, and competition between producers, keep it from happening. The impossibility of specific forces—companies or countries—to command thoroughly oil supply and prices clearly emerged also when the world oil market was characterized by powerful oligopolies.

The Seven Sisters were only partially successful in restricting production in order to sustain prices from 1950 to 1970, because they did not succeed in avoiding either the destructive competition of independent companies or the rising tide of producing countries' reactions. By the same token, OPEC was incapable of managing its apparent success after the 1973 shock, which was by no means the outcome of a plan devised by the organization. The cartel's producers only exploited consumers' anxiety, which boosted oil prices higher and higher even if supply was wide and growing. The final result of that mismanagement was the countershock of 1986. Since then, the era of oil oligopolies has ended, and this has further made the control of oil an impossible dream.

In 2005, the eleven OPEC countries controlled about 31 million barrels per day of oil production out of the more than 83 million barrels that the world consumed daily. What is more, the organization is far from being a monolithic body. Its members have different ideologies, policies, and economic targets, so that their discipline tends to be driven by self-interest, rather than by a sense of common purpose. In their turn, international oil companies are driven by targets that structurally collide with OPEC's. Indeed, as a whole they control nearly 8 percent of global oil reserves, and about 25–30 percent of oil production. Every year, they need to replace reserves in order to sustain their future production, so that their main interest is to open new frontiers and experiment with new technologies in their quest for survival. Naturally, the more OPEC limits its own production to raise oil prices, the more international oil companies have an incentive to spend money to develop new resources outside the OPEC realm. In other words, extended shortages breed high prices, which in turn lead to new investment in high-cost production areas that would be shunned in a low price environment. The consequence, then, is that oil *can* be a weapon, but only for brief periods. And its use as such can backfire, inflicting serious damage on the weapon wielder—as the Arab oil producers learned the hard way in the 1980s.

Yet, neither explanations rooted in economics nor historical evidence have ever succeeded in countering the powerful mythology of oil insecurity.

Today the latter is gaining new strength from the sweeping cultural and religious militancy that has been highlighted in Samuel Huntington's *The Clash of Civilizations and the Reshaping of World Order*. Is it not true, pessimists say, that oil reserves are highly concentrated in a handful of unstable countries rife with anti-Western feelings, and which could be hit by political crises leading to major disruptions in oil supply? And is it not true, they add, that most of those countries, particularly the Arab ones, have already demonstrated a propensity to use oil as a weapon?

Unfortunately, today's public perception of oil issues is colored by the oil shocks of the 1970s and the selective embargo promoted by Arab producers in 1973. The gloomy shadow of those events always looms over any discussion of oil, despite the fact that those shocks and the embargo were largely the product of a distorted collective psychology, "a classic case of buyers' panic"[2] resulting from poor information about the actual oil supply. They also drew strength from catastrophic predictions spread by the media and think tanks, misguided interventionist policies by the United States, and many other factors. For those whose opinions are rooted in uninformed knowledge, however, little purpose is served by repeating that the effective oil shortage was small, and could have been easily managed, and that even those peculiar shocks were nothing but an extraordinary exception in the history of oil.

I realize fully that any attempt to ease the fear of oil insecurity against this background, and to put it the right perspective, could be seen as whistling in the wind. Nonetheless, given the importance of the issue, it is worth trying.

It is true, as we have seen, that 65 percent of the world's proven oil reserves are now concentrated in five Persian Gulf countries: Saudi Arabia, Iraq, Iran, United Arab Emirates, and Kuwait, and that while today they produce less than 30 percent of global oil, their role is likely to expand in the future.

It is equally true, however, that all of these countries, and other OPEC members as well, need decent oil prices and steady oil revenues to ensure the survival of their economic and social systems. For them it is essential that oil remain the world's fuel of choice in the decades to come, since their economies are overwhelmingly oil based, and their budgets and spending capacity critically oil dependent. On average, oil represents 40 percent of the GDP of the Persian Gulf countries, and more than 85 percent of their exports. It is their lifeblood.

The suffocating correlation between oil and economic survival has been dramatically increasing because of the region's demographic explosion. Saudi Arabia, Iran, Iraq, and many others in the area have witnessed a three-fold population increase since the 1970s, and today more than 60 percent of their population is less than twenty-four years old. This demographic revolution has created a completely new set of expectations and frustrations for which stagnant and monocultural economies offer no credible answers. The absence of industrial and economic development in sectors other than oil limits the Gulf countries' ability to provide for the self-realization of their growing population.

In Saudi Arabia, for example, only one young man out of three may be able to find a job when he reaches working age. The oil industry cannot respond to this critical challenge because it is not labor intensive. Even the mighty Saudi Aramco, driver of the country's economy, formal owner of all of the kingdom's oil reserves, and the number one oil producing company in the world, employs only 54,000 people.[3] By the same token, the giant Saudi petrochemical corporation Sabic has a workforce of only 16,000 people, and 2,000 of them are Europeans.[4] Yet the Saudi population numbers about 21 million, and keeps growing.

As a consequence, while only sustained oil revenues will allow these countries to control unrest by preserving their huge social welfare programs, their oil wealth is much lower today than it was in the late 1970s and the early 1980s. And cheap oil, coupled with population growth, means a dramatic dip in per capita oil income. In fact, oil earnings have fallen in both real and nominal terms after the memorable windfall profits of the 1970s, and today's Middle Eastern oil producers have a smaller cake on the table that must feed a much larger family. Saudi Arabia, for example, had a 1981 per capita GDP of $28,600—equal to that of the United States. By 2000, it had fallen to $7,000.[5] As a result, a collapse or a stagnation of global oil consumption threatens both the minimum needs of the people, and the stability of the most oil-rich countries.

For all of these reasons, great oil producers are much more dependent on consuming countries than the latter are dependent on them—at least in the long term. By the same token, whereas industrial countries worry about energy security, producing countries are preoccupied by the security of oil demand. That is why it is critical to their long-term economic and social performance that they are perceived as reliable suppliers of oil.

Actually, all the great oil producers have learned a lesson from the 1970s: an oil shock can be a terrible experience for the industrial countries, but it is not a fatal blow. As soon as they perceive the long-term nature of such a shock they react, and their reaction can turn into a permanent nightmare for any producer. Any structural reaction implies not only a reduction in demand, but also much more money devoted to research and development of alternative sources of energy or investments in new oil-producing areas.

Only the relatively low prices of oil we have experienced so far have prevented a major shift to other sources of energy like the one begun in the late 1970s. Nonetheless, as we have previously observed, oil's supremacy as a source of energy has already been eroded in the last thirty years, and while oil is bound to retain an extraordinary economic advantage in the transportation sector, it could be replaced in most other human activities. The speed and size of that replacement are only a question of price and reliability of supply.

That is why so far, whatever their problems and in spite of uninformed knowledge, Arab governments and non-Arab, fundamentalist Iran have always made every possible effort to ensure the stability of oil prices and supplies, and they are still trying to pursue this objective. This does not exclude that—for short periods of time—one of these countries may be tempted to use oil as a weapon, especially when the global oil supply is tight, prices are high, and political tensions rose to unprecedented levels. But this scenario represents an exception, not the rule. And above all, it cannot last for long.

In sum, oil is both a blessing and a curse for those countries that depend heavily on its revenues, a kind of Sword of Damocles threatening their own future much more than that of the West.

The situation would not be much different for a radical, religious Arab government installed in the Persian Gulf by a violent takeover—the real nightmare that obsesses the mind of many Western strategists.

So far, the only experience we have of a fundamentalist regime installed in a great oil-producing country is that of Iran. And what we know for sure is that even at the height of Khomeini's Islamic revolution, Iran did not use oil as a weapon or consider using it. To the contrary, in the early 1980s the country rejected OPEC oil quotas and strove to produce as much as possible.

There are several further considerations that counterbalance the prevailing view that radical Islam per se is a mortal threat to oil supplies and to the West in broader terms. To start with, a profound misinterpretation of Islam's attitude towards the West persists. As *Newsweek*'s Fareed Zakaria has perceptively observed:

> The trouble with thundering declarations about "Islam's nature" is that Islam, like any other religion, is not what books make it but

what people make it. Forget the ranting of the fundamentalists, who are a minority. Most Muslims' daily lives do not confirm the idea of a faith that is intrinsically anti-Western or anti-modern.[6]

At its core, fundamentalism is an attempt to find in the Koran the guidelines for shaping a civil and political model of the modern Islamic society. In this perspective, it is the revival of an unfulfilled quest for autonomy and independence that has swept and bloodied the Middle East since World War I.

The most extreme fringe of fundamentalism derives from the *Salafiyya*, a religious movement that dates back to the very early days of Islam, and that considers most forms of current Muslim practice a corruption of the Prophet's original message. But beyond its strict observance of a rigid religious orthodoxy, as Michael Scott Doran has pointed out, even "the *Salafiyya* is not a unified movement, and it expresses itself in many forms, most of which do not approach the extremism of Osama bin Laden or the Taliban."[7]

Thus, like pan-Arabism, fundamentalism is not a monolithic doctrine but the expression of a highly fragmented universe with many faces and political opinions. The violent or terrorist fringes of fundamentalism are only a small part of this universe, and their appeal to Arab and Islamic societies is probably already declining after peaking in the mid-1990s. At that time, many Arab countries were militarized in response to frequent terrorist attacks that alienated a frightened civil society. The most striking example of such failure was probably Algeria, whose population in the 1990s was the first victim of radicals' fury.

The terrorist attacks against the United States in 2001 were driven partly by the desperate need of religious extremists to increase the visibility of their efforts, after their failure to penetrate the Middle East. As much as it was spectacular and tragic, it was not necessarily a sign of strength. Rather, it revealed an incredible underestimation and lack of preparation by U.S. authorities. By the same token, later terrorist attacks by supposed al-Qa'ida cells in many parts of the world have been certainly frightening, but they also reveal an erratic organization. There is nothing as simple as putting a bomb on a train, in the metro, or in a holiday resort, all places that are virtually uncontrollable due to the concentration of hundreds of people and the ease of access. As a consequence, one should expect that a sound military organization may attack places like these once a week, not once a year. If a large part of the Islamic population were to support al-Qa'ida's vision, the situation would be much different, and the world would really face a global war. But things are not like that. As Doran put it:

> The war between extremist *Salafis* and the broader populations around them is only the tip of the iceberg. The fight over religion among Muslims is but one of a number of deep and enduring regional struggles that originally had nothing to do with the United States and even today involve it only indirectly.[8]

Within the various Arab societies, divisions nourished by national interests, tribal constituencies, ethnic juxtapositions, and religion partisanships have so far doomed any effort to achieve a cohesive political architecture, leaving autocracy to prevail over any other form of government. Despite the hyped Western perception, the reality is that there is neither one single Arab identity nor a single Islamic one. Furthermore, with the exception of Iran, 90 percent of the Muslim world is devoted to Sunni Islam, which provides for neither a single institutional religious authority nor any specific hierarchy of religious leaders. In Zakaria's prose:

> The decision to oppose the state on the grounds that it is insufficiently Islamic belongs to anyone who wishes to exercise it. This much Islam shares with Protestantism. Just as any Protestant with just a little training—Jerry Falwell, Pat Robertson—can declare himself a religious leader, so also any Muslim can opine on issues of faith. . . . The problem, in other words, is the absence of religious authority in Islam, not its dominance.[9]

Such features of Islam and Arab societies make it hard for a single fundamentalist leader to emerge as the undisputed authority in one country, unless he forges alliances with the most important tribal forces of that country, to whom he must concede a great degree of self-government and autonomy. That is what happened in Afghanistan, where local warlords were free to rule their areas of influence and to practice their economic activities—including the not-very-Islamic cultivation and trade of opium. Thus, while extremist Islam can be "profoundly effective in mounting a protest movement," and may "produce a cadre of activists whose devotion to the cause knows no bounds," its absence of hierarchy and organization underlies its inability to establish long-lasting political and civil institutions.[10]

If the rise to power of a fundamentalist government is difficult to envisage, the preservation of power by such a government would be even more difficult, and its ability to use oil as a weapon highly complicated.

Having to deal with a highly fragmented population, powerful tribal forces, and ethnic and religious divides within its own borders, such a government would find it hard to preserve its legitimacy and authority if it were to impose draconian economic hardships. The very consolidation and development of its military and repressive apparatus would require huge sums of money, forcing it to reckon with economic imperatives for its own survival.

Because oil is the predominant source of income for all Arab countries of the Persian Gulf, all of this calls into question the idea that a fundamentalist regime would resort to an oil embargo to hurt the West. For a very short time, perhaps, this might be possible. But over the long term, results would not differ from those provoked by the oil price hikes of the 1970s.

In this context, it is interesting to also note that even Osama bin Laden appears to have taken a somewhat cautions view on oil prices. In an audiotape posted on an Islamic website on December 16, 2004, a speaker identified by the U.S. Department of State as bin Laden urged Islamic militants to stop Westerners from obtaining Middle Eastern oil. "Try your best to stop the biggest theft in history," he said, accusing the West of buying oil too cheaply. He then underlined that the price of oil had fallen several times while the cost of other commodities had doubled. "Today its price should be at least $100 per barrel," he said, adding that a price like that would make it a "fair and legitimate trade."

If confirmed, such a declaration would have deep political significance. It would mean that even the al-Qa'ida leader's vision of the world's future oil order is not based on banning sales of oil to Western countries, but rather on efforts to increase its price. And although worrisome by our standards, that target is almost moderate if compared with the claims of OPEC and the pro-Western Shah Reza Pahlavi, who in the 1970s called for a price of 60 dollars per barrel—equivalent to more than 250 dollars in today's purchasing power!

In any case, any unilateral action by a single producer aimed at reducing or curtailing production will, in the medium-to-long term, only advantage other producers. What is more, a selected embargo directed against "enemy countries" will only marginally affect the global oil market. Given the nature of the latter, the countries that are not blacklisted and continue to receive oil could always resell their oil to the countries under embargo. The truth, as Morris Adelman put it, is that

> Whether a supplier loves or hates a customer (or vice versa) does not matter because, in the world oil market, a seller cannot isolate any customer and a buyer cannot isolate any supplier.[11]

Thus there is reason to believe that the Western obsession with Persian Gulf oil security is largely overstated. By the same token, its main corollaries, the perceived need for a soft or hard form of control of the Persian Gulf and the parallel quest for "oil independence" from it, are the result of a dangerous and even naïve miscalculation, both in political and economic terms.

The only real centripetal force in Islamic societies is the rejection of any threat to their existence by external powers, a feeling shared by all societies. For most Arabs, moreover, the disappointing lack of improvement in their lives, which has marked their modern history, is cushioned in part by the comforting and reassuring view that foreign hands have conspired against their progress. Although only partially confirmed by history, the evocative power of such a view is so strong in the collective psychology of the Arab masses that any direct or indirect Western involvement in the shaping of a future Middle East is bound to fail.

In particular, stationing Western troops in an Islamic Arab country is anathema to any Muslim, an act of impiety and an offence against the sacredness of the Holy Land of the believers. It is something that will lead to further alienating the indigenous populations. By embarking on a long-term military and political presence in the Middle East to secure its oil supply, the West would provide a boost to the radical Islamists' most effective weapon, the leveraging of people's frustration and hopelessness, which is the mainstay of every revolutionary movement. The result would be a permanent state of guerrilla warfare, where the overwhelming high-tech forces of the West would confront the zeal of hundreds of small groups of militant, impoverished people, to whom the alternative between life and death is a choice of no great significance.

In plainly economic terms, then, no Islamic oil embargo–led shock can justify the massive spending required to deploy troops across a vast war theater, when just keeping two divisions engaged in "stability operations" in Iraq for one week costs \$1 billion, and keeping them engaged for a full year would cost the entire GDP of New Zealand.[12]

If no kind of military action is an option, the supposed need for consuming countries to devise policies aimed at ensuring "oil independence" from the Middle East, or to undertake broader actions to shift to energy sources other than oil, is simply nonsense.

Past efforts along these lines have always turned into a sterile exercise. In the real world, the replacement of a resource with another is driven by economics, not by politics. The latter can only favor the rise of a source of energy that is already close to fill the gap that distances it from the dominant one. Above all, politics should promote more responsible and environmental friendly consuming habits, both by devising a different architecture for our societies, our cities, our systems of transportation, and by imposing higher taxes on the most polluting sources of energy. Politics can also promote research and development on realistic new sources of energy and more efficient systems to consume it, but cannot impose them by law.

Coal replaced wood in the eighteenth century because it was cheaper and much more efficient than wood. Great Britain decided to convert its navy to oil at the beginning of the twentieth century because it was cheaper and much more efficient than coal, despite the fact that the country had no oil reserves on its territory at that time. And oil's supremacy has been eroded in the last thirty years because other forms of energy have proved to be more efficient and less expensive.

Only totalitarian systems maintain the illusion that political decisions can ignore the harsh reality of economics. The Nazi attempt to replace oil with synthetic fuels did not fail from a technical point of view: it failed because synthetic fuels were much more expensive than gasoline and diesel. Thus, any attempt by consuming countries to shift away from Middle Eastern oil—or from oil in general—would only weaken their own economies, driving up their energy costs. And it would do nothing to combat terrorism, despite the assertions of those who argue that oil imports from the region help finance it. Even if the West were to implement such a program, the Arab countries will continue selling their oil and have no trouble finding buyers.

In sum, like the Greek god Proteus, the oil market is escaping control by constantly assuming different forms, which makes political manipulation of oil difficult, indeed useless—for both producers and consumers. In this framework, the notion of oil security is simply a confusing myth when referred to oil supply. Western governments must explain clearly to their constituencies that oil—like many other goods—is prone to price volatility, which makes occasional high prices, disruptions, and temporary shortages unavoidable. Furthermore, they must disabuse their citizens of bonanza oil expectations—particularly when oil prices fall—and promote more careful consumption habits, while leaving market forces to work.

As to the issue of the Middle East, of course there is no easy or immediate solution to its several dilemmas. Throughout history, the shaping and consolidation of national identities has been a prolonged process fraught with considerable suffering. The countries of the Middle East are relatively new, forged mainly after World War I, and Western states must ready themselves for the long road ahead, on which they must avoid either underestimating the strength of Middle Eastern states or exaggerating the threats that they pose. Above all, Western governments must overcome their misguided obsession with oil security so that they can begin to cope more impartially with the Middle East's problems because so long as that obsession remains the dominant paradigm of Western politics in the region, we will be doomed to repeat the mistakes and dramas that characterized relations between the West and the Arab peoples in the twentieth century.

CHAPTER 22

The Long Wave of Resource Nationalism

The two-year period of 2006 and 2007 has had significant impact on the world of oil. It has seen prices—at least in nominal terms—reach record heights, peaking at $77 per barrel in July and August 2006, and closing the year at an average of $66 per barrel, while continuing to oscillate between $60 and $77 during the first half of 2007. Even more significantly, however, it has been a period in which a new wave of resource nationalism has shaken up the world of oil and gas.

Although resource nationalism had already manifested itself strongly at the beginning of the decade, it gained new strength during 2006 and early 2007, causing serious new concerns and casting a shadow over the reliability of some major oil and gas producing countries.

From Latin America to Iran, and from Russia to Africa, a wide range of nationalist impulses have come into play, combined with some common elements similar to what prevailed during the 1970s: an oil and gas supply crisis seized upon by governments to exert greater control over production and generate extra profits, while at the same time flexing their political muscles to enhance their domestic and international stature.

This wave of resource nationalism has been especially visible in Venezuela and Russia. As early as April 2005, the Venezuelan Energy and Mines Ministry had ordered the national oil company, PDVSA, to review the thirty-two Operating Service Agreements it had signed with several international oil companies between 1992 and 1997, and transform them into joint-venture agreements, creating joint-companies in which PDVSA would hold at least 60 percent ownership. After a long period of uncertainty and extended negotiations, some foreign companies accepted the new formula, but others chose to sell their assets and leave the country. Some refused to accept the compensation offered by the Venezuelan government for their assets, which was based on PDVSA's calculation of their book value (instead of market value), and called for international arbitration.

PDVSA took a similar approach toward the foreign companies that had entered into strategic partnerships to develop and exploit the resources in the Orinoco River basin. This river basin stretches for hundreds of miles north of the river and contains huge reserves of ultra-heavy oil, estimated at around 270 billion barrels, at par with Saudi Arabia's 260 billion barrels of conventional oil. During the 1990s, with WTI selling at $18 or less a barrel, this heavy, costly-to-refine crude was worth only a few dollars a barrel, and the Western oil companies held a virtual monopoly on the technology to extract and refine it. As a result, the Venezuelan government had agreed to let the foreign companies back into the country, offering them favourable fiscal conditions. But the surge in oil prices had drastically changed the situation. With crude at $60 to $70 a barrel, even the less valuable extra-heavy oils from the Orinoco were now much more financially attractive. Now there were Chinese and Russian companies offering to develop the region in place of the major Western companies. Between 2006 and 2007, Venezuela made clear that the operating and ownership conditions in the Orinoco belt had to change. In June 2007, PDVSA almost doubled its stake from an average of 40 percent to around 80 percent in four major oil projects in the area, involving several larger Western majors such as Conoco-Phillips, Total, Exxon-Mobil, and BP. Once again, most of them accepted the new rules—that is, they become junior partners of PDVSA.

Buoyed by high prices and strong popular support, Venezuela's Hugo Chavez hasn't blinked in the face of international criticism of his increasingly extreme resource nationalism. On the contrary, he has used his growing oil revenues to launch new social initiatives at home. With PDVSA's oil representing around 30 percent of Venezuelan gross domestic product (GDP) and around 50 percent of the government's revenues, Chavez channeled more than $13 billion in state company funds to social programs in 2006, almost doubling the amount channeled in 2005, while heavily subsidizing domestic oil consumption at a cost of $9 billion in 2006 alone.

Chavez also used his financial strength to poke a sharp stick in the eye of his perceived "enemy," the United States, by financing governments and alliances at odds with Washington. He has traveled extensively to countries hostile to the United States, including Iran, and cemented his friendship with Cuba's Fidel Castro, whose mantle he apparently hopes to inherit. His anti-Western policies culminated in May 2007 with Venezuela's announcement of withdrawal from the International Monetary Fund (IMF) and World Bank, which he labeled as tools of United States foreign and economic policy.

At the same time, Chavez expanded Venezuela's political and economic relations with Russia and particularly China. In March 2007, the two countries established a $6 billion fund to finance development projects in Venezuela and other Latin American countries. China will provide two-thirds of the fund, and Venezuela one-third.

Chavez's activities have resonated throughout Latin America, but especially in Bolivia and Ecuador. In Bolivia, in May 2006, just six months after his election, President Evo Morales re-nationalized the country's oil and natural gas exploration and production facilities, which had been partly privatized during the 1990s, and turned them over to the state-owned Yacimentos Petroliferos Fiscales Bolivianos (YPFB). In Ecuador, President Rafael Correa announced his intention to renegotiate contracts signed with foreign energy companies by previous governments, and at the same time tightened oversight over all energy operations in the country.

It's still too early to predict if the Chavez approach will turn into a model adopted by other major energy-producing countries. But history has shown that Latin America has been the pacesetter for some of the more dramatic twists in the oil world. The first nationalization of the oil industry—not including the USSR—took place in Argentina in 1922. From there, the phenomenon spread through the region, culminating in the 1938 nationalization of the industry in Mexico. In 1948, Venezuela was the first to impose a "fifty-fifty" profit sharing formula on the international oil companies, a formula that quickly spread to the Middle East and the rest of the world. And it was Venezuela again that spearheaded the creation of the Organization of Petroleum Exporting Countries (OPEC) in 1960 as a counterweight to the power of the "Seven Sisters."

Today, Chavez and Morales are simply the latest standard-bearers of an extreme strain of implacable energy irredentism, legitimate and understandable in a region beset by staggering inequalities, but easily subject to populist extremism, and thus prepared to sacrifice oil for the sake of political ends beyond the comprehension of Westerners. Nevertheless, it would be a serious mistake to ignore this reality.

Vladimir Putin's efforts to use Russia's energy resources as instruments of political power has been equally determined, but with far greater international impact.

President Putin used his first five-year term to gain complete control over the energy sector, and then embarked on a double-barrelled strategy to control both distribution to the markets and imports from former members of the USSR. This involved the transportation of gas to consumers, mainly those in Europe, and the signing of new import contracts with oil and gas producers in countries like Turkmenistan and Kazakhstan.

Implementation of this strategy produced a sudden crisis centered in Ukraine, which was the critical link in the pipeline network that transports gas from Siberia and Central Asia to Europe. Eighty percent of this network passes through Ukraine. The crisis was set off in late December 2005, as a consequence of a disagreement between Ukraine and Russia, which had unilaterally set the price for natural gas to Ukraine. Gazprom reacted to Ukrainian objections by cutting off supplies to the country on January 1, 2006, and Urkraine countered by diverting gas destined for Europe for its own use, causing serious problems for the Europeans, who import nearly 40 percent of their requirements from Russia. However, the story behind this crisis is complex and merits a closer look.

Like other former members of the Soviet Union, Ukraine had long benefited from a low "political" price for the gas it bought from Gazprom. By the end of 2005, it was about a quarter of what Gazprom charged the Europeans, and half of what Russians paid. But by then tensions were building between Russia and Ukraine. Moscow had been accusing the Ukrainians for some time of illegally siphoning off gas destined for Europe and of failing to pay for gas directly delivered to them. Concurrently, Ukraine had gone through its "orange revolution," replacing a pro-Russian regime with one more closely aligned to the West. Friction within the new regime had already resulted in new elections being scheduled for March 2006. In the eyes of many analysts, the Russian decision to cut off supplies to Ukraine was part of an effort to influence the political outcome. But Gazprom's action had unexpected consequences.

In the middle of an especially cold winter, several European countries found themselves suddenly short of natural gas. As tension grew, charges and counter-charges flew. The Russians accused Ukraine of diverting Europe-bound supplies, while the Ukrainians said Gazprom had reduced supplies too drastically. The situation rapidly escalated. Many European governments, the European Union, and the White House condemned Russia's unilateral action and accused the Kremlin of neo-imperialism and of using its gas as an instrument of blackmail. Gazprom was forced to back off, even as the fierce Russian winter strained its ability to maintain a high level of exports to Europe.

Once the bitter cold ended, Gazprom reasserted its position, and this time it went beyond just Ukraine: there was no longer a need to maintain "political" prices for *any* former member of the USSR, it argued, since that country no longer existed, and therefore all would have to accept market prices. The extension of this principle to all countries once united into the USSR seemed to reveal once again an aggressive stance by the Kremlin, aimed at reconquering past influence over the same countries. On May 4, 2006, while taking part in an international pro-democracy conference in Vilnius, the capital of Lithuania, United States vice president Dick Cheney strongly attacked Moscow's energy policy, accusing it of using oil and gas supplies as "tools of intimidation and blackmail," a hard accusation that soon seemed to herald the dawn of a new cold war era between the United States and Russia.[1] The Russian answer arrived soon. On May 8, 2006, the energy minister of the Russian Federation, Viktor Khristenko, wrote an article for the *Financial Times* where he rebuffed all Cheney's arguments and stressed that as far as the energy policy of Moscow was concerned:

". . . Russia has moved away from Soviet-era arrangements of subsidising energy prices to our neighbours and turned to market-based pricing mechanisms. We continue to push for co-operative and practical collaboration among nations to establish a balanced and equitable energy security system."[2]

But even more clouds have since appeared on the horizon. By October 2006, Russia and Ukraine had agreed on a price of gas for 2007 and laid out plans for a Russian-Ukranian joint venture company to manage the pipeline network in Ukraine. But four months later, in February 2007, the Ukrainian parliament cast a shadow over that project by forbidding any sub-contracting of the national pipeline network to third parties.

While the Ukrainian situation was playing out, Russia opened a new front in the oil and gas wars, issuing an ultimatum to Belarus: accept the new market prices imposed by Gazprom or face a supply cutoff on January 1, 2007. A repeat of the Ukrainian crisis was apparently averted at the eleventh hour, shortly before the dawn of the new year, increasing Blarus's price by around 60 percent, which still left it more than 50 percent below the prices paid by the European Union. But the apparent solution soon vanished because of eventual recriminations and additional requests by the Russian side, in particular, that by 2010 Belarus would sell to Gazprom half the company that manages the piplines that pass through Belarus on their way to Western Europe. In a rapid deterioration of the second stage of the crisis, on January 8, 2007, Moscow closed down the Druzhba ("Friendship") pipeline taking oil from Russia through Belarus to Poland, Germany, and other European countries, accusing Belarus of illegally stealing oil. In support of Belarus, Azerbaijian stopped exporting its own oil through the Russian port of Novorossysk, on the Black Sea. A major crisis seemed at that point just around the corner. But finally, Belarus and Russia entered an agreement with Moscow getting most of its requests.

The Kremlin's strategy went beyond simple efforts to control the existing gas distribution networks and the upward adjustment of gas prices for the former Soviet republics. In May 2007, presidents Putin, Nursultan Nazarbayev of Kazakhstan, and Gurbanguly Berdimuhammedov of Turkmenistan signed a multifaceted agreement to modernize the creaking Soviet-era pipeline network that brings natural gas to Russia from Turkmenistan via Uzbekistan and Kazakhstan, and to build a new pipeline to carry additional gas from Turkmenistan to Russia along the Kazakh coast of the Caspian Sea. When completed, the new and improved pipelines could deliver between 70 and 80 billion cubic meters of gas a year from Turkmenistan to the Russian Federation, a vast increase over the current 42 billion cubic meters. In the process, the Russians dealt a major blow to those—led by the United States—who had been promoting an alternative route for the delivery of natural gas from Turkmenistan to Europe, by way of the Caspian Sea, Azerbaijan, Georgia, and Turkey. At the same time, they have succeeded in retaining control over a large part of the future energy resources of the Caspian region.

Moscow has also moved ahead with efforts to regain control over some oil and gas development projects that were turned over to foreign companies during the 1990s, foremost among them the Sakhalin 2 and Kovykta projects.

The Sakhalin 2 project was managed by Shell and was intended to develop and begin extraction of some of the vast reserves of oil and natural gas discovered on Sakhalin Island, in the Sea of Okhotsk, north of Japan. The project, by Shell's own admission, had been facing significant cost overruns, nearly doubling from original estimates of $12 billion to $22 billion. The increased costs implied lower revenues for the Russians, and they moved to block the project, alleging that Shell had caused unspecified environmental damage to the area. Ultimately, in December 2006, Shell and its Japanese partners—Mitsui and Mitsubishi—were forced to accept a new agreement with Gazprom in which the Russian company became the majority shareholder in the project.

A similar fate befell the project to develop the giant natural gas field at Kovykta, in the Irkutsk region of Eastern Siberia, originally assigned to RUSIA Petroleum, with 63 percent controlled by the British-Russian joint venture BP-TNK. This time the Russian authorities took advantage of project delays to invoke "violations of licensing terms," which called for the delivery of 9 billion cubic meters of natural gas to Gazprom during 2007, and presented a three-month ultimatum to BP-TNK to resolve the issue or face the loss of its rights in the field. In June 2007, BP-TNK agreed to surrender a majority share in Kovykta to Gazprom. This would allow Kovykta to access Gazprom's Russian pipeline network and open the door to exporting the gas to China. Without access to the network, BP-TNK couldn't have exported the gas beyond the Irkutsk region.

Looking beyond Latin America and Russia, resource nationalism has followed different models in Africa. As one of the only regions still open to international oil companies, the continent has been attracting massive investments in a hyper-competitive environment that has been described as a "scramble for African oil," with Chinese companies playing an especially active role in the process. In recent years, the China National Petroleum Corporation (CNPC) and its subsidiary PetroChina, the China National Offshore Oil Company (CNOOC), and the China Petroleum & Chemical Corporation (Sinopec) have vastly expanded their presence in the region. Today China depends on Africa for 45 to 50 percent of its petroleum imports. CNPC/PetroChina, for example, is the biggest producer in Sudan and also owns a share in the 1,500-kilometer (900-mile) pipeline that links the oilfields to the Marsa al-Bashair terminal near Port Sudan on the Red Sea. The company is also active in Algeria, Angola, Mauritania, Niger, and Nigeria. CNOOC is present in Equatorial Guinea and Nigeria, while Sinopec has participations in Algeria, Angola, Gabon, and the Republic of the Congo, as well as Sudan.

African governments have taken advantage of the feverish competition between the Chinese companies, as well as Russian, Malaysian, and Indonesian companies, albeit on a lesser scale, and their reserves-hungry Western counterparts to fashion better terms for themselves in new oil and gas exploration and production contracts. In the 2005–2007 bid rounds for exploration blocks in countries like Algeria, Angola, and Libya, the producing countries managed to push the government take past the 90 percent threshold. At the beginning of the decade, the average worldwide government take (excluding the Western oil producing countries that generally have much more moderate rates) hovered within a range of 70 to 80 percent.

This shift in the balance of power between producing countries and the Western oil companies has been carried out in a less visible and quieter way than the more extreme cases of resource nationalism that we have seen in Venezuela and Russia, or other countries where government-controlled companies have simply expropriated the largest part of underground resources. As a matter of fact, between 2006 and 2007 a general trend toward host governments' re-appropriation of a larger share of oil economic rent has been generalized, also outside the borders of so-called under-developed or developing countries. Even in Western countries, the call for a major taxation of oil companies has taken place, bringing countries and states, such as Great Britain, Alaska, and California, to increase their taxes on oil activity, in the midst of a strong popular support to deprive oil companies of most of their "windfall profits."

In any event, and in whatever form it may take, it is clear that re-source nationalism is making a significant impact on the world of hy-drocarbons and foretells a serious identity crisis for the Western oil companies. Above all, it's not a new phenomenon linked to a particular and temporary situation of high prices and limited supplies, but part of a long-term historical trend.

In fact, the process of re-appropriation of underground resources by producing countries has been underway since the end of World War II. It began with the fifty-fifty formula in Venezuela, was later applied in the Middle East and elsewhere, and has continued ever since, sometimes at a rapid pace and sometimes more slowly. Whatever its speed, however, the trend has been consistently in the direction of less access to resources by foreign companies, and except for a handful of marginal cases, it has never taken a step back. The paradigm of the 1970s is instructive.

After the two oil shocks, sky-high prices, and nationalization of West-ern oil companies' activities that marked those years and seemed to give the producing countries virtually limitless power, the latter made no change whatsoever in their policies once the price of crude crashed in 1986. And although prices remained in the basement through the 1990s and led to social and economic difficulties for the producers, they held their ground and stayed the course even though their technical and man-agerial abilities were far less developed than they are today.

It is of little importance that the process of reasserting control over re-sources changed from time to time, ranging from tighter financial and contractual formulas to outright expropriations. And even when it led to inefficient practices, or major problems for the producing countries, it retained its legitimacy, whether foreigners liked it or not, as a manifestation of the local population's deeply felt desire to control its own strategic resources, especially oil and gas, which are seen as an integral element of national identity and self-sufficiency.

Two other major countries have dominated the fragile equilibrium of the world's oil market in the last two years: Nigeria and Iran. In Nigeria, an explosive mix of causes has provoked a dramatic escalation of violence in the country's most oil-endowed region, the Niger Delta area, leading to a major drop in oil production and to a quite constant situation of insecurity that has caused several oil companies to periodically stop their activities in the area. The main causes behind such a chaos derived from two long-terms problems, in addition to occasional causes that may have led to their explosion. On the one hand is the growing resentment of the population of the Niger Delta not to have enjoyed any fruit of the oil bonanza of the last few years, as well as probably not to have ever enjoyed the fruit of oil since the beginning of the Nigerian oil saga in the late 1950s. On the other hand is the growing strength on an ethnic-based new movement, the *Movement for the Emancipation of the Niger Delta* (MEND), which leveraging on the arrest of its leader for corruption and fraud, launched a campaign of destabilization of all oil activities in the Niger Delta, trying to have a major effect on the May 2007 presidential elections, which led to the departure of former president Olesung Obasanjo and the rise to power of his heir, Umaru Yar'adua.

However, in spite of the openings of the new president to the problems of the Niger Delta region, violence, killings, kidnappings, and attacks to oil installations continued, making it difficult to envision how the forces in motion could influence the future of the major oil-producing country in Africa. While the Nigerian government take on oil activities is quite high, amounting to an average 75 percent of oil revenues from companies' operation, a problem of re-distribution of such money still exists. At the same time, the dramatic ethnic divide and fragmentation of Nigeria—rifted among more than 300 ethnic groups, each one striving to get a larger slice of the oil wealth—makes it difficult to find a common ground for re-distribution and local development. It has also facilitated over time the rapacious attitudes of several leaders, both national and local, who deprived the population of funds and money.

The case of Iran is different. Its race to enter the nuclear club has given rise to fears of an international crisis of global proportions against the background of the already tenuously balanced world oil market. The specter of United States military action to destroy Iran's nuclear facilities, and the prospect of a chain-reaction global uprising in much of the Islamic world, cast a long shadow. In an Islamic world where many have been disappointed by the failure of local leaders to stand up to perceived American ambitions, Iranian President Mahmoud Ahmadinejad has won support from a wide range of radical groups with multiple agendas.

The deteriorating relationship between Iran and the United States has been weighing heavily on the price of crude. Fear of an interruption in the flow of Iranian exports as a result of a possible U.S. attack, or an Iranian reaction to sanctions based on the "oil weapon," has pushed market makers to hedge or speculate against a possible crisis, keeping crude prices at a high level. This climate of fear has obscured the real crux of the problem: the Iranians' determination to independently develop their own nuclear program.

It may seem counter-intuitive, but oil-rich Iran may indeed need such a program for peaceful commercial use. Because of a false perception about Iran's vast oil and gas resources, the world at large assumes that the only goal of its nuclear development program must be to obtain atomic weapons. In fact, and despite the unquestioned abundance of its oil and natural gas resources, Iran stands at the edge of an energy and economic precipice.

Ninety percent of its petroleum comes from fields discovered between 1929 and 1969 that rely on outdated technology for the extraction of only a fraction, about 27 percent, of their content. Because the Iranian constitution prohibits foreigners from controlling hydrocarbon reserves or production, there are limited opportunities for cooperation between Tehran and foreign companies with more advanced technologies. Also, like many other major producers, Iran has made only very limited efforts to find new sources of oil and has had difficulty in developing newly discovered fields.

International isolation and U.S. sanctions worsen the situation. To prevent a rapid decline in production from existing fields, Iran pumped almost 40 billion cubic meters of natural gas back into the ground in 2005 to force more oil to the surface. That's equivalent to roughly half the annual output of a major gas producer like Algeria. Re-injecting gas into the ground to boost oil production makes financial good sense for the

Iranians. At a price of $28 per barrel—less than half of today's going rate—the re-injected gas is worth 80 times what it would fetch on the regulated domestic market, and four times what its export would yield.

But this scenario is no longer sustainable. Domestic consumption of energy and electricity in Iran is skyrocketing, and natural gas is the only current option for generating power. No Iranian government can afford to deprive the country of that gas, which is sold at politically driven prices below the cost of production. But giving gas to the people means taking it away from the oil fields, setting off a downward production spiral that has reached 13 percent a year in 2006—or 500,000 barrels per day.

The country now produces with difficulty around 4 million barrels per day (mbd), in contrast to its historical peak of 6 mbd in 1974. Putting it another way, without drastic changes, Iranian oil production risks extinction in less than ten years. Thus, it makes perfect sense for the Iranians to build a nuclear-based electricity production system as partial replacement for gas-fired plants. Even a pro-Western regime would find it necessary to support such a program, and in fact today's efforts are simply a continuation of a program begun by the Shah in the 1970s. Of course, the economic need for a peaceful nuclear program does not negate the possibility that Iran, despite its constant denials, may also secretly seek to develop nuclear weapons.

Nevertheless, it seems futile and dangerous to base policy on assumed Iranian intentions. At this time Iran is only able to produce enriched uranium, the fuel needed to generate electricity, with a purity level of less than 5 percent. The production of nuclear weapons requires purity close to 100 percent, and large quantities. It will take years for Iran to attain either target. Simply having enriched uranium is not enough for a nuclear weapon, and simply having a nuclear weapon is not enough to challenge a superpower like the United States.

Many ask themselves why Iran doesn't prove its good faith by giving up its own uranium enrichment program and buying needed supplies from third parties. Given the country's recent experience with international sanctions, isolation, and Western hostility, such dependence on the goodwill of others would probably not be psychologically acceptable to the Iranian public.

For the West, Iran's development of nuclear capabilities also raises the possibility of proliferation. Many other energy producers in the Middle East are also looking at peaceful nuclear energy to help stretch out their oil and gas reserves. This is a real phenomenon, but it should be noted that it is not unique to the region. Even in the industrialized

West the nuclear energy debate has been revived, and many governments now see nuclear energy as the only realistic alternative to continued dependence on imported petroleum and gas. And if nuclear energy turns out to be the best option for the industrialized world, why can't it be an option for the developing world? Or are we going to go back to a double standard by which only *we* (the West) have a right that is denied to *them* (the developing world)?

Appendixes

APPENDIX 1

Proven Reserves, Production, and Reserves Life Index of the First Twenty Oil Countries in the World and World Totals (2004)

Country	Proven Reserves (Billion Barrels)	Daily Production (Million Barrels)	Reserves Life Index
Saudi Arabia	262	10.136	69
Iran	126	4.167	83
Iraq	115	2.010	157
Kuwait	101	2.171	110
United Arab Emirates	98	2.748	98
Venezuela	77	2.964	72
Russia	60	9.227	18
Libya	39	1.614	61
Nigeria	35	2.505	27
United States	30	7.675	11
China	18	3.492	14
Qatar	15	1.030	40
Mexico	14.6	3.825	11
Algeria	11.8	1.930	16
Brazil	10.6	1.767	13
Kazakhstan	9	1.175	21
Norway	8.5	3.158	9
Azerbaijan	7	0.317	60
Oman	5.5	0.988	20
Angola	5.4	0.764	15
World Total	**1,111**	**82**	**38**

Source: Eni, World Oil&Gas Review, 2005.

APPENDIX 2

Consumption Trends of the
First Twenty Countries
in the World (1980–2004)
(Million Barrels per Day)

Country	2004	2000	1995	1990	1980
United States	20.85	20	18	17.2	17.5
China	6.4	4.5	3.3	2.3	1.7
Japan	5.4	5.6	5.7	5.2	5
Germany	2.7	2.8	2.9	2.7	3
Russia	2.6	2.6	2.8	4.3	4.6
India	2.4	2.2	1.6	1.2	0.65
Canada	2.3	2	1.8	1.7	1.9
South Korea	2.1	2.1	2	1	0.52
Brazil	2.1	2.1	1.8	1.4	1.2
Saudi Arabia	2.1	1.7	1.3	0.9	0.6
France	2	2	1.9	1.8	2.2
Mexico	2	2	1.8	1.7	1.2
Italy	1.85	1.85	1.9	1.9	1.9
United Kingdom	1.85	1.75	1.8	1.8	1.65
Spain	1.6	1.4	1.2	1	1
Iran	1.5	1.3	1.2	1	0.57
Indonesia	1.2	1	0.8	0.65	0.4
Taiwan	1	0.8	0.7	0.6	0.4
Netherlands	0.96	0.85	0.77	0.73	0.77
Thailand	0.95	0.73	0.7	0.4	0.2
World Total	**82.2**	**76.6**	**70**	**66**	**62**

APPENDIX 3

Main Features of Some Qualities of Crude Oil
(Benchmarks in **Bold**)

Name	Origin	Daily Production (thousand barrels)	API degree	Sulphur Content (%)
Brent Blend	**United Kingdom**	300	**38.7°**	**0.31**
Forties	United Kingdom	650	37.3°	0.40
Ekofisk	Norway	500	37.8°	0.22
Statfjord	Norway	480	37.7°	0.29
WTI Blend	**United States**	300	**38.7°**	**0.45**
Alaskan North Slope	United States	950	31°	1
Light Louisiana	United States	400	38.7°	0.13
West Texas Sour	United States	775	34.2°	1.30
BCF-17	Venezuela	800	16.5°	2.5
Maya	Mexico	2,450	21.6°	3.6
Isthmus	Mexico	500	32.8°	1.4
Olmeca	Mexico	400	39.3°	0.8
Urals	Russia	3,200	32°	1.30
Siberian Light	Russia	100	35.6°	0.46
Arabian Light	**Saudi Arabia**	**5,000**	**33.4°**	**1.80**
Arabian Extra Light	Saudi Arabia	1,200	37°	1.33
Arabian Medium	Saudi Arabia	1,500	30.3°	2.45
Arabian Heavy	Saudi Arabia	800	28.7°	2.8
Basrah Light	Iraq	1,600	30.2°	2.6
Kirkuk	Iraq	350	33.3°	2.3

Name	Origin	Daily Production (thousand barrels)	API degree	Sulphur Content (%)
Iran Heavy	Iran	1,700	30°	2
Iran Light	Iran	1,300	33.4°	1.6
Kuwait	Kuwait	2,000	31°	2.63
Dubai	**Dubai**	**100**	**31.4°**	**2**
Qua Iboe	Nigeria	500	36°	0.11
Bonny Light	Nigeria	450	34.3°	0.15
Forcados	Nigeria	400	30.4°	0.18
Escavros	Nigeria	300	34.4°	0.15
Cabina	Angola	300	32°	0.12
Palanca	Angola	200	37°	0.17
Brega	Libya	120	42°	0.20
Bu Attifel	Libya	100	43°	0.03
Es Sider	Libya	300	36.6°	0.42
Saharan Blend	Algeria	350	47°	0.11
Tapis	**Malaysia**	**300**	**45.2°**	**0.03**
Daquing	China	1,000	32.2°	0.09
Shengli	China	550	26°	0.76

APPENDIX 4

The Largest National Oil Companies
(As of December 31, 2004)

Company	Oil & Gas Reserves (1)	Oil Reserves (2)	Oil Production (3)	Total Revenues (4)
Saudi Aramco—S. Arabia	299.8	259.4	8.9	122
NIOC—Iran	287.9	125.8	3.9	32.5
Gazprom*—Russia	189.6	13.6	0.2	26.9
QGPC—Qatar	172.1	15.2	0.7	12.6
INOC—Iraq	134.0	115.0	2	12.1
ADNOC—Abu Dhabi	126.0	92.2	2	17.5
PDVSA—Venezuela	104.8	79	3.1	67.5
KPC—Kuwait	99	99	2.3	25.2
NNPC—Nigeria	65.6	35.3	2.3	21.6
NOC—Libya	48	39	1.6	16

1. Billion barrels of oil equivalent
2. Billion barrels of oil
3. Million barrels per day
4. Billion dollars
*Without Yuganskneftegaz and Sibneft

APPENDIX 5

The Largest International Oil Companies

(As of December 31, 2004)

Company	Oil & Gas Reserves (1)	Oil Reserves (2)	Oil Production (3)	Total Revenues (4)*	Market Capitalization (4)
ExxonMobil	21.3	10.9	2.5	264.0	328.1
BP	18.3	9.9	2.5	285.1	209.5
Shell	11.9	4.9	2.2	265.2	201.9
Chevron	11.4	8.0	1.7	142.9	109.9
Total	10.9	7.0	1.7	152.6	139.5
ConocoPhillips	8.6	5.5	1.0	118.7	60.3
Eni	7.2	4.0	1.0	72.6	94.9
RepsolYPF	4.8	1.7	0.6	45.0	29.1
Occidental	2.5	2.0	0.5	11.4	23.1
Anadarko	2.4	1.1	0.2	6.1	15.4

1. Billion barrels of oil equivalent
2. Billion barrels of oil
3. Million barrels per day
4. Billion dollars
*Net of excise taxes

Notes

Preface

1. Maugeri, Leonardo. *Not in Oil's Name*. In: Foreign Affairs, Vol. 82, No. 4, July/August 2003.
2. Maugeri, Leonardo. *Oil, Never Cry Wolf: Why the Petroleum Age Is Far From Over*. In: Science, Vol. 304, No. 5674, May 21, 2004.
3. Yergin, Daniel. *The Prize: The Epic Quest for Oil, Money and Power*. New York: Simon & Schuster, 1991.
4. Ibidem, p. 222.
5. Adelman, Morris A. *The Genie Out of the Bottle: World Oil since 1970*. Cambridge (MA): MIT Press, 1995, p. 178.
6. On this issue see: Gordon, Richard L. *Viewing Energy Prospects*. In: The Energy Journal, Vol. 26, No. 3, 2005, pp. 122–133.

Chapter 1

1. The word kerosene derives from the Greek words *keros* and *elaion*, which respectively mean *wax* and *oil*. Gesner altered the word *elaion* in *ene* in order to make it more similar to the then popular *camphrene*, another illuminating substance. See: Yergin, *The Prize*, p. 23.
2. Giddens, Paul. *Early Days of Oil*. Princeton (NJ): Princeton University Press, 1948, p. 1.
3. Knowles, Ruth Sheldon. *The Greatest Gamblers: The Epic of the American Oil Exploration*. Norman: University of Oklahoma Press, 1978 (2nd Ed.), p. 3.
4. Ibidem, p. 5.
5. As brilliantly explained by Ruth Sheldon Knowles, the difference between Drake's method and rotary drilling was like the difference between driving a nail and putting in a screw. Drake's method pounded a hole. Rotary drilling had

been devised as a faster means of going through soft formations for salt and water. Later on, rotary drilling became the dominant technique whatever the features of the subsoil. See: Knowles, p. 31.

6. American Petroleum Institute (API). *Petroleum Facts and Figures: Centennial Edition, 1959.* New York: API, 1959, p. 1.

7. Williamson, Harold F., and Arnold R. Daum. *The American Petroleum Industry: The Age of Illumination, 1859–1899.* Evanston (IL): Northwestern University Press, 1959, p. 118.

8. For a comprehensive and detailed analysis of Rockefeller's personality and achievements see: Chernow, Ron. *Titan: The Life of John D. Rockefeller, Sr.* New York: Random House, 1998.

9. Ibidem, p. 130.

10. Ibidem, p. 156.

11. On Flagler's life see: Akin, Edward. *Flagler: Rockefeller Partner and Florida Baron.* Kent (OH): Kent State University Press, 1988.

12. For example, Standard Oil was able to transport a barrel of oil from Cleveland to New York City at a cost of $1.65—against an official fee for its competitors of $2.40. See: Chernow, p. 113; Yergin, p. 39.

13. Yergin, p. 43.

14. Chernow, p. 224.

15. Yergin, p. 56.

16. The history of the Nobels and the development of Russian oil are described by: Tolf, Robert W. *The Russian Rockefellers: The Saga of the Nobel Family and the Russian Oil Industry.* Stanford (CA): Hoover Institution Press, Stanford University, 1976.

17. On the history and oil venture of the Rothschild family see: Cowles, Virginia. *The Rothschilds: A Family of Fortune.* London: Weidenfeld and Nicolson, 1973.

18. Standard Oil could afford such a policy abroad with little damage because it offset lower revenues in one area by increasing prices in others—notably the United States and Asia.

19. On Marcus Samuel's life and accomplishments see: Henriques, Robert. *Marcus Samuel: First Viscount Bearsted and Founder of the "Shell" Transport and Trading Company, 1853–1927.* London: Barrie and Rocklift, 1960.

20. Standard did not hesitate to spread the suspicion among Arabs that behind Samuel's project there was a Jewish conspiracy.

21. API. *Petroleum Facts and Figures: Centennial Edition, 1959.*

22. On the origins of Royal Dutch see: Gerretson, E. C. *History of the Royal Dutch.* Leiden: E. J. Brill, 1955 (4 vols.).

23. Yergin, pp. 126–127.

24. Chernow, p. 148.

25. See: Tarbell, Ida. *The History of the Standard Oil Company.* New York: McClure, Phillips & Co., 1904 (2 vols.).

26. Chernow, p. 438.

27. Flynn, John T. *Men of Wealth: The Story of Twelve Significant Fortunes from the Renaissance to the Present Day*. New York: Simon & Schuster, 1941, p. 449.

28. Williamson, Harold F., Ralph L. Andreano, Arnold R. Daum, and Gilbert C. Klose. *The American Petroleum Industry: The Age of Energy, 1899–1959*. Evanston (IL): Northwestern University Press, 1963, p. 728.

29. In his excellent biography of John D. Rockefeller, Ron Chernow presents detailed evidence that the founder of Standard Oil was aware of the corrupt practices his men resorted to in running the business, and even that he had in some instances directly financed politicians and public officials.

30. Nevins, Allan. *Study in Power: John D. Rockefeller, Industrialist and Philanthropist*. New York: Charles Scribner's Sons, 1953, Vol. 1, pp. 91–92.

31. Chernow, p. 117.

Chapter 2

1. Yergin, p. 79.

2. Flink, James. *The Automobile Age*. Cambridge (MA): MIT Press, 2001 (Sixth Printing, paperback ed.), pp. 10–14.

3. Bardou, Jean-Pierre, Jean Jacques Chanaron, Patrick Fridenson, and James M. Laux. *The Automobile Revolution: The Impact of an Industry*. Chapel Hill: University of North Carolina Press, 1982, p. 5.

4. Flink, p. 33.

5. Rae, John B. *The American Automobile*. Chicago: University of Chicago Press, 1965, pp. 58–60.

6. Bardou et al., p. 45.

7. The assembly line was behind this staggering success. Before its appearance, a chassis assemblage required twelve and a half hours. Thanks to the assembly line, that time fell to one hour and a half. See: Tedlow, Richard S. *New and Improved: The Story of Mass Marketing in America*. Cambridge (MA): Harvard Business School Press, 1996, p. 127.

8. McCraw, Thomas K., and Richard S. Tedlow. "Henry Ford, Alfred Sloan, and the Three Phases of Marketing." In: McCraw, Thomas K. (Ed.). *Creating Modern Capitalism: How Entrepreneurs, Companies, and Countries Triumphed in Three Industrial Revolutions*. Cambridge (MA): Harvard University Press, 1997, p. 274.

9. Ibidem, pp. 275–276.

10. Ibidem, p. 267.

11. Ibidem, p. 556.

12. Lesser, Ian O. *Resources and Strategy: Vital Materials in International Conflict, 1600–The Present*. New York: St. Martin's Press, 1989, p. 25.

13. That provision implied a total concession area of 500,000 square miles. For details about the D'Arcy concession in Persia, see: Stocking, George W. *Middle East Oil: A Study in Political and Economic Controversy*. Knoxville (TN): Vanderbilt University Press, 1970, pp. 9–10.

14. Ibidem.

15. Meyer, Karl E., and Shareen Blair Brysiac. *Tournament of Shadows: The Great Game and the Race for Empire in Central Asia*. Washington (DC): Counterpoint, 1999, p. xx.

16. Yergin, p. 140.

17. Ibidem, p. 160.

18. Stocking, p. 19.

19. The following table shows the main world oil producers during World War I.

Output of Crude—Total World and Major Producing Areas: 1914–1919
(thousands of barrels per day)

Country	1914	1915	1916	1917	1918	1919
United States	728	770	824	919	975	1,037
Russia	184	188	180	173	75	87
Rumania	35	33	24	10	24	18
Poland (Galicia)	18	15	18	17	16	16
East Indies*	31	33	34	36	35	42
Mexico	72	90	111	152	175	239
Persia	11	10	12	19	24	28
Others	11	16	27	34	56	53
World	**1,090**	**1,155**	**1,230**	**1,360**	**1,380**	**1,520**

Sources: Energy Information Administration (EIA); DeGolyer and McNaughton, Twentieth Century Petroleum Statistics.

*Includes Indonesia, Myanmar, and Brunei

20. Mejcher, Helmut. *Imperial Quest for Oil: Iraq 1910–1928*. London: Ithaca Press, 1976, p. 37.

21. Yergin, p. 194.

22. Williamson et al., *The Age of Energy*, p. 311.

23. Sobel, Robert. *Coolidge: An American Enigma*. Washington (DC): Regnery Publishing, 1998, p. 343.

24. Yergin, p. 188.

25. Shwadran, Benjamin. *The Middle East, Oil and the Great Powers*. New York: Praeger, 1955, p. 208.

26. Germany looked at the Ottoman Empire as an area of its own influence. One of the most revealing signs of this attitude was the construction of the Berlin-Baghdad railway, undertaken by Germany at the end of the nineteenth century.

27. See: Klieman, Arnold. *The Foundations of British Policy in the Arab World*. Baltimore: Johns Hopkins University Press, 1972, pp. 77, 87; Longrigg, Steve. *Iraq 1900–1950*. London: Oxford University Press, 1953, p. 123.

28. Ferguson, Niall. *Empire: The Rise and Demise of the British World Order and the Lessons for Global Power*. London: Allen Lane, 2002, p. 313.

29. Mejcher, p. 76.

30. Stocking, p. 52; Stork, p. 16.

31. Shwadran, p. 237.

32. Williamson et al., pp. 515–519.

33. Actually, in 1928 the Western oil venture in Mesopotamia was still named Turkish Petroleum Company. The latter was renamed Iraq Petroleum Company in 1929, after Mesopotamia assumed its modern name of Iraq.

34. Such a clause already existed in the original bylaw of TPC, written in 1914. Also in that case, the proponent had been Gulbenkian.

35. After his Mexican venture, Weetman Pearson devoted himself to creating an economic empire. Even today, the group bearing his name owns—among other things—50 percent of *The Economist* and the *Financial Times*.

36. At the core of the confrontation was the will of the former to regain sovereignty over the country's resources, which had been sold to foreign interests by the corrupt men of the Diaz regime, all of whom took cuts of the deal. The revolutionary spirit also translated in the newly approved constitution, which stated that subsurface resources belonged to the State and could not be transferred to private parties—whatever their nationality. In this context, the controversy was seemingly irresolvable, as both parties were unwilling to accept any compromise; consequently, it affected the development of Mexican oil for many years.

37. Philip, George. *Oil and Politics in Latin America: Nationalist Movements and State Companies*. Cambridge (UK): Cambridge University Press, 1982, p. 32.

38. Yergin, p. 233.

39. On the process that led to the suppression of the new oil laws see: Liewen, Edwin. *Petroleum in Venezuela: A History*. Berkeley: University of California Press, 1954, pp. 20–22.

40. Betancourt, Romulo. *Venezuela: Oil & Politics*. Boston: Houghton Mifflin Co., 1979, p. 26.

41. DeGolyer and McNaughton. *Twentieth Century Petroleum Statistics*. Dallas: DeGolyer and McNaughton, 1994 (50th Ed.).

42. Liewen, p. 85.

43. For the Venezuelan version of the Dutch Disease see: Karl, Terry Lynn. *The Paradox of Plenty: Oil Booms and Petro-States*. Berkeley: University of California Press, 1997.

Chapter 3

1. Philby, John B. *Arabian Oil Ventures*. Washington (DC): Middle East Institute, 1964, p. 68.

2. Shwadran, pp. 370–385.

3. Holden, David, and Richard Johns. *The House of Saud: The Rise and Rule of the Most Powerful Dynasty in the Arab World.* New York: Holt, Rinehart and Winston, 1981, pp. 79–80.

4. Yergin, p. 281.

5. Philby, *Arabian Oil Ventures,* p. 88.

6. For a detailed history of the Saudi monarchy see: Holden and Johns.

7. Brown, Anthony Cave. *Oil, God, and Gold: The Story of Aramco and the Saudi Kings.* New York: Houghton Mifflin Co., 1999, p. 15.

8. Holden and Johns, p. 84.

9. Ibidem, p. 90.

10. Ibidem, pp. 91–92.

11. The salary Great Britain allowed to ibn Saud was part of the British strategy aimed at conserving special relations and influence upon many emirs and kings of the Arabian Peninsula. Hussein had also been granted a salary that was much higher than that of ibn Saud.

12. Philby, John. *Arabian Jubilee.* London: Robert Hale, 1952, p. 176.

13. For a detailed reconstruction of France's entry into the international oil arena see: Melby, Eric D. K. *Oil and the International System: The Case of France, 1918–1969.* New York: Arno Press, 1981.

14. Williamson et al., *The Age of Energy,* p. 524.

15. Ibidem.

16. Coal hydrogenation consists of determining a reaction by mixing coal and hydrogen under high pressures and temperatures. The process permits the molecular breaking and fusing of both, thereby obtaining liquid organic compositions. The German chemist Bergius was the first scientist to demonstrate the possibility of carrying out the process, which he patented in 1913.

17. During World War II, IG Farben was one the main firms responsible for the ignominious use of enslaved workforce—particularly of Jewish origin—coming from Nazi concentration camps. One of its most important synthetic gasoline producing plants was just outside the Auschwitz camp. After the end of the war, the Allies dismembered the Group, and some of its main components—among them current BASF—survived.

18. On the German plans for development of synthetic fuels see: Goralski, Robert, and Russell W. Freeburg. *Oil and War: How the Deadly Struggle for Fuel in WWII Meant Victory or Defeat.* New York: William Morrow and Company, 1987.

Chapter 4

1. The U.S. Geological Survey was established in 1879, and over the first two decades of its activity, it did not deal with hydrocarbons. Only at the

beginning of the twentieth century did it begin to study oil and natural gas, at the urging of its new chief, George Otis Smith (1907–1930). For more information see: www.usgs.gov.

2. *Oil&Gas Journal*'s predecessor—*The Oil Investors' Journal*—was first published in Beaumont, Texas, in 1902. Sold by its founder in 1910, it was renamed *Oil&Gas Journal* by its new publisher, which transformed it into a weekly publication devoted to technical issues concerning the hydrocarbon industry. For more information see: www.ogj.com.

3. Williamson et al., *The Age of Energy*, p. 45.

4. Scuola Superiore Enrico Mattei. *Glossario dell'industria petrolifera*. Milano: Biblioteca Eni, 2002 (2nd Ed.), p. 15.

5. Williamson et al., *The Age of Energy*, pp. 313–314.

6. Yergin, pp. 220–223.

7. Ibidem, p. 219.

8. Hyne, Norman J. *Non-technical Guide to Petroleum Geology, Exploration, Drilling, and Production*. Tulsa (OK): PennWell Corporation, 2001 (2nd Ed.), p. 213.

9. American Petroleum Institute. *Petroleum Facts and Figures: Centennial Edition, 1959*. New York: API, 1959, p. 251.

10. Williamson et al., *The Age of Energy*, p. 146.

11. DeGolyer and McNaughton, *Twentieth Century Petroleum Statistics* (1994 ed.).

12. Yergin, p. 551.

13. Ibidem.

14. For a detailed account of the "As-Is Agreement" see: Blair, John. *The Control of Oil*. New York: Pantheon Books, 1976, pp. 54–63; Yergin, pp. 260–265.

15. Sampson, Anthony. *The Seven Sisters: The Great Oil Companies and the World They Shaped*. New York: Viking Press, 1975, pp. 73–74.

16. DeGolyer and McNaughton (1994).

17. Ibidem.

18. Yergin, p. 254.

19. DeGolyer and McNaughton (1994).

20. For a detailed description of Texas Railroad Commission's role as arbiter to the U.S. oil production rationing see: Blair, pp. 159–164.

21. In particular, it was Juan Perez Alfonzo—the future oil minister of Venezuela and the "brain" behind the creation of OPEC in 1960—who was influenced by the American production regulatory system of the 1930s.

22. Yergin, p. 274.

23. Sampson, p. 85.

24. Grayson, George W. *Oil and Mexican Foreign Policy*. Pittsburgh (PA): University of Pittsburgh Press, 1988, p. 3.

Chapter 5

1. DeGolyer and McNaughton (1994).

2. Anderson, Irvine H. *Aramco, the United States, and Saudi Arabia: A Study of the Dynamics of Foreign Oil Policy 1933–1950.* Princeton (NJ): Princeton University Press, 1981, pp. 36–37.

3. Ibidem, pp. 37–57.

4. Yergin, p. 395.

5. Ibidem.

6. API (1959).

7. Yergin, p. 395.

8. Among the leading supporters of Ickes's vision were War Secretary Henry Stimson, Navy Secretary Frank Knox, and many others.

9. Miller, Aron David. *Search for Security: Saudi Arabian Oil and the American Foreign Policy, 1939–1949.* Chapel Hill: University of North Carolina Press, 1980, p. 44.

10. Yergin, p. 395.

11. Kapstein, Ethan B. *The Insecure Alliance: Energy Crises and Western Politics since 1944.* New York-Oxford: Oxford University Press, 1990, p. 62.

12. Ibidem.

13. Ibidem, p. 65.

14. Ibidem, p. 68.

15. The "fifty-fifty" formula was based on an equal sharing of operating profits—i.e., the company's gross profit less all costs and depreciation but before taxes.

16. Through Getty Oil, Paul Getty granted the Saudis a fifty-five-cent royalty per barrel, more than doubling the amount paid by Aramco, which was about twenty cents. Moreover, Getty paid a lump sum of 9.5 million dollars as the concession agreement was signed, and committed his company to paying the Saudi government 1 million dollars yearly, independent of production results. See: Yergin, pp. 439 and following.

17. U.S. Congress. Senate. Committee on Foreign Relations. Subcommittee on Multinational Corporations. *Multinational Oil Corporations and United States Foreign Policy.* Hearings. [Hereafter referred to as *Oil Multinational Hearings*] 93rd Congress, 1st Session. Washington (DC): Government Printing Office, 1975, Vol. 4, p. 89.

18. *Oil Multinational Hearings*, Vol. 4, pp. 89–91; Blair, pp. 47 and following. Thanks to the "fifty-fifty" system, in 1951, for instance, Saudi Arabia's share of profits was $110 million, while U.S. taxes on Aramco operations in the country were a modest $6 million—against $43 million in 1949, before the application of the new system and on the basis of a much lower production.

19. *Oil Multinational Hearings*, Vol. 4, pp. 12, 95.

20. Ibidem, pp. 84–85.

21. Yergin, p. 428.

22. There is a vast literature proving the indifference toward the Jewish question and the opposition to the birth of Israel. See in particular: Clifford, Clark (with Richard Holbrooke). *Counsel to the President: A Memoir.* New York: Random House, 1991, pp. 3–25. For a more general review of the issue see: Levantrosser, William F. (Ed.). *Harry S. Truman: The Man from Independence.* Westport (CT): Greenwood Press, 1986. (In particular, see essays by Charles J. Tull, pp. 17–20, and Herbert Drunks, pp. 56–67.)

23. On the whole issue, see: McCullough, David. *Truman.* New York: Simon & Schuster, 1992, pp. 595–607.

24. The Clifford memo is quoted in: Snetsinger, John. *Truman, the Jewish Vote, and the Creation of Israel.* Stanford (CA): Stanford University Press, 1974, p. 78.

25. McCullough, pp. 596–597.

26. Ibidem, p. 598.

27. Ibidem, p. 616.

28. Ibidem, pp. 617–618.

29. Levantrosser, p. 386.

Chapter 6

1. Kinzer, Stephen. *All the Shah's Men: An American Coup and the Roots of Terror in the Middle East.* Hoboken (NJ): John Wiley & Sons, 2003, pp. 1–16.

2. Farmanfarmaian, Manucher, and Roxane Farmanfarmaian. *Blood and Oil: Inside the Shah's Iran.* New York: Modern Library, 1999, pp. 184–185.

3. Abramson, Rudy. *Spanning the Century: The Life of W. Averell Harriman, 1891–1986.* New York: William Morrow, 1992, p. 479.

4. The following table shows BP's profits (gross and net) along with taxes paid by the company to the British government and overall [total] payments to the Iranian government.

Year	Gross Profits	British Taxes	Payments to Iran	Net Profits
1947	40,469	14,800	7,104	18,565
1948	61,547	28,310	9,172	24,065
1949	54,359	22,480	13,489	18,390
1950	99,842	50,707	16,032	33,103

5. For a detailed analysis of the 1949 Supplemental Agreement see: Elm, Mostafa. *Oil, Power and Principle: Iran's Oil Nationalization and Its Aftermath.* Syracuse (NY): Syracuse University Press, 1992, pp. 54–56.

6. Kinzer, p. 77.

7. Blair, p. 79.

8. Bill, James A., and William Roger Louis (Eds.). *Mussadiq, Iranian Nationalism and Oil.* Austin: University of Texas Press, 1988, pp. 235 and following.

9. This conspiratorial behavior is evidenced by a wide literature based on recently opened archives. See: Bill and Louis, *Mussadiq*; Kinzer; Woodhouse, Christopher M. *Something Ventured.* London: Granata, 1982. Woodhouse was the British secret agent who convinced the CIA and the U.S. Department of State to defenestrate Mossadegh.

10. McLellan, David S. *Dean Acheson: The State Department Years.* New York: Dodd, Mead, 1976, p. 387.

11. Woodhouse, p. 117.

12. Bill and Louis, *Mussadiq*, p. 278.

13. Ambrose, Stephen. *Eisenhower: The President.* New York: Simon & Schuster, 1984, p. 111.

14. The documents published by *NYT* were part of an official CIA account of the events in Iran entitled *The Overthrow of Premier Mossadegh of Iran, November 1952–August 1954*, written but not published in 1954 by Donald M. Wilber. See also: Kinzer, which made extensive use of the documents published by the *NYT*. See: *New York Times*, April 16, 2000. For the full account see: www.nytimes.com.

15. *The Overthrow of Premier Mossadegh of Iran.* In: New York Times, April 16, 2000. The two senior officers who prepared the blueprint for Operation Ajax were Donald Wilber (CIA) and Norman Darbyshire (SIS). See also: Kinzer, pp. 162–163.

16. Eventually, the majors' share was reduced by 5 percent in order to make room for small independent U.S. companies—a stratagem devised to avoid future antitrust actions.

17. Report to the National Security Council by the Departments of State, Defense, and the Interior: National Security Problems Concerning Free World Petroleum Demands and Potential Supplies. Top Secret, NSC 138/1. Washington, January 6, 1953. FRUS, 1952–1954, Vol. 1, Part 2, pp. 1317 and following. Quoted in: Maugeri, Leonardo. *L'Arma del Petrolio: Questione Petrolifera Globale, Guerra Fredda e Politica Italiana nella Vicenda di Enrico Mattei.* Firenze: Loggia dè Lanzi, 1994. [Translation—*The Oil Weapon: The Global Oil Issue, Cold War, Italian Politics, and the Case of Enrico Mattei*], pp. 47–48.

18. Ibidem.

19. Ibidem.

20. Sampson, p. 106.

21. Ibidem.

22. It was the Economic Cooperation Administration (ECA), presided over by Paul Hoffman.

23. U.S. Congress. Senate. Select Committee on Small Business. Subcommittee on Monopoly. *The International Petroleum Cartel: Staff Report to, and submitted by, the Federal Trade Commission.* 82nd Congress, 2nd Session. Washington (DC): Government Printing Office, 1952, p. 114.

24. Ibidem.

25. Stork, Joe. *Middle East Oil and the Energy Crisis.* New York: Monthly Review Press, 1975, p. 55.

26. Yergin, p. 472.

Chapter 7

1. DeGolyer and McNaughton (1994).

2. In 1970, oil accounted for 43 percent of global energy consumption, as against 34 percent for coal.

3. Motor Vehicle Association of the United States. *World Motor Vehicle Data*, 1990, p. 35.

4. Bardou et al., p. 180.

5. Ibidem, p. 183.

6. Ibidem, p. 182.

7. U.S. oil production passed from 5.9 million barrels per day in 1948 to 11.3 in 1970. See: DeGolyer and McNaughton (1994).

8. DeGolyer and McNaughton (1994).

9. Adelman, *The Genie*, p. 56.

10. Jacoby, Neil. *Multinational Oil: A Study in Industrial Dynamics.* New York: Macmillan, 1974, pp. 68–69.

11. Sampson, p. 131.

12. The workings of the Texas Railroad Commission and its compliance with U.S. government targets were revealed by a U.S. Senate investigation in 1969. Also by its own initiative, however, the Commission had constantly reduced production in the United States throughout the 1950s—except in periods of international supply disruptions. After the Suez crisis (1956), that policy tightened; in 1962, for example, the Commission authorized production of oil for only 91 days (against 261 in 1952 or 191 in 1954). See: U.S. Congress. Senate. Committee on the Judiciary. Subcommittee on Antitrust and Monopoly. *Hearings on Governmental Intervention in the Market Mechanism.* 91st Congress, 1st Session. Washington (DC): Government Printing Office, 1969, Part 1, pp. 137 onward.

13. Jacoby, p. 89.

14. Ibidem.

15. Adelman, *The Genie*, p. 96.

16. Oil production of the Seven Sisters in relation to total world production, 1950–1966 is represented in the table below; figures represent millions of barrels per day.

Company	1950	1955	1960	1966
Exxon	1.5	2.2	2.5	4.1
Shell	0.9	1.55	2	3
BP	0.8	0.9	1.5	2.5
Gulf	0.55	0.95	1.6	2.3
Texaco	0.55	0.9	1.35	2.25
Chevron	0.5	0.7	1	1.73
Mobil	0.3	0.6	0.8	1.32
Seven Sisters	*5.1*	*7.8*	*10.75*	*17.2*
Total World	*10.2*	*15.6*	*21.5*	*34.4*

17. Jacoby, p. 153.

18. Maugeri, *L'Arma del Petrolio*, pp. 230–234.

19. For a detailed analysis of the effects of Eisenhower's Mandatory Oil Import Program see: Morse, Edward. *The US and the International Petroleum System: Rogue Elephant in the Jungle of Geopolitics*. A Presentation for the Oxford Energy Seminar. In: Petroleum Intelligence Weekly (PIW), August 31, 1995, pp. 2–4.

20. DeGolyer and McNaughton (1994). In 1959, the Soviet Union produced 2.6 million barrels per day. In 1960, it outstripped Venezuela with slightly less than 3 mbd as against 2.85 mbd for Venezuela.

21. For Tariki's background see: Yergin, pp. 513–515.

22. As to the origins of OPEC see: Skeet, Ian. *OPEC: Twenty-Five Years of Prices and Politics*. Cambridge (UK): Cambridge University Press, 1988.

23. DeGolyer and McNaughton (1994).

24. Jacoby, p. 65.

25. Blair, p. 212.

26. At that time, independent companies accounted for around 50 percent of the country's total production, in contrast to less than 10 percent on average in the other OPEC countries. Ibidem.

27. For a complete account of IPC and Iraq government confrontation see: Stocking, pp. 212–239.

28. Issawi, Charles, and Mohammed Yeganeh. *The Economics of the Middle East Oil*. New York: Praeger Publishers, 1962, p. 93.

29. Stocking, pp. 214–255.

30. What IPC was left with was essentially the relatively small area surrounding the Kirkuk field plus the newly discovered North Rumaila field in the south—or about 2,000 square kilometers: however, the latter was also

confiscated in 1962, due to IPC's unwillingness to come to terms with the Iraqi government. See: Stocking, pp. 247–253.

31. Eni assumed the function of a public holding company charged with the management of existing state companies, such as Agip, Snam—a distributor and transporter of natural gas—and Anic, a petrochemical firm. Quite soon, Agip was divided in two parts, one upstream, Agip Mineraria, and the other dedicated to refining and marketing oil products, which kept the original name. After its transformation into a joint stock company in 1992, the acronym Eni lost its original significance of Ente Nazionale Idrocarburi and became a simple name.

32. Maugeri, *L'Arma del Petrolio*, pp. 71–77.

33. His creative intuition in this area was decisive in catalyzing the growth of industry in central northern Italy and with it the economic boom that brought Italy the highest rates of growth in Europe from 1957 to 1962.

34. The importance accorded natural gas and its integration into Eni's core business activities brought the company, from its foundation, a distinctive character with regard to the world oil industry, which virtually ignored gas, considering it a bothersome by-product of oil.

35. Maugeri, *L'Arma del Petrolio*, pp. 143–154.

36. Elbrick to Dulles. *Excerpts from Italian Progress Report (Mattei Iranian Oil Deal)*, elaborated and approved on September 3, 1957, by the Office for Coordinating Board (OCB). Top Secret. Washington (DC): National Archives, 36// 765.13/5–2455. In: Maugeri, *L'Arma del Petrolio*, pp. 155–156.

37. Memorandum of Discussion at the 337th Meeting of the National Security Council, September 23, 1957. Top secret. Abilene (KS): Eisenhower Library, Whitman File, NSC Records. In: Maugeri, *L'Arma del Petrolio*, pp. 156–157.

38. See: Foreign Office, Circular 029, March 31, 1961. Confidential. London: Foreign Office, 371/160305 RT 153/3. In: Maugeri, *L'Arma del Petrolio*, pp. 259–261.

39. Ibidem, pp. 266–273.

40. Ibidem, pp. 274–282.

41. Ibidem, p. 279.

42. In 1996, the Italian authorities reopened investigations into Mattei's death, long thought to have been orchestrated by the Seven Sisters. Investigations proved Mattei's airplane had been sabotaged but failed to reveal the organizers of the "accident." The file was closed, but many findings of the Italian magistrates indicated that Mattei's homicide had probably been masterminded in Italy.

Chapter 8

1. For a profile of Nasser see: Nutting, Anthony. *Nasser*. New York: E. P. Dutton, 1972.

2. For the origins of Nasser's visions of Pan-Arabism, see: Dawisha, Adeed. *Arab Nationalism in the Twentieth Century: From Triumph to Despair.* Princeton (NJ): Princeton University Press, 2003, pp. 135–159.

3. Nasser, Gamal Abdel. *The Philosophy of the Revolution.* Buffalo (NY): Smith, Keynes, and Marshall, 1959, pp. 60–61.

4. Cordesman, Anthony. *Saudi Arabia Enters the Twenty-First Century: The Political, Foreign Policy, Economic, and Energy Dimension.* Westport (CT): Praeger Publishers and CSIS, 2003, p. 107.

5. On the Suez Crisis see: Neff, Donald. *Warriors at Suez: Eisenhower Takes America into the Middle East.* New York: Simon & Schuster, 1981; Heikal, Mohammed H. *Cutting Through the Lion's Tale: Suez Through Egyptian Eyes.* London: Andre Deutsch, 1986.

6. Dawisha, pp. 234–237.

7. Quoted in: Brown, p. 252.

8. Kepel, Gilles. *Jihad: The Trial of Political Islam.* Cambridge (MA): Harvard University Press, 2002, pp. 52–54.

9. Ibidem, pp. 25–27.

10. On Hammer's oil venture, see: Weinberg, Steve. *Armand Hammer: The Untold Story.* Boston: Little, Brown & Co., 1989.

11. Blair, p. 221.

12. The company's production was halved (from 800,000 to 440,000 barrels daily), its assets put under the threat of expropriation. In a desperate attempt to find a way out, Hammer met Exxon chairman and chief executive Kenneth Jamieson and asked him to supply Occidental with Persian Gulf oil at cost-plus-tax, "plus a reasonable profit, such as ten percent," so that the company could withstand Libyan ultimatum and honor its contracts. But the largest of the Seven Sisters remained unmoved by Hammer's problems and two weeks later offered him oil at market price with no discount. See: Hammer, Armand (with Neil Lyndon). *Hammer.* New York: Putnam's Sons, 1987, p. 345. See also: Sampson, p. 253.

Chapter 9

1. Yergin, p. 589.

2. Author's calculations from different sources.

3. This policy had been formulated by President Lyndon Johnson; Nixon continued the course set by his predecessor.

4. On the effects of Nixon's decision to impose price controls on oil products see: Morse, Edward. *The US and the International Petroleum System Sector,* pp. 5–8.

5. Yergin, p. 590.

6. In the late 1960s, Akins had prepared a study indicating the risks of the world's energy situation, and particularly the end of the "buyers market" that had characterized the postwar era. See: Yergin, p. 590.

7. Sampson, p. 211.

8. On James Akins see: Yergin, pp. 590–591.

9. See: Middle East Economic Journal (MEES), January 1, 1971.

10. In an attempt to withstand, twenty-one companies formed a common front and agreed to help each other with oil supply, in case Libya acted on its *aut-aut*. As later revealed by U.S. Senate investigations, the secret agreement was called "Libyan Safety Net," and it consisted exactly of what Hammer had asked of Exxon only two years before, with no success.

11. Stork, p. 184.

12. A graduate of Harvard University, he had been chosen by King Faisal in 1962 to replace Abdullah Tariki, the first Saudi oil minister and the radical enemy of the Seven Sisters, who had cofounded OPEC. Quiet, intellectually brilliant, and rational and moderate in judgment, Yamani was well acquainted with the boom-and-bust nature of the oil market, a consequence of the free-for-all of blind greed among the oil protagonists. See: Robinson, Jeffrey. *Yamani: The Inside Story*. New York: The Atlantic Monthly Press, 1988.

13. Sampson, p. 232.

14. The lifting of price controls was based on an intricate system as well. The price of "old oil" remained fully controlled, while stripper oil was fully decontrolled. In the middle, new oil and released oilfields were partially decontrolled.

15. For a detailed analysis of the Nixon administration's oil regulation system see: Morse, pp. 5–8.

16. Ibidem, p. 7.

17. PIW, August 13, 1973, p. 3.

18. Yergin, p. 594.

19. For the complete text of Yamani's speech see: MEES, September 22, 1972.

20. Sampson, p. 244; Yergin, p. 594.

21. *Oil Multinational Hearings*, Vol. 7, p. 504.

22. Ibidem, pp. 509, 246.

23. At that time, OAPEC was formed by Saudi Arabia, Iraq, Kuwait, United Arab Emirates, Syria, Libya, Algeria, Egypt, and Qatar. The Organization had been founded in January 1968 by Saudi Arabia, Kuwait, and Libya, as a forum to discuss Arab policies and plans made possible by oil revenues.

24. For a detailed reconstruction of the decisions taken by OPEC and OAPEC in October 1973, see: Skeet, p. 100.

25. For a precise reconstruction of data concerning Arab cutbacks and rising output from other countries see: Adelman, *The Genie*, pp. 110–111.

26. Calculations by the author, on the basis of different sources.

27. *New York Times*, January 7, 1974. Quoted in: Stork, p. 231.

28. Yergin, pp. 621–625.

29. Skeet, p. 100.

30. Ibidem, pp. 99–105.

31. The Club of Rome was a private, nonprofit association formed by several international academics and experts whose main purpose was to address the issues of sustainable growth and environmental protection. The report "The Limits to Growth" was prepared for a meeting of the association. It received worldwide recognition particularly after the first oil shock. See: Meadows, Donella, Dennis Meadows, Jorgen Rahders, and William Bahrens III. *The Limits to Growth: A Report for the Club of Rome's Project on the Predicament of Mankind.* New York: Signet Books, 1972.

32. Eni estimates.

33. Akins, James. *The Oil Crisis: This Time the Wolf Is Here.* In: Foreign Affairs, April 1973, Vol. 11, No. 1, pp. 462–490.

34. Yergin, pp. 617–618.

35. DeGolyer and McNaughton (1994).

36. The Senate Subcommittee that investigated the oil corporations was chaired by the Democrat Frank Church, a staunch, longtime critic of multinationals who had already investigated the connections between International Telephone and Telegraph Corporation (ITT) and the violent coup that brought Augusto Pinochet to power in Chile. The hearings on oil multinationals were published in several volumes by the Government Printing Office (see: *Oil Multinational Hearings*). As to the final report of the subcommittee see: U.S. Congress. Senate. Committee on Foreign Relations. Subcommittee on Multinational Corporations. *Multinational Oil Corporations and United States Foreign Policy.* Final Report. 93rd Congress, 1st Session. Washington (DC): Government Printing Office, 1975.

37. Sampson, p. 274.

38. The final deal between the Saudis and the American partners of Aramco also established a role for the companies: they would continue to provide technical assistance in the extraction of crude, receiving twenty-one cents per barrel produced; in addition, they would market 80 percent of Aramco's oil, acquiring it according to a particularly advantageous formula.

Chapter 10

1. Sampson, p. 301.

2. U.S. Central Intelligence Agency. *The International Energy Situation: Outlook to 1985.* Washington (DC): Government Printing Office, 1977.

3. *New York Times*, October 6, 1977, p. 67. Quoted in: Adelman, *The Genie*, p. 164.

4. The IEA's study was *A Comparison of Energy Projections to 1985*. See: Parra, Francisco. *Oil Politics: A Modern History of Oil.* London: I.B. Tauris, 2004, p. 218.

5. Adelman, *The Genie*, p. 160.

6. Rockefeller Foundation. *Working Paper on International Energy Supply.* New York: Rockefeller Foundation, 1978.

7. Bill, James A. *The Eagle and the Lion: The Tragedy of American-Iranian Relations*. New Haven (CT): Yale University Press, 1988, p. 200.

8. On these problems see: Ghods, Mohammed Reza. *Iran in the Twentieth Century: A Political History*. Boulder (CO): Lynne Rienner, 1989, pp. 199–202.

9. For a vivid account of Khomeini's life see: Moin, Baquer. *Khomeini: Life of the Ayatollah*. New York: Thomas Dunne Books, 2004.

10. Ibidem, p. 195.

11. Bill, pp. 236–237.

12. Estimates by the author.

13. See: Cordesman, p. 173.

14. For an insight of the genesis of the "Carter Doctrine" see: Christopher, Warren, et. al. The *American Hostages in Iran: The Conduct of a Crisis*. New Haven (CT): Yale University Press, 1985, pp. 112 and following.

15. Skeet, p. 168.

16. For a short account of those predictions, see: Adelman, *The Genie*, pp. 189, 205–206.

17. Exxon. Annual Report 1979.

18. BP. *Oil Crisis . . . Again*. London: BP, 1979.

19. Adelman, *The Genie*, p. 178.

20. Adelman, Morris. *The Real Oil Problem*. In: Regulation, Spring 2004, p. 16.

Chapter 11

1. Yergin, p. 704.

2. BP's Statistical Review of World Energy, June 1987 and earlier editions.

3. In 1983, the United States produced 10.2 million barrels daily—including Alaska—as against 11.3 mbd in 1970. See: DeGolyer and McNaughton (1994).

4. Ibidem.

5. Skeet, p. 183.

6. The decision was taken during the OPEC extraordinary meeting held in Vienna, March 19–21. See: Skeet, pp. 184–185.

7. It is noteworthy that OPEC had assigned both Iran and Iraq the same production quota of 1.2 million barrels per day—a fact that exacerbated Iran's reaction against OPEC itself: Iran's oil potential, at that time, was indeed higher than that of Iraq. See: Skeet, p. 187.

8. The decision was taken at the OPEC's extraordinary meeting in London, on March 21.

9. Skeet, p. 202.

10. Ibidem.

11. For a detailed account of the Iran-Iraq oil war see: Chubin, Shahram, and Charles Tripp. *Iran and Iraq at War*. London: I.B. Tauris, 1988.

12. Mabro, Robert (Ed.). *OPEC and the World Oil Market: The Genesis of the 1986 Price Crisis.* Oxford (UK): Oxford University Press-Oxford Institute for Energy Studies, 1986, p. 166.

13. The following table shows OPEC's production (million barrels per day) and revenues (in current U.S. dollars) between 1973 and 1983.

OPEC Production and Revenues

Year	Production	Revenues
1973	30.9	22.6
1974	30.5	87
1975	27.2	92.4
1976	30.7	108
1977	31.3	122.5
1978	29.8	114.3
1979	30.9	192.6
1980	26.9	275
1981	22.6	247.7
1982	19	192.6
1983	17	153.9
1984	16.3	158.6
1985	15.4	132
1986	18.3	77.1

Source: OPEC Statistical Bulletin, various editions.

14. Parra, p. 287.

15. *New York Times*, April 3, 1986, p. A1.

16. Parra, p. 290.

17. Based on unproven declarations by former Reagan administration top officials, this thesis is reported in: Schweitzer, Peter. *Victory: The Reagan Administration's Secret Strategy That Hastened the Collapse of the Soviet Union.* New York: Atlantic Monthly Press, 1994.

18. For an account of Reagan's energy deregulation see: Morse, pp. 9–11.

19. Actually, it was BP itself that rejected as unnecessary the government proposal to take a "golden share." See: Graham, Cosmo, and Tony Prosser. *Privatizing Public Enterprises: Constitutions, the State, and Regulation in Comparative Perspective.* Oxford (UK): Clarendon Press, 1991, p. 162.

20. The British government floated its last 31.5 shares of BP, an operation of about 5.5 billion pounds at time. Unfortunately, the offer coincided with the world stock market collapse of October 1987. The majority of BP shares were soon traded well below the offer price, which favored KIO's progressive takeover.

21. The Kuwait Investment Office partial takeover of BP and the reaction of BP's government is well analyzed in: Graham and Cosmo, pp. 162–163.

22. International Petroleum Encyclopedia 2003. Tulsa (OK): PennWell Corporation, 2003. Statistical Tables, p. 229.

23. For a detailed account of oil industry overspending in the 1970s and 1980s, and its drive towards diversification, see: Grant, Robert M. *The Oil Companies in Transition, 1970–1987*. Milano: Franco Angeli, 1991.

24. The following table shows the oil companies' main mergers and acquisitions, 1981–1984.

Oil Companies' Main Mergers and Acquisitions, 1981–1984

Buyer	Target	Year	Value*
DuPont	Conoco	1981	7.8
U.S. Steel	Marathon Oil	1983	–
Phillips Petroleum	General America	1983	1.1
Mobil	Superior Oil	1984	5.7
Texaco	Getty Oil	1984	10.2
Chevron	Gulf	1984	13.2
Shell	Shell USA	1984	5.7
BP	Standard Ohio	1984	7.6

Billion Dollars

Chapter 12

1. Marr, Phebe. *The Modern History of Iraq*. Boulder (CO): Westview Press, 2003 (2nd Ed.).

2. Ibidem, p. 220.

3. Freedman, Lawrence, and Efriam Karsh. *The Gulf Conflict, 1990–1991*. Princeton (NJ): Princeton University Press, 1993, p. 41.

4. Ibidem, pp. 47–49.

5. Marr, p. 221.

6. The list of formal charges made by Iraq against Kuwait was contained in a thirty-seven-page letter sent by Iraqi Foreign Minister, Tariq Aziz, to the Arab League, on July 16, 1990. See: Jentleson, Bruce. *With Friends Like These: Reagan, Bush, and Saddam, 1982–1990*. New York: W. W. Norton & Company, 1994, p. 168.

7. Schwarzkopf, Norman (with Peter Petre). *It Doesn't Take a Hero*. New York: Bantam Books, 1992, p. 316.

8. Baker, James (with Thomas M. DeFrank). *The Politics of Diplomacy*. New York: G. P. Putnam's Sons, 1995, p. 273

9. Quoted in: Yergin, p. 773.

10. Freedman and Karsh, pp. 281–282.

11. Ibidem, p. 380.

12. Parra, p. 305.

13. Ibidem.

14. Ibidem, p. 306.

15. For the complex evolution of United Nations resolutions and sanctions against Iraq see: Melby, Eric D. K. *Iraq*. In: Haas, Richard N. (Ed.). *Economic Sanctions and American Diplomacy*. New York: Council on Foreign Relations, 1998, pp. 107–128.

16. For a detailed and well-documented reconstruction of the mismanagement thesis, see Jentleson.

17. Ibidem, p. 15.

18. Ibidem, p. 19.

19. Ibidem.

Chapter 13

1. Perestroika and glasnost were launched by Gorbachev during the XXVII Congress of the Soviet Communist Party, in February 1986. For a detailed account see: Boffa, Giuseppe. *Dall'URSS alla Russia: Storia di Una Crisi Non Finita. (From USSR to Russia: History of an Unfinished Crisis)*. Roma-Bari: Editori Laterza, 1995, pp. 186–191.

2. Adelman, *The Genie*, p. 315.

3. Eni. World Oil and Gas Review 2001 and 2004. Rome: Eni, 2001, 2004. See also: www.eni.it.

4. Ibidem.

5. For a good account of the Russian privatization process see: Blasi, Joseph R., Maya Douglas, and Kruse Kroumova. *Kremlin Capitalism: Privatizing the Russian Economy*. Ithaca (NY): Cornell University Press, 1997; Aslund, Anders. *Building Capitalism: The Transformation of the Former Soviet Bloc*. Cambridge (UK): Cambridge University Press, 2002.

6. International Energy Agency (IEA). *Energy Policies of the Russian Federation*. Paris: IEA, 1995, p. 36.

7. For details see: Blasi et al.; Freeland, Chrystia. *Sale of the Century: Russia's Wild Ride from Communism to Capitalism*. New York: Crown Business, 2000; Hoffman, David. *The Oligarchs*. New York: Public Affairs, 2001.

8. The first version of the loans-for-shares scheme was presented to the Russian government by Potanin himself, together with Khodorkovsky and Alexander Smolensky (another leader of the new Russian finance industry), at the end of March 1995. See: Freeland, pp. 172–181; Hoffman, pp. 302–308.

9. The Russian government reduced the number of companies involved in the "loans for share" scheme to sixteen in October 1995, following Yeltsin's decree that had launched the scheme in August.

10. In detail, Menatep gave the Russian government a $159 million loan in exchange for 45 percent of Yukos shares held by the government. The remaining 33 percent of the oil company's shares were acquired by Menatep in a bid for $150 million. Menatep made a commitment with the state to invest $200 million in Yukos over the next three years, bringing its overall spending commitment for closing the deal to $509 million.

11. Hoffman, p. 304.

12. Ibidem, p. 315.

13. Poussenkova, Nina. *From Rags to Riches: Oilmen Vs. Financiers in the Russian Oil Sector*. In: *The Energy Dimension in Russian Global Strategy*, a study prepared for the James Baker Institute for Public Policy of Rice University, October 2004. See: www.rice.edu/energy/publications/russianglobalstrategy.html.

14. See: Freeland's book, *Sale of the Century*, quoted in note 7 of this chapter.

15. See: *Azerbaijani Deal Looks Not All That Different from Tengiz*. In: PIW, September 26, 1994.

16. See: *Two Giant Kazakh Contracts Signed in US*. In: Platt's Oilgram News, Vol. 75, No. 225, p. 3.

17. See: *Caspian Export Line Neither Comes Nor Goes Away As Issue*. In: PIW, September 19, 1994.

18. See in particular: Garnett, Sherman W., Alexander Rahr, and Koji Watanabe. *The New Central Asia: In Search of Stability*. A report to the Trilateral Commission. New York: The Trilateral Commission, 2000, pp. 22–24.

19. McGuinn, Bradford R., and Mohiaddin Mesbahi. *America's Drive to the Caspian*. In: Amirahmadi, Hoosang (Ed.). *The Caspian Region at a Crossroad: Challenges of a New Frontier of Energy and Development*. New York: St. Martin's Press, 2000, pp. 187–191. See also: Garnett et al., pp. 22–24.

20. Tsepkalo, Valery V. *The Remaking of Eurasia*. In: Foreign Affairs, Vol. 77, No. 2, March/April 1998, p. 123.

21. For a detailed analysis of the Chechen Wars and their motivations see: Evangelista, Matthew. *The Chechen Wars: Will Russia Go the Way of the Soviet Union?* Washington (DC): Brookings Institution Press, 2003.

22. As to the debate concerning the legal status of the Caspian Sea see: Horton, Scott, and Natik Mamedov. *Legal Status of the Caspian Sea*. In: Amirahmadi, Hoosang (Ed.). *The Caspian Region*, pp. 265–271.

23. Garnett et al., p. 15.

24. The Caspian Pipeline Consortium was set up in its final structure in 1996. The Kazakh government held a 50 percent share of it, with eight international companies holding the other share. Among these are Chevron, Exxon-Mobil, Eni, Lukoil, Oman Oil Company, and others.

Chapter 14

1. Keddie, Nikki R. *Modern Iran: Roots and Results of Revolution.* New Haven (CT): Yale University Press, 2003, p. 265.

2. On Sosa Pietri's policy see: Mommer, Bernard. *Global Oil and the Nation State.* Oxford (UK): Oxford Institute for Energy Studies, 2002, pp. 212–215.

3. See: *Venezuelan Capacity Keeps on Growing and Growing.* Electronic PIW, October 28, 1996.

4. *Giusti: OPEC Must Shift.* Platt's Oilgram News Online, November 21, 1996.

5. For a review of flawed expectations concerning Asia see: Manning, Robert A. *The Asian Energy Factor.* New York: Palgrave-Council on Foreign Relations, 2000.

6. The document was IEA's *Oil Market Report* of April 1998. See: www.IEA.org.

7. Rhodes, Ann. *IPAA: Independents Seek Relief from Low Crude Oil Prices.* In: Oil&Gas Journal, November 23, 1998, pp. 36–37.

8. *Supply-Demand Disparity Tops 1998 Oil Trends.* In: Oil&Gas Journal, March 8, 1999.

9. Williams, Bob. *Oil Producers Face Key Question: How Long Will Prices Stay Low?* In: Oil&Gas Journal, December 28, 1998.

10. For the complete text of the speech see: Yamani, Zaki Ahamad. *Past and Future Trends in the Oil Industry.* Speech held at World Oil Prices Conference, London, September 8, 1998. In: Middle East Economic Survey, September 14, 1998.

11. *The Economist,* March 6, 1999, pp. 21–23.

12. *GCC Summit Extends Oil Cuts to End-1999. Saudi Crown Prince Says Oil Boom Is Over.* In: MEES, Vol. 41, No 59, December 14, 1999.

13. Brown, pp. 315–317.

14. Specifically, ROCE indicates the profitability of capital employed coming from operating activities for a company that is fully financed with equity. It is the ratio between the unleveraged after-tax income before minority interests and the net capital employed.

15. Adelman, *The Genie,* pp. 152–153.

16. Probably, the best essay on oil companies' strategic and financial problems in the 1990s and at the onset of the new century is: Antill, Nick, and Robert Arnott. *Oil Company Crisis: Managing Structure, Profitability, and Growth.* Oxford (UK): Oxford Institute for Energy Studies, 2003.

17. The Kyoto Protocol obliges its signatories to ensure that their average annual carbon dioxide equivalent emissions of six greenhouse gases will be at least 5.2 percent below 1990 levels in the commitment period 2008 to 2012. In December 1997 only thirty-eight countries (later joined by Kazakhstan) out of eighty-four that signed the Protocol explicitly committed to pursue this target (these countries are part of the Protocol's Annex B).

18. For a detailed and brilliant reconstruction of Enron's origins, rise, and fall see: McLean, Bethany, and Peter Elkind. *The Smartest Guys in the Room: The Amazing Rise and Scandalous Fall of Enron.* New York: Portfolio-Penguin Books, 2003. A more concise history of the Texas company is also given by: Fusaro, Peter, and Ross Miller. *What Went Wrong at Enron: Everyone's Guide to the Largest Bankruptcy in U.S. History.* Hoboken (NJ): John Wiley & Sons, 2002.

19. McLean and Elkind, p. 34.

20. Ibidem, p. 39.

21. Ibidem, pp. 30–40.

22. Ibidem, p. xxv.

23. Ibidem.

24. The following table shows the main merger and acquisition deals of the oil sector between 1998 and 2000.

Main Merger and Acquisition Deals of the Oil Sector, 1998–2000

Buyer	Target	Year	Value*
BP	Amco	1998	56
Exxon	Mobil	1998	77
Total	Petrofina	1998	12
Repsol	YPF	1999	15
BP-Amoco	Arco	1999	27
TotalFina	Elf	1999	52
Chevron	Texaco	2000	36

*Billion Dollars

Chapter 15

1. Kepel, Gilles. *The War for Muslim Minds: Islam and the West.* Cambridge (MA): Harvard University Press, 2004, p. 10.

2. See: Maggie, Michael. *Tape Attributed to bin Laden Praises Attack on U.S. Consulate in Saudi Arabia.* Associated Press Newswires, December 16, 2004.

3. West, J. Robinson. *Five Myths about the Oil Industry.* In: The International Economy, Summer 2003.

4. Eni Database. Brent quotations, on average annual basis, were: $12.7 per barrel in 1998, $17.9 in 1999, $28.4 in 2000, $24.5 in 2001, $25 in 2002, $28.8 in 2003, $38.2 in 2004, and $54.4 in 2005.

5. Cambridge Energy Research Associates (CERA). *The Brave New world of Oil Prices: A Permanent Shift Upward?* Cambridge (MA): CERA, 2004, p. 8.

6. Campbell, Colin J., and Jean Laherrére. *The End of Cheap Oil.* In: Scientific American, March 1998.

7. This ample literature featured melodramatic titles that left little hope for the future. Among them: Deffeyes, Kenneth S. *Hubbert's Peak: The Impending World Oil Shortage*. Princeton (NJ): Princeton University Press, 2001; Heinberg, Richard. *The Party's Over: Oil, War and the Fate of Industrial Societies*. Gabriola (BC): New Society Publishers, 2003; Goodstein, David. *Out of Gas: The End of the Age of Oil*. New York: W. W. Norton & Company, 2004; Roberts, Paul. *The End of Oil: On the Edge of a Perilous New World*. Boston-New York: Houghton Mifflin, 2004; Austin Fitts, Catherine. *Crossing the Rubicon: The Decline of the American Empire at the End of the Age of Oil*. Gabriola (BC): New Society Publishers, 2004.

8. Matthew Simmons made his remarks on Saudi Arabia's oil depletion during a conference hosted by the Center for Strategic & International Studies, in Washington (DC) on February 24, 2004. For a detailed report of Simmons's address see: *US Critic Predicts End of Saudi Arabia's Oil Miracle*. In: PIW, March 1, 2004, p. 6. In 2005, he exposed his ideas in a book. See: Simmons, Matthew. *Twilight in the Desert: The Coming Saudi Oil Shock and the World Economy*. Hoboken (NJ): John Wiley, 2005.

9. A protagonist of the conquest of Sibneft in 1995, Berezovsky, who also became a telecommunications magnate, had been the closest magnate to Yeltsin, so much so that he was first named vice president of the National Security Council and then executive general secretary of the Community of Independent States (CIS—the federation of the former countries of the USSR). Berezovsky was not arrested but threatened with arrest; he sold all his properties and took refuge abroad.

10. The *dominus* of Gazprom, Rem Vyakhirev, was retired and replaced by Aleksei Miller, an economist close to Putin, who in 2000 had been named Deputy Minister of Energy.

11. The whole procedure followed by the government was strongly opposed by Yukos's top management, which branded it unlawful and threatened lawsuits at the international level. Still behind bars, Khodorkovsky had consistently refused to reach agreements with the Kremlin, making the matter even more intricate. But the sale of Yugyskneftegaz delivered a deadly blow to the whole of Yukos. The most important production unit of the group, Yugyskneftegaz had arrived at producing nearly one million barrels a day, and accounted for 60 percent of Yukos's entire production. Without it, Khodorkovsky's group could hardly survive.

12. Putin, Vladimir. *Strategicheskoe Planirovanie Vosproizvodstva Mineralno-Syrevoy Bazy Regiona v Vusloviyach Formirovaniya Rynochnych Otnosheniy.* (Mineral Raw Materials in the Strategy for Development of the Russian Economy.) St. Petersburg: Zapiski Gornogo Institura (Note of the Mining Institute), January 1999, Vol. 144. A good account of such dissertation is contained in: Brill Olcott, Martha. *Vladimir Putin and the Geopolitics of Oil*. In: *The Energy*

Dimension in Russian Global Strategy, a study prepared for the James Baker Institute for Public Policy of Rice University, October 2004. See: www.rice.edu/energy/publications/russianglobalstrategy.html.

13. *Financial Times*, February 11, 2005.

14. During the 1990s, the rate of oil recovery of the Russian oil industry averaged a very poor 8–9 percent, which was not so far from the one registered during the Soviet era. In 2004, that rate had jumped to 16 percent, that, albeit low with respect to the world average (35 percent), was the major reason behind the Russian oil production recovery.

15. Eni. World Oil and Gas Review 2001 and 2004. Rome: Eni, 2001, 2004. See also: www.eni.it.

16. Eni. World Oil and Gas Review 2004. The European per capita consumption takes into account only EU-15 members, before the enlargement of the Union in 2004.

17. See: *Naimi Says Markets Balanced, But Saudis Will Hike Output If Asked*. In: Platt's Oilgram News Online, November 30, 2004.

18. Goldman Sachs. *US Energy: Oil Super Spike Period May Be Upon Us*. Goldman Sachs Global Investment Research Report, March 30, 2005.

19. They were both sons of ibn Saud, but from different mothers.

20. See: Khalaf, Roula. *Saudis Reassure Oil Markets as Fahd Dies*. In: Financial Times, August 2, 2005, p. 1.

21. Posner, Gerald. *The Kingdom and the Power*. In: New York Times, August 2, 2005, p. A21.

22. See: Hoyos, Carola, and Javier Blas. *Riyadh Closer to Pressing for Higher Oil Prices*. In: Financial Times, August 2, 2005, p. 2.

23. See: Yergin, Daniel. *The Katrina Crisis*. In: The Wall Street Journal, September 5, 2005, p. 6.

24. IEA. *Oil Market Report*. Paris: IEA, December 13, 2005.

25. Nakamura, David. *Refiners Add 2.7 Million b/d of Crude Refining Capacity in 2005*. In: Oil&Gas Journal, December 19, 2005, pp. 60–64.

26. *The Pusher-in-Chief*. In: The Economist, February 4, 2006, p. 40.

Chapter 16

1. The theory was first outlined in: Hubbert, Marion King. *Nuclear Energy and the Fossil Fuels*. In: *Drilling and Production Practice—Proceedings of the Spring Meeting, San Antonio*. Washington (DC): American Petroleum Institute, 1956, pp. 7–25.

2. "The world has been very extensively explored with the help of sophisticated technology and great advances in petroleum geology. It is accordingly almost inconceivable that any large provinces—that is to say, with a potential to supply the world for more than, say, a year or two—have been missed." See:

Campbell, Colin J. *Depletion Patterns Show Change Due for Production of Conventional Oil.* In: Oil&Gas Journal, December 29, 1997, p. 33.

3. Miller, Keith. *Worldwide Oil Reserve Estimates and the Decline in Oil Field Development.* Paris: International Energy Agency, 1998, p. 67.

4. Eni Database.

5. Cavallo, Alfred J. *Hubbert's Model: Uses, Meanings, and Limits—2.* In: Oil&Gas Journal, June 13, 2005, p. 21. (Second of two articles. The first one was published in OGJ, June 6, 2005.)

6. Deffeyes, *Hubbert's Peak*, p. 139.

7. See: Downey, Marlan W., Jack C. Threet, and William A. Morgan (Eds.). *Petroleum Provinces of the Twenty-first Century.* Tulsa (OK): American Association of Petroleum Geologists, 2002.

8. See: Lynch, Michael. *Oil Scarcity, Energy Security and Long-term Oil Prices: Lessons Learned (and Unlearned).* In: International Association of Energy Economists' Newsletter, First Quarter 1998, p. 6.

9. Campbell summed up his views in *"The Coming Oil Crisis"* (Multi-Science Publishing Company & Petroconsultants, 1997); his most successful essay—*"The End of Cheap Oil"*—was written with Jean H. Laherrére and published by Scientific American Magazine in 1998.

10. For a detailed analysis of Campbell's mistakes, see: Lynch, Michael. *The Analysis and Forecasting of Petroleum Supply: Sources of Error and Bias.* In: Energy Watchers VII, International Research Center for Energy and Economic Development, 1996. See also: Adelman, Morris, and Michael Lynch. *Natural Gas Supply to 2010.* Hoersholm: International Gas Union, 2002. The following table—concerning the main revisions of URR made by Hubbert, Campbell, and the U.S. Geological Survey—is presented in the latter work.

Billion Barrels (between parentheses, year of estimate)

Hubbert	Campbell	U.S. Geological Survey
1,350 (1969)	1,578 (1989)	1,796 (1987)
2,000 (1973)	1,650 (1990)	2,079 (1991)
	1,750 (1995)	2,272 (1994)
	1,800 (1996)	3,021 (2000)
	1,950 (2002)	

Chapter 17

1. See: Odell, Peter. *Why Carbon Fuels Will Dominate the 21st Century's Global Energy Economy.* Brentwood (Essex): Multi-Science Publishing

Co., 2004, pp. 112–120; Smil, Vaclav. *Energy at the Crossroads: Global Perspectives and Uncertainties.* Cambridge (MA): MIT Press, 2003, pp. 200–203.

2. See: Scott, Henry, Russell J. Hemley, Ho-kwang Mao, Dudley R. Herschbach, Laurence E. Fried, W. Michael Howard, and Sorin Bastean. *Generation of Methane in the Earth's Mantle: In Situ High Pressure-Temperature Measurements of Carbonate Reduction.* In: PNAS, Vol. 101, No. 39, September 28, 2004, 14023–14026, www.pnas.org.

3. Highfield, Roger. *Are We Down to Our Last Drop of Oil?* In: The Daily Telegraph, October 13, 2004, p. 16.

4. See, for example, the excellent analysis of Gulf of Mexico Hydrocarbon reservoir properties in: Haeberle, Frederick R. *Gulf of Mexico Reservoir Properties Are Helpful Parameters for Explorers.* In: Oil&Gas Journal, June 27, 2005, pp. 34–37.

5. Knowles, pp. 142–143.

6. Adelman, *The Genie,* p. 15.

7. IEA. *World Energy Outlook 2004.* Paris: IEA, 2004, pp. 99–100.

8. Maugeri, *Oil, Never Cry Wolf.*

9. Ibidem.

10. See: www.spe.org.

11. See: www.world-petroleum.org.

12. There are several sources that every year estimate the amount of global proven oil reserves. Here are their conclusions:

- *Oil&Gas Journal* (December 2004): 1.277 trillion barrels T/b. This is the only source that includes "nonconventional oils";
- *Eni—World Oil and Gas Review* (May 2005): 1.111 T/b;
- *BP's Statistical Review of World Energy* (June 2005): 1.188 T/b;
- *World Oil* (August 2005): 1.082 T/b.

13. U.S. Geological Survey. *World Petroleum Assessment 2000.* Washington (DC): U.S. Geological Survey, 2000.

14. IEA. *World Energy Outlook 2001.* Paris: OECD/IEA, 2001, pp. 62–64.

15. Ibidem.

16. See: *Oil Reserves' Worldwide Report.* In: Oil&Gas Journal, December 23, 2002, p. 62.

17. See: Williams, Bob. *Heavy Hydrocarbon Playing Key Role in Peak-Oil Debate, Future Energy Supply.* In: Oil&Gas Journal, July 28, 2003.

18. Klett, Thomas, and James Schmoker. *Reserve Growth of the World's Giant Oil Fields.* In: AAPG Memoir 78. Tulsa (OK): The American Association of Petroleum Engineers, 2003, pp. 107–122.

19. Ibidem.
20. IEA. *World Energy Outlook 2001*, p. 56.
21. Giles, Jim. *Every Last Drop*. In: Nature Magazine, Vol. 429, June 17, 2004, pp. 694–695.
22. See: Adelman, Morris. *The World Petroleum Market*. Baltimore: Johns Hopkins University Press, 1972. By the same author, see also: *The Economics of Petroleum Supply*. Cambridge (MA): MIT Press, 1993.

Chapter 18

1. CERA. *Fact or Fiction: Non-OPEC's Potential for Exceptional Production Growth*. Cambridge (MA): CERA, December 2004.
2. In particular, in 1984 Kuwait increased its oil reserves by 26 billion barrels; in 1986, both the United Arab Emirates and Iran increased their proven reserves by 64 billion barrels and by 34 billion barrels respectively; in 1987 Iraq made its own increase (28 billion barrels); finally, in 1988 Saudi Arabia increased its oil reserves by 85 billion barrels.
3. As we have seen, the policy of limiting the development of oil reserves in the Persian Gulf emerged during the Hearings on Multinational Petroleum Corporations held by the U.S. Senate in 1974–1975. A summary and an accurate explanation of the secret agreement among the Seven Sisters are given by: Blair, John M. *The Control of Oil*. New York: Pantheon Books, 1976.
4. Maugeri, Leonardo. *Two Cheers for Expensive Oil*. In: Foreign Affairs, Vol. 2, No. 85, March–April 2006, pp. 149–161.
5. IEA. *World Energy Outlook 2004*. Paris: IEA, 2004, p. 98.
6. Rach, Nina M. *Drilling Boom Continues Worldwide*. In: Oil&Gas Journal, April 18, 2005, pp. 38–39.
7. Author's estimates on the basis of IHS (formerly Petroconsultant) and WoodMcKenzie's databases.
8. U.S. Energy Information Administration. *Iraq Country Analysis Brief*. August 2003 (www.eia.doe.gov).
9. Estimates of the number of Iraq's productive fields frequently vary because several fields are grouped together. For example, the deposits at Amara, Abu Ghirba, Buzurgan, and Jabal Fauqui are often lumped together as the Missan fields.
10. Maugeri, Leonardo. *The Virgin Oilfields of Iraq*. In: Newsweek International, July 5, 2004, p. 25.
11. See: Petroleum Economist, April 2004, p. 24. This estimate has been produced by Saudi Aramco.
12. Williams, Bob. *Saudi Oil Minister Al-Naimi Sees Kingdom Sustaining Oil Supply Linchpin Role for Decades*. In: Oil&Gas Journal, April 5, 2005, pp. 19.

13. Author's estimates. According to Saudi Aramco, the figure is 136. See: Saudi Aramco. *Fifty Years Crude Oil Supply Scenarios: Saudi Aramco Perspective.* Informational material distributed on behalf of Saudi Aramco at the U.S.-Saudi Relations and Global Energy Security conference held by the Center for Strategic and International Studies, Washington, 2004.

14. These figures are mainly based on IHS (formerly Petroconsultants) database and other sources analyzed by the author.

15. See: Petroleum Intelligence Weekly, March 1, 2004, p. 7.

16. IEA. *World Energy Outlook 2005: Middle East and North Africa Insights.* Paris: IEA/OECD, 2005, p. 512.

17. Hoyos, Carola. *Field Work: Why Kuwait's Rulers Are Being Forced to Ponder a New Pact With Big Oil.* In: Financial Times, January 14, 2006, p. 9.

18. IEA. *World Energy Outlook 2004.* Paris: IEA, 2004, p. 301.

19. For a comprehensive treatment of the SEC's regulation, see: CERA. *In Search of Reasonable Certainty: Oil and Gas Reserves Disclosure.* Cambridge (MA): CERA, February 2005.

20. In the United States, there are two institutions that establish accounting standards and specify general information that listed companies are obliged to disclose: they are the Securities and Exchange Commission (SEC) and the Financial Accounting Standards Board (FASB). Upon a specific request by SEC, FASB in 1982 established the 10 percent discount rate to calculate the net present value of oil companies' cash flows. That figure was the result of a survey among the members of the "Society of Petroleum Evaluation Engineers." FASB made its decision official through its *Statement of Financial Accounting Standard no. 69* (1982). See: www.fasb.org/pdf/fas69.pdf.

21. Malkiel, Burton G. *A Random Walk Down Wall Street.* New York: W. W. Norton & Company, 1999 (7th Rev. Ed.), p. 29.

22. CERA. *In Search of Reasonable Certainty*, p. 8.

23. For a detailed analysis of fiscal regimes' influence on a company's proven reserves, see: CERA. *In Search of Reasonable Certainty*, pp. 50–59.

24. CERA. *Worldwide Liquids Capacity Outlook to 2010: Tight Supply or Excess of Riches?* Cambridge (MA): CERA, June 2005.

Chapter 19

1. CERA. *Worldwide Liquids Capacity Outlook to 2010.* Cambridge (MA): CERA, 2005.

2. Author's evaluations.

3. On today's oil price regime and its origins, see the excellent essay: Mabro, Robert. *The International Oil Price Regime: Origins, Rationale, and Assessment.* In: Journal of Energy Literature—Oxford Institute for Energy Studies, Vol. XI, No. 1, June 2005.

4. Spot market price information is reported by specialized news agencies (such as Platt's and Reuters), who canvass the market on a daily basis and obtain transaction figures for crude oil and petroleum products from individual traders.

5. Knapp, David. *The ABCs of US Gasoline*. In: Energy Compass, June 18, 2004, p. 3.

6. Author's estimates.

7. IEA. *World Energy Outlook 2004*, pp. 84–85.

8. Author's estimates.

9. CERA. *China's Oil Market Is Distorted by Its Pricing System*. Cambridge (MA): CERA, May 2005.

10. See the excellent analysis of this issue made by *The Economist* on October 2, 2004 ("Oil In Asia: Pump Priming"), pp. 55–56.

11. For a detailed account of oil subsidies' costs for the main Asian countries see: Schmollinger, Christian. *Governments Grapple With Fuel Subsidies*. In: Energy Compass, April 29, 2005, pp. 6–7.

12. Eni. *World Oil and Gas Review 2004*.

13. IEA. *World Energy Outlook 2004*. Paris: IEA, 2004, p. 86.

14. There are two main loopholes in the United States right now. One offers a tax exemption allowing small business owners to write off $25,000 or more when they purchase large SUVs, pickups, and vans. The second stems from U.S. corporate average fuel economy (CAFE) standards, which de facto require less efficient mileage per gallon (mpg) from SUVs, pick-ups, and similar vehicles relative to other cars.

15. Author's estimates.

16. IEA. *Oil Market Report*. Paris: IEA, December 13, 2005.

Chapter 20

1. On these arguments, see: Mabro, Robert. *The International Oil Price Regime*.

2. Cacchione, Nicholas. *Is Gasoline Still a Great Bargain?* Norwalk (CT): John S. Herold, May 2005.

3. Ibidem.

4. Ibidem.

5. Center for Global Energy Studies. *Global Oil Report*. London: CGES, July–August 2004, p. 6.

6. See, for example: CERA. *Do High Oil Prices Matter? $1.6 Trillion Say Yes*. Cambridge (MA): CERA, Summer 2005.

7. U.S. Department of Energy, Office of Energy Efficiency and Renewable Energy—Hydrogen, Fuel Cells and Infrastructure Technologies Program. In: www.eere.energy.gov, 2004. See also: Romm, Joseph. *The Hype about Hy-*

World's Hydrogen Production (2003)

Source	Millions M3*	%
Natural Gas	240	48
Oil	150	30
Coal	90	18
Water	20	4
Total	500	100

Source: U.S. Department of Energy.
*Cubic meters

drogen: Fact and Fiction in the Race to Save the Climate. Washington (DC): Island Press, 2004, pp. 71–72.

8. Elaboration by the author based on data reported in Romm, p. 74. Romm uses estimates contained in: Simbeck, Dale, and Elaine Chang. Hydrogen Supply: Cost Estimates for Hydrogen Pathways—Scoping Analysis. Golden (CO): U.S. Department of Energy, National Renewable Energy Laboratory, 2002. Dale and Chang use a long-term price reference for natural gas of $3.50 per million British Thermal Unit (Btu) for centralized production and $6 per million Btu for forecourt production. Those prices are roughly equivalent to a price of $30 per barrel of oil, even if the relation may change from time to time.

9. Ealey, Lance, and Glenn Mercer. Tomorrow's Cars, Today's Engines. In: The McKinsey Quarterly, No. 3, 2002.

10. All these figures are nominal, because official vehicle mileage tests released by the U.S. Environmental Protection Agency (EPA) are outmoded, being performed at 47 miles per hour, with no wind resistance and no air conditioning. Nonetheless, they are the standard used for all EPA vehicle tests, so the relative advantages remain. See: Nakamura, David. It Never Fails. In: Oil&Gas Journal, May 24, 2004.

11. Romm, p. 150.

12. Honda released this figure at the 24th Annual North American Conference of the USAEE/IAEE, Washington, July 8–10, 2004.

13. See: Lowry, Karen, and Keith Naughton. Green & Mean. In: Newsweek International, November 29, 2004, p. 50.

14. Maugeri, Leonardo. The Price Is Wrong. In: Newsweek International, September 6, 2004, p. 72.

15. Moreover, while in the mid-1980s the cost of an engine represented about 15 percent of a vehicle's total costs, now the figure is below 8 percent.

16. For instance, according to U.S. EPA, the most fuel-efficient diesel model in the United States, the Volkswagen New Beetle TDI, gets 47 mpg on highways,

Ranking of Most Fuel-Efficient Cars Sold in the U.S.—2003

Vehicles	City mpg*	Highway mpg*	Price (US $)**
Honda Insight Hybrid	60	66	21,300
Toyota Prius Hybrid	60	51	
Honda Civic Hybrid	48	47	29,500
VW New Beetle TDI	38	46	
VW Golf TDI	38	46	
VW Jetta TDI	38	46	
VW Jetta Wagon TDI	36	47	
Toyota Echo	35	43	
Toyota Corolla	32	40	
Scion xA	32	38	

Source: U.S. Environment Protection Agency.

*Miles per gallon

**Base vehicle

as against 66, 51, and 47 mpg respectively by Honda Insight, Toyota Prius, and Honda Civic hybrid vehicles.

17. Lowry and Naughton (Newsweek International), p. 52.

Chapter 21

1. Thomas, Hugh. *The Suez Affair*. Middlesex (UK): Penguin Books, 1970, p. 32.

2. Adelman, Morris. *The Real Oil Problem*. In: Regulation, Spring 2004, p. 20.

3. See: www.saudiaramco.com (Our People).

4. See: www.sabic.com (About Sabic-Human Resources).

5. Kepel, *The War for Muslim Minds*.

6. Zakaria, Fareed. *The Future of Freedom: Illiberal Democracy at Home and Abroad*. New York: W. W. Norton & Company, 2003, p. 126.

7. See: Doran, Michael Scott. *Somebody Else's Civil War: Ideology, Rage, and the Assault on America*. In: Hoge, James F., Jr., and Gideon Rose (Eds.). *How Did This Happen? Terrorism and the New War*. New York: Public Affairs—Council on Foreign Relations, 2001, p. 34.

8. Doran, pp. 49–50.

9. Zakaria, pp. 124–125.

10. Doran, p. 44.

11. Adelman, *The Real Oil Problem*, p. 19.

12. See: Peterson, Peter G. *Riding for a Fall*. In: Foreign Affairs, Vol. 83, No. 5, September–October 2004, p. 112.

Chapter 22

1. Wagstyl, Stefan. "US Warns Putin over 'Blackmail' on Energy." *Financial Times*, May 5, 2006, p.1

2. Khristenko, Viktor. "Energy Collaboration Is Free From Soviet Ghosts." *Financial Times*, May 8, 2006, p.13.

Bibliography

Selected Books and Articles

Abramson, Rudy. *Spanning the Century: The Life of W. Averell Harriman, 1891–1986.* New York: William Morrow, 1992.

Adelman, Morris A. *The World Petroleum Market.* Baltimore: Johns Hopkins University Press, 1972.

———. *The Economics of Petroleum Supply.* Cambridge (MA): MIT Press, 1993.

———. *The Genie Out of the Bottle: World Oil since 1970.* Cambridge (MA): MIT Press, 1995.

———. *The Real Oil Problem.* In: Regulation, Spring 2004.

Adelman, Morris, and Michael Lynch. *Natural Gas Supply to 2010.* Hoersholm: International Gas Union, 2002.

Akin, Edward. *Flagler: Rockefeller Partner and Florida Baron.* Kent (OH): Kent State University Press, 1988.

Akins, James. *The Oil Crisis: This Time the Wolf Is Here.* In: Foreign Affairs, Vol. 11, No. 1, April 1973.

Ambrose, Stephen. *Eisenhower: The President.* New York: Simon & Schuster, 1984.

American Petroleum Institute (API). *Petroleum Facts and Figures: Centennial Edition, 1959.* New York: API, 1959.

Amirahmadi, Hoosang (Ed.). *The Caspian Region at a Crossroad: Challenges of a New Frontier of Energy and Development.* New York: St. Martin's Press, 2000.

Anderson, Irvine H. *Aramco, the United States, and Saudi Arabia: A Study of the Dynamics of Foreign Oil Policy 1933–1950.* Princeton (NJ): Princeton University Press, 1981.

Antill, Nick, and Robert Arnott. *Oil Company Crisis: Managing Structure, Profitability, and Growth.* Oxford (UK): Oxford Institute for Energy Studies, 2003.

Aslund, Anders. *Building Capitalism: The Transformation of the Former Soviet Bloc.* Cambridge (UK): Cambridge University Press, 2002.

Associated Press Newswires (Maggie, Michael). *Tape Attributed to bin Laden Praises Attack on U.S. Consulate in Saudi Arabia.* December 16, 2004.

Austin Fitts, Catherine. *Crossing the Rubicon: The Decline of the American Empire at the End of the Age of Oil.* Gabriola (BC): New Society Publishers, 2004.

Baker, James (with Thomas M. DeFrank). *The Politics of Diplomacy.* New York: G. P. Putnam's Sons, 1995.

Bardou, Jean-Pierre, Jean Jacques Chanaron, Patrick Fridenson, and James M. Laux. *The Automobile Revolution: The Impact of an Industry.* Chapel Hill: University of North Carolina Press, 1982.

Betancourt, Romulo. *Venezuela: Oil & Politics.* Boston: Houghton Mifflin Co., 1979.

Bill, James A. *The Eagle and the Lion: The Tragedy of American-Iranian Relations.* New Haven (CT): Yale University Press, 1988.

Bill, James A., and William Roger Louis (Eds.). *Mussadiq, Iranian Nationalism and Oil.* Austin: University of Texas Press, 1988.

Blair, John M. *The Control of Oil.* New York: Pantheon Books, 1976.

Blasi, Joseph R., Maya Douglas, and Kruse Kroumova. *Kremlin Capitalism: Privatizing the Russian Economy.* Ithaca (NY): Cornell University Press, 1997.

Boffa, Giuseppe. *Dall'URSS alla Russia: Storia di Una Crisi Non Finita. (From USSR to Russia: History of an Unfinished Crisis).* Roma-Bari: Editori Laterza, 1995.

BP. *Oil Crisis . . . Again.* London: BP, 1979.

Brill Olcott, Martha. *Vladimir Putin and the Geopolitics of Oil.* In: *The Energy Dimension in Russian Global Strategy*, a study prepared for the James Baker Institute for Public Policy of Rice University, October 2004. www.rice.edu/energy/publications/russianglobalstrategy.html.

Bronson, Rachel. *Rethinking Religion: The Legacy of the US-Saudi Relationship.* In: The Washington Quarterly, Autumn 2005.

Brown, Anthony Cave. *Oil, God, and Gold: The Story of Aramco and the Saudi Kings.* New York: Houghton Mifflin Co., 1999.

Cacchione, Nicholas. *Is Gasoline Still a Great Bargain?* Norwalk (CT): John S. Herold, May 2005.

Cambridge Energy Research Associate. *The Brave New World of Oil Prices: A Permanent Shift Upward?* Cambridge (MA): CERA, 2004.

———. *Fact or Fiction: Non-OPEC's Potential for Exceptional Production Growth.* Cambridge (MA): CERA, 2004.

———. *China's Oil Market Is Distorted by Its Pricing System.* Cambridge (MA): CERA, 2005.

————. *Do High Oil Prices Matter? $1.6 Trillion Say Yes.* Cambridge (MA): CERA, 2005.

————. *In Search of Reasonable Certainty: Oil and Gas Reserves Disclosure.* Cambridge (MA): CERA, 2005.

————. *Refining Capacity Additions Lag behind Demand Growth.* Cambridge (MA): CERA, 2005.

————. *Staying Wilde: The New Paradigm for Light-Heavy Differentials?* Cambridge (MA): CERA, 2005.

————. *Worldwide Liquids Capacity Outlook to 2010: Tight Supply or Excess of Riches?* Cambridge (MA): CERA, 2005.

Campbell, Colin J. *Depletion Patterns Show Change Due for Production of Conventional Oil.* In: Oil&Gas Journal, December 29, 1997.

Campbell, Colin J., and Jean Laherrére. *The End of Cheap Oil.* In: Scientific American, March 1998.

Cavallo, Alfred J. *Hubbert's Model: Uses, Meanings, and Limits.* (Two articles.) In: Oil&Gas Journal, June 6 and 13, 2005.

Center for Global Energy Studies. *Global Oil Report.* London: CGES, July–August 2004.

Central Intelligence Agency. *The Overthrow of Premier Mossadegh of Iran, November 1952–August 1954.* Summary published in *New York Times,* April 16, 2000.

Chernow, Ron. *Titan: The Life of John D. Rockefeller, Sr.* New York: Random House, 1998.

Christopher, Warren, et al. *The American Hostages in Iran: The Conduct of a Crisis.* New Haven (CT): Yale University Press, 1985.

Chubin, Shahram, and Charles Tripp. *Iran and Iraq at War.* London: I.B. Tauris, 1988.

Clark, Judy. *Geopolitics, Unconventional Fuels to Reshape Industry.* In: Oil&Gas Journal, April 25, 2005.

Clifford, Clark (with Richard Holbrooke). *Counsel to the President: A Memoir.* New York: Random House, 1991.

Cordesman, Anthony. *Saudi Arabia Enters the Twenty-first Century: The Political, Foreign Policy, Economic, and Energy Dimension.* Westport (CT): Praeger Publishers and CSIS, 2003.

Cowles, Virginia. *The Rothschilds: A Family of Fortune.* London: Weidenfeld and Nicolson, 1973.

Dawisha, Adeed. *Arab Nationalism in the Twentieth Century: From Triumph to Despair.* Princeton (NJ): Princeton University Press, 2003.

Deffeyes, Kenneth S. *Hubbert's Peak: The Impending World Oil Shortage.* Princeton (NJ): Princeton University Press, 2001.

DeGolyer and McNaughton. *Twentieth Century Petroleum Statistics.* Dallas: DeGolyer and McNaughton, 1994 (50th Ed.).

Doran, Michael Scott. *Somebody Else's Civil War: Ideology, Rage, and the Assault on America.* In: Hoge, James F., Jr., and Gideon Rose (Eds.). *How Did This Happen? Terrorism and the New War.* New York: Public Affairs—Council on Foreign Relations, 2001.

Downey, Marlan W., Jack C. Threet, and William A. Morgan (Eds.). *Petroleum Provinces of the Twenty-first Century.* Tulsa (OK): American Association of Petroleum Geologists, 2002.

Ealey, Lance, and Glenn Mercer. *Tomorrow's Cars, Today's Engines.* In: The McKinsey Quarterly, No. 3, 2002.

Elm, Mostafa. *Oil, Power and Principle: Iran's Oil Nationalization and its Aftermath.* Syracuse (NY): Syracuse University Press, 1992.

Evangelista, Matthew. *The Chechen Wars: Will Russia Go the Way of the Soviet Union?* Washington (DC): Brookings Institution Press, 2003.

Exxon. Annual Report 1979.

Farmanfarmaian, Manucher, and Roxane Farmanfarmaian. *Blood and Oil: Inside the Shah's Iran.* New York: Modern Library, 1999.

Ferguson, Niall. *Empire: The Rise and Demise of the British World Order and the Lessons for Global Power.* London: Allen Lane, 2002.

Flink, James. *The Automobile Age.* Cambridge (MA): MIT Press, 2001 (Sixth Printing, paperback ed.).

Flynn, John T. *Men of Wealth: The Story of Twelve Significant Fortunes from the Renaissance to the Present Day.* New York: Simon & Schuster, 1941.

Freedman, Lawrence, and Efriam Karsh. *The Gulf Conflict, 1990–1991.* Princeton (NJ): Princeton University Press, 1993.

Freeland, Chrystia. *Sale of the Century: Russia's Wild Ride from Communism to Capitalism.* New York: Crown Business, 2000.

Fusaro, Peter, and Ross Miller. *What Went Wrong at Enron: Everyone's Guide to the Largest Bankruptcy in U.S. History.* Hoboken (NJ): John Wiley & Sons, 2002.

Garnett, Sherman W., Alexander Rahr, and Koji Watanabe. *The New Central Asia: In Search of Stability.* A report to the Trilateral Commission. New York: The Trilateral Commission, 2000.

Gerretson, E. C. *History of the Royal Dutch.* Leiden: E. J. Brill, 1955 (4 Vols.).

Ghods, Mohammed Reza. *Iran in the Twentieth Century: A Political History.* Boulder (CO): Lynne Rienner, 1989.

Giddens, Paul. *Early Days of Oil.* Princeton (NJ): Princeton University Press, 1948.

Giles, Jim. *Every Last Drop.* In: Nature Magazine, Vol. 429, June 17, 2004.

Goldman Sachs. *US Energy: Oil. Super Spike Period May Be Upon Us.* Goldman Sachs Global Investment Research Report, March 30, 2005.

Goodstein, David. *Out of Gas: The End f the Age of Oil.* New York: W. W. Norton & Company, 2004.

Goralski, Robert, and Russell W. Freeburg. *Oil and War: How the Deadly Struggle for Fuel in WWII Meant Victory or Defeat.* New York: William Morrow and Company, 1987.

Gordon, Richard L. *Viewing Energy Prospects.* In: The Energy Journal, Vol. 26, No. 3, 2005.

Graham, Cosmo, and Tony Prosser. *Privatizing Public Enterprises: Constitutions, the State, and Regulation in Comparative Perspective.* Oxford (UK): Clarendon Press, 1991.

Grant, Robert M. *The Oil Companies in Transition, 1970–1987.* Milano: Franco Angeli, 1991.

Grayson, George W. *Oil and Mexican Foreign Policy.* Pittsburgh (PA): University of Pittsburgh Press, 1988.

Haas, Richard N. (Ed.). *Economic Sanctions and American Diplomacy.* New York: Council on Foreign Relations, 1998.

Haeberle, Frederick R. *Gulf of Mexico Reservoir Properties Are Helpful Parameters for Explorers.* In: Oil&Gas Journal, June 27, 2005.

Hammer, Armand (with Neil Lyndon). *Hammer.* New York: Putnam's Sons, 1987.

Heikal, Mohammed H. *Cutting Through the Lion's Tale: Suez Through Egyptian Eyes.* London: Andre Deutsch, 1986.

Heinberg, Richard. *The Party's Over: Oil, War and the Fate of Industrial Societies.* Gabriola (BC): New Society Publisher, 2003.

Henriques, Robert. *Marcus Samuel: First Viscount Bearsted and Founder of the "Shell" Transport and Trading Company, 1853–1927.* London: Barrie and Rocklift, 1960.

Highfield, Roger. *Are We Down to Our Last Drop of Oil?* In: The Daily Telegraph, October 13, 2004.

Hoffman, David. *The Oligarchs.* New York: Public Affairs, 2001.

Hoge, James F., Jr., and Gideon Rose (Eds.). *How Did This Happen? Terrorism and the New War.* New York: Public Affairs—Council on Foreign Relations, 2001.

Holden, David, and Richard Johns. *The House of Saud: The Rise and Rule of the Most Powerful Dynasty in the Arab World.* New York: Holt, Rinehart and Winston, 1981.

Hoyos, Carola. *Field Work: Why Kuwait's Rulers Are Being Forced to Ponder a New Pact with Big Oil.* In: Financial Times, Januray 14, 2006.

Hubbert, Marion King. *Nuclear Energy and the Fossil Fuels.* In: Drilling and Production Practice—Proceedings of the Spring Meeting, San Antonio. Washington (DC): American Petroleum Institute, 1956.

Hyne, Norman J. *Non-technical Guide to Petroleum Geology, Exploration, Drilling, and Production.* Tulsa (OK): PennWell Corporation, 2001 (2nd Ed.).

International Energy Agency (IEA). *Energy Policies of the Russian Federation.* Paris: IEA, 1995.

———. *World Energy Outlook 2001.* Paris: IEA, 2001.

————. *World Energy Outlook 2004.* Paris: IEA, 2004.

————. *World Energy Outlook 2005: Middle East and North Africa Insights.* Paris: IEA, 2005.

International Petroleum Encyclopedia 2003. Tulsa (OK): PennWell Corporation, 2003.

Issawi, Charles, and Mohammed Yeganeh. *The Economics of the Middle East Oil.* New York: Praeger Publishers, 1962.

Jacoby, Neil. *Multinational Oil: A Study in Industrial Dynamics.* New York: Macmillan, 1974.

Jentleson, Bruce. *With Friends Like These: Reagan, Bush, and Saddam, 1982–1990.* New York: W. W. Norton & Company, 1994.

Kapstein, Ethan B. *The Insecure Alliance: Energy Crises and Western Politics since 1944.* New York-Oxford: Oxford University Press, 1990.

Karl, Terry Lynn. *The Paradox of Plenty: Oil Booms and Petro-States.* Berkeley: University of California Press, 1997.

Keddie, Nikki R. *Modern Iran: Roots and Results of Revolution.* New Haven (CT): Yale University Press, 2003.

Kepel, Gilles. *Jihad: The Trial of Political Islam.* Cambridge (MA): Harvard University Press, 2002.

————. *The War for Muslim Minds: Islam and the West.* Cambridge (MA): Harvard University Press, 2004.

Kinzer, Stephen. *All the Shah's Men: An American Coup and the Roots of Terror in the Middle East.* Hoboken (NJ): John Wiley & Sons, 2003.

Klare, Michael T. *Blood and Oil: the Dangers and Consequences of America's Growing Petroleum Dependency.* New York: Metropolitan Books/Henry Holt & Co., 2004.

Klett, Thomas, and James Schmoker. *Reserve Growth of the World's Giant Oil Fields.* In: *AAPG Memoir 78.* Tulsa (OK): The American Association of Petroleum Engineers, 2003.

Klieman, Arnold. *The Foundations of British Policy in the Arab World.* Baltimore: Johns Hopkins University Press, 1972.

Knapp, David. *The ABCs of US Gasoline.* In: Energy Compass, June 18, 2004.

Knowles, Ruth Sheldon. *The Greatest Gamblers: The Epic of the American Oil Exploration.* Norman: University of Oklahoma Press, 1978 (2nd Ed.).

Lesser, Ian O. *Resources and Strategy: Vital Materials in International Conflict, 1600–The Present.* New York: St. Martin's Press, 1989.

Levantrosser, William F. (Ed.). *Harry S. Truman: the Man from Independence.* Westport (CT): Greenwood Press, 1986.

Liewen, Edwin. *Petroleum in Venezuela: A History.* Berkeley: University of California Press, 1954.

Longrigg, Steve. *Iraq 1900–1950.* London: Oxford University Press, 1953.

Lowry, Karen, and Keith Naughton. *Green & Mean.* In: Newsweek International, November 29, 2004.

Lynch, Michael. *The Analysis and Forecasting of Petroleum Supply: Sources of Error and Bias.* In: Energy Watchers VII, International Research Center for Energy and Economic Development, 1996.

————. *Oil Scarcity, Energy Security and Long-term Oil Prices: Lessons Learned (and Unlearned).* In: International Association of Energy Economists' Newsletter, First Quarter 1998.

Mabro, Robert (Ed.). *OPEC and the World Oil Market: The Genesis of the 1986 Price Crisis.* Oxford (UK): Oxford University Press—Oxford Institute for Energy Studies, 1986.

————. *The International Oil Price Regime: Origins, Rationale, and Assessment.* In: Journal of Energy Literature—Oxford Institute for Energy Studies, Vol. XI, No. 1, June 2005.

Malkiel, Burton G. *A Random Walk Down Wall Street.* New York: W. W. Norton & Company, 1999 (7th Rev. Ed.).

Manning, Robert A. *The Asian Energy Factor.* New York: Palgrave-Council on Foreign Relations, 2000.

Marr, Phebe. *The Modern History of Iraq.* Boulder (CO): Westview Press, 2003 (2nd Ed.).

Maugeri, Leonardo. *L'Arma del Petrolio: Questione Petrolifera Globale, Guerra Fredda e Politica Italiana nella Vicenda di Enrico Mattei.* Firenze: Loggia dè Lanzi, 1994. (Translation—*The Oil Weapon: The Global Oil Issue, Cold War, Italian Politics, and the Case of Enrico Mattei.*)

————. *Not in Oil's Name.* In: Foreign Affairs, Vol. 82, No. 4, July/August 2003.

————. *Oil, Never Cry Wolf: Why the Petroleum Age is Far from Over.* In: Science, Vol. 304, No. 5674, May 21, 2004.

————. *The Virgin Oilfields of Iraq.* In: Newsweek International, July 5, 2004.

————. *The Price Is Wrong.* In: Newsweek International, September 6, 2004.

————. *Two Cheers for Expensive Oil.* In: Foreign Affairs, Vol. 85, No. 2, March/April 2006.

McCraw, Thomas K. (Ed.). *Creating Modern Capitalism: How Entrepreneurs, Companies, and Countries Triumphed in Three Industrial Revolutions.* Cambridge (MA): Harvard University Press, 1997.

McCullough, David. *Truman.* New York: Simon & Schuster, 1992.

McLean, Bethany, and Peter Elkind. *The Smartest Guys in the Room: The Amazing Rise and Scandalous Fall of Enron.* New York: Portfolio-Penguin Books, 2003.

McLellan, David S. *Dean Acheson: The State Department Years.* New York: Dodd, Mead, 1976.

Meadows, Donella, Dennis Meadows, Jorgen Rahders, and William Bahrens III. *The Limits to Growth: A Report for the Club of Rome's Project on the Predicament of Mankind.* New York: Signet Books, 1972.

Mejcher, Helmut. *Imperial Quest for Oil: Iraq 1910–1928.* London: Ithaca Press, 1976.

Melby, Eric D. K. *Oil and the International System: The Case of France, 1918–1969*. New York: Arno Press, 1981.

Meyer, Karl E., and Shareen Blair Brysiac. *Tournament of Shadows: The Great Game and the Race for Empire in Central Asia*. Washington (DC): Counterpoint, 1999.

Middle East Economic Survey. *GCC Summit Extends Oil Cuts to End-1999. Saudi Crown Prince Says Oil Boom Is Over*. December 14, 1999.

Miller, Aron David. *Search for Security: Saudi Arabian Oil and the American Foreign Policy, 1939–1949*. Chapel Hill: University of North Carolina Press, 1980.

Miller, Keith. *Worldwide Oil Reserve Estimates and the Decline in Oil Field Development*. Paris: International Energy Agency, 1998.

Moin, Baquer. *Khomeini: Life of the Ayatollah*. New York: Thomas Dunne Books, 2004.

Mommer, Bernard. *Global Oil and the Nation State*. Oxford (UK): Oxford Institute for Energy Studies, 2002.

Morse, Edward. *The US and the International Petroleum System: Rogue Elephant in the Jungle of Geopolitics*. A Presentation for the Oxford Energy Seminar. In: Petroleum Intelligence Weekly (PIW), August 31, 1995.

Motor Vehicle Association of the United States. *World Motor Vehicle Data*, 1990.

Nakamura, David. *It Never Fails*. In: Oil&Gas Journal, May 24, 2004.

———. *Refiners Add 2.7 Million b/b of Crude Refining Capacity in 2005*. In: Oil&Gas Journal, December 19, 2005.

Nasser, Gamal Abdel. *The Philosophy of the Revolution*. Buffalo (NY): Smith, Keynes, and Marshall, 1959.

National Energy Policy Development Group. *National Energy Policy. Reliable, Affordable, and Environmentally Sound Energy for America's Future*. Washington (DC): U.S. Government Printing Office, 2001.

Neff, Donald. *Warriors at Suez: Eisenhower Takes America into the Middle East*. New York: Simon & Schuster, 1981.

Nevins, Allan. *Study in Power: John D. Rockefeller, Industrialist and Philanthropist*. New York: Charles Scribner's Sons, 1953, Vol. 1.

Nutting, Anthony. *Nasser*. New York: E. P. Dutton, 1972.

Odell, Peter. *Why Carbon Fuels Will Dominate the 21st Century's Global Energy Economy*. Brentwood (Essex): Multi-Science Publishing Co., 2004.

Parra, Francisco. *Oil Politics: A Modern History of Oil*. London: I.B. Tauris, 2004.

Peterson, Peter G. *Riding for a Fall*. In: Foreign Affairs, September–October, 2004, Vol. 83, No.5.

Philby, John B. *Arabian Jubilee*. London: Robert Hale, 1952.

———. *Arabian Oil Ventures*. Washington (DC): Middle East Institute, 1964.

Philip, George. *Oil and Politics in Latin America: Nationalist Movements and State Companies*. Cambridge (UK): Cambridge University Press, 1982.

Posner, Gerald. *The Kingdom and the Power*. In: New York Times, August 2, 2005.

Poussenkova, Nina. *From Rags to Riches: Oilmen Vs. Financiers in the Russian Oil Sector.* In: *The Energy Dimension in Russian Global Strategy*, a study prepared for the James Baker Institute for Public Policy of Rice University, October 2004. www.rice.edu/energy/publications/russianglo balstrategy.html.

Rach, Nina M. *Drilling Boom Continues Worldwide.* In: Oil&Gas Journal, April 18, 2005.

Rae, John B. *The American Automobile.* Chicago: University of Chicago Press, 1965.

Rhodes, Ann. *IPAA: Independents Seek Relief From Low Crude Oil Prices.* In: Oil&Gas Journal, November 23, 1998.

Roberts, Paul. *The End of Oil: On the Edge of a Perilous New World.* Boston-New York: Houghton Mifflin, 2004.

Robinson, Jeffrey. *Yamani: The Inside Story.* New York: The Atlantic Monthly Press, 1988.

Rockefeller Foundation. *Working Paper on International Energy Supply.* New York: Rockefeller Foundation, 1978.

Romm, Joseph. *The Hype about Hydrogen: Fact and Fiction in the Race to Save the Climate.* Washington (DC): Island Press, 2004.

Sampson, Anthony. *The Seven Sisters: The Great Oil Companies and the World They Shaped.* New York: Viking Press, 1975.

Schmollinger, Christian. *Governments Grapple with Fuel Subsidies.* In: Energy Compass, April 29, 2005.

Schwarzkopf, Norman (with Peter Petre). *It Doesn't Take a Hero.* New York: Bantam Books, 1992.

Schweitzer, Peter. *Victory: The Reagan Administration's Secret Strategy That Hastened the Collapse of the Soviet Union.* New York: Atlantic Monthly Press, 1994.

Scott, Henry, Russell J. Hemley, Ho-kwang Mao, Dudley R. Herschbach, Laurence E. Fried, W. Michael Howard, and Sorin Bastean. *Generation of Methane in the Earth's Mantle: In Situ High Pressure-Temperature Measurements of Carbonate Reduction.* In: PNAS, Vol. 101, No. 39, September 28, 2004, 14023–14026. www.pnas.org.

Scuola Superiore Enrico Mattei. *Glossario dell'industria petrolifera.* Milano: Biblioteca Eni, 2002 (2nd Ed.).

Shwadran, Benjamin. *The Middle East, Oil and the Great Powers.* New York: Praeger, 1955.

Simbeck, Dale, and Elaine Chang. *Hydrogen Supply: Cost Estimates for Hydro-gen Pathways—Scoping Analysis.* Golden (CO): U.S. Department of Energy, National Renewable Energy Laboratory, 2002.

Simmons, Matthew. *Twilight in the Desert: The Coming Saudi Oil Shock and the World Economy.* Hoboken (NJ): John Wiley, 2005.

Skeet, Ian. *OPEC: Twenty-five Years of Prices and Politics.* Cambridge (UK): Cambridge University Press, 1988.

Smil, Vaclav. *Energy at the Crossroads: Global Perspectives and Uncertainties.* Cambridge (MA): MIT Press, 2003.

Snetsinger, John. *Truman, the Jewish Vote, and the Creation of Israel.* Stanford (CA): Stanford University Press, 1974.

Sobel, Robert. *Coolidge: An American Enigma.* Washington (DC): Regnery Publishing, 1998.

Stocking, George W. *Middle East Oil: A Study in Political and Economic Controversy.* Knoxville (TN): Vanderbilt University Press, 1970.

Stork, Joe. *Middle East Oil and the Energy Crisis.* New York: Monthly Review Press, 1975.

Tarbell, Ida. *The History of the Standard Oil Company.* 2 vols. New York: McClure, Phillips & Co., 1904.

Tedlow, Richard S. *New and Improved: The Story of Mass Marketing in America.* Cambridge (MA): Harvard Business School Press, 1996.

Thomas, Hugh. *The Suez Affair.* Middlesex (UK): Penguin Books, 1970.

Tolf, Robert W. *The Russian Rockefellers: The Saga of the Nobel Family and the Russian Oil Industry.* Stanford (CA): Hoover Institution Press, Stanford University, 1976.

Tsepkalo, Valery V. *The Remaking of Eurasia.* In: Foreign Affairs, Vol. 77, No. 2, March/April 1998.

U.S. Central Intelligence Agency. *The International Energy Situation: Outlook to 1985.* Washington (DC): U.S. Government Printing Office, 1977.

U.S. Congress. Senate. Committee on Foreign Relations. Subcommittee on Multinational Corporations. *Multinational Oil Corporations and United States Foreign Policy.* Hearings. 93rd Congress, 1st Session. Washington (DC): Government Printing Office, 1975.

U.S. Congress. Senate. Committee on Foreign Relations. Subcommittee on Multinational Corporations. *Multinational Oil Corporations and United States Foreign Policy.* Final Report. 93rd Congress, 1st Session. Washington (DC): Government Printing Office, 1975.

U.S. Congress. Senate. Committee on the Judiciary. Subcommittee on Antitrust and Monopoly. *Hearings on Governmental Intervention in the Market Mechanism.* 91st Congress, 1st Session. Washington (DC): Government Printing Office, 1969.

U.S. Congress. Senate. Select Committee on Small Business. Subcommittee on Monopoly. *The International Petroleum Cartel: Staff Report to, and submitted by, the Federal Trade Commission.* 82nd Congress, 2nd Session. Washington (DC): Government Printing Office, 1952.

U.S. Geological Survey. *World Petroleum Assessment 2000.* Washington (DC): U.S. Geological Survey, 2000.

Weinberg, Steve. *Armand Hammer: The Untold Story.* Boston: Little, Brown & Co., 1989.

West, J. Robinson. *Five Myths about the Oil Industry*. In: The International Economy, Summer 2003.

Williams, Bob. *Oil Producers Face Key Question: How Long Will Prices Stay Low?* In: Oil&Gas Journal, December 28, 1998.

———. *Heavy Hydrocarbon Playing Key Role in Peak-Oil Debate, Future Energy Supply*. In: Oil&Gas Journal, July 28, 2003.

———. *Saudi Oil Minister Al-Naimi Sees Kingdom Sustaining Oil Supply Linchpin Role for Decades*. In: Oil&Gas Journal, April 5, 2005.

Williamson, Harold F., and Arnold R. Daum. *The American Petroleum Industry: The Age of Illumination, 1859–1899*. Evanston (IL): Northwestern University Press, 1959.

Williamson, Harold F., Ralph L. Andreano, Arnold R. Daum, and Gilbert C. Klose. *The American Petroleum Industry: The Age of Energy, 1899–1959*. Evanston (IL): Northwestern University Press, 1963.

Woodhouse, Christopher M. *Something Ventured*. London: Granata, 1982.

Yamani, Zaki Ahamad. *Past and Future Trends in the Oil Industry*. In: MEES, September 14, 1998.

Yergin, Daniel. *The Prize: The Epic Quest for Oil, Money and Power*. New York: Simon & Schuster, 1991.

———. *The Katrina Crisis*. In: The Wall Street Journal, September 5, 2005.

Zakaria, Fareed. *The Future of Freedom: Illiberal Democracy at Home and Abroad*. New York: W. W. Norton & Company, 2003.

Data Sources

American Petroleum Institute. www.api.org

BP's Statistical Review of World Energy. www.bp.org

Eni. World Oil and Gas Review, various editions. www.eni.it

Fuel Cells and Infrastructure Technologies Program. www.eere.energy.gov

IHS Energy (formerly Petroconsultant). www.ihs.com.

International Energy Agency. www.IEA.org

Oil&Gas Journal. www.ogj.com

OPEC Statistical Bulletin. www.opec.org

Society of Petroleum Engineers. www.spe.org

U.S. Energy Information Administration. www.doe.gov

U.S. Geological Survey. www.usgs.gov

WoodMcKenzie. www.woodmacresearch.com

World Oil. www.worldoil.com

World Petroleum Congress. www.world-petroleum.org

Index

About the Author

LEONARDO MAUGERI is Group Senior Vice President for Corporate Strategies and Planning for the Italian energy company, Eni, the sixth-largest publicly listed oil company in the world. His articles have appeared in *Newsweek*, *Foreign Affairs*, *Science*, and *The Wall Street Journal*, among other publications.